LANGUAGE, QUANTUM, MUSIC

SYNTHESE LIBRARY

STUDIES IN EPISTEMOLOGY,

LOGIC, METHODOLOGY, AND PHILOSOPHY OF SCIENCE

VOLUME 281

LANGUAGE, QUANTUM, MUSIC

Selected Contributed Papers of the Tenth International Congress of Logic, Methodology and Philosophy of Science, Florence, August 1995

Edited by

MARIA LUISA DALLA CHIARA
University of Florence, Italy

ROBERTO GIUNTINI
University of Cagliari, Italy

FEDERICO LAUDISA
University of Florence, Italy

KLUWER ACADEMIC PUBLISHERS
DORDRECHT / BOSTON / LONDON

Library of Congress Cataloging-in-Publication Data is available.

ISBN 0-7923-5727-2

Published by Kluwer Academic Publishers,
P.O. Box 17, 3300 AA Dordrecht, The Netherlands.

Sold and distributed in North, Central and South America
by Kluwer Academic Publishers,
PO Box 358, Accord Station, Hingham, MA 02018-0358, U.S.A.

In all other countries, sold and distributed
by Kluwer Academic Publishers, Distribution Center,
P.O. Box 322, 3300 AH Dordrecht, The Netherlands

Printed on acid-free paper

Printed and bound in Great Britain by MPG Books, Bodmin, Cornwall.

TABLE OF CONTENTS

PART IV: HISTORY OF SCIENCE

PART V: SCIENCE AND MUSIC

EDITORIAL

The *10th International Congress of Logic, Methodology and Philosophy of Science*, which took place in Florence in August 1995, offered a vivid and comprehensive picture of the present state of research in all directions of Logic and Philosophy of Science. The final program comprised 51 invited lectures and around 700 contributed papers, distributed into 15 sections.

Following the tradition of previous LMPS meetings, some authors whose papers aroused particular interest were invited to submit their works for publication in a collection of selected contributed papers. Due to the large number of interesting contributions, it was decided to split the collection into two distinct volumes: one covering the areas of Logic, Foundations of Mathematics and Computer Science, the other focusing on the general Philosophy of Science and the Foundations of Physics.

As a leading criterion for the present volume, we tried to combine papers containing relevant technical results in the field of foundations of science with papers devoted to conceptual analyses, deeply rooted in advanced present-day research. After all, we believe this is part of the genuine spirit underlying the whole enterprise of LMPS studies.

Our initial selection addressed 40 papers taken from Congress sections 6, 7, 11, 13 (*General Philosophy of Science*: 'Methodology', 'Probability, Induction and Decision Theory'; *Philosophical and Foundational Problems of the Sciences*: 'Physical Sciences', 'Cognitive Sciences and AI'). It has been our intention to reflect not only the value of the single contributions, but also the relative size and the richness of the cited sections.

For a number of reasons, the original list reduced to the present 29 papers (including two papers that had been presented in the affiliated meeting *Science and Music*). A few papers appear more or less as they were presented at the Congress; others contain additional results, or provide a survey of a specific research issue.

We believe that the volume is attractive both for specialists in foundational questions and for scholars interested in general epistemology.

ACKNOWLEDGMENTS

We thank all the referees who assisted us in the difficult task of selecting among many interesting contributions. We wish to express our gratitude to the editors of the *Synthese Library* for making available this collection of papers and for contributing so much to the circulation of logical and epistemological ideas.

Maria Luisa Dalla Chiara
Roberto Giuntini
Federico Laudisa

ERMANNO BENCIVENGA

REALISM, IDEALISM, AND GENERAL TERMS[1]

Bencivenga (1990) pointed out that classical and free quantification theories are naturally associated with the frameworks of the transcendental realist and the transcendental idealist, respectively.[2]

The former's conceptual starting point is objects, and a singular term can only *be* a term if it refers to an object: non-denoting singular terms are an absurdity. The latter's conceptual starting point, on the other hand, is language, and singular terms are expressions fulfilling certain grammatical roles. Thus it is perfectly possible that some such expressions refer to nothing (real): non-denoting singular terms are just part of life.

Lambert (1967) brought out an analogy between the 'problem' addressed by free logic and the traditional problem of the existential import of general terms. The analogy is best illustrated by considering the traditional square of opposition: relations of contrariety, subcontrariety, and subordination only hold around the square (for a given statement form '*S* are *P*') if the general term in subject position has instances. Since 'unicorn' has no instances, 'All unicorns are white' (*A*) and 'No unicorns are white' (*E*) are both true, 'Some unicorns are white' (*I*) and 'Some unicorns are not white' (*O*) are both false, and *A* does not entail *I* (nor does *E* entail *O*). Limiting the square (and hence traditional logic) to general terms having instances is for Lambert the analogue of the treatment of singular terms in classical quantification theory, whereas inserting a 'filter' statement explicitly requiring subject terms to have instances is the analogue of the strategy adopted by free logicians.

When these two contributions are brought together, the following reflection emerges. Non-denoting singular terms no longer seem a problem for which free logic provides a better solution than its classical counterpart. In the conceptual framework most appropriate to classical quantification theory, there simply cannot be non-denoting singular terms, and hence there is no relevant problem. The transcendental realist is forced to say that there can be at most an *appearance* of non-denoting singular terms, and hence his problem is at most that of dispelling the appearance.[3] Does this mean that in the realist's case the analogy between singular and general terms breaks down? For the existential import of general terms does seem to be a problem no matter what one's conceptual framework is: it seems undeniable that there are general terms with no instances, and that the relations of contrariety, subcontrariety, and subordination fail to hold for statements containing such terms in subject position

M.L. Dalla Chiara et al. (eds.), Language, Quantum, Music, 1–5.

in the traditional square. For a realist, it seems, the traditional strategy of limiting the validity of the square to general terms having instances is one way of dealing with this problem, whereas the 'analogous' strategy concerning singular terms is not a *strategy* at all, but a conceptual necessity. In the present note I intend to argue that the analogy can be saved: the 'strategy' is a conceptual necessity in the case of general terms, too.

For the realist, I said, a singular term *is* a term – as opposed to, say, an inkblot – *because* it refers to an object. What about a general term now? What makes *that* a term? A simple answer would be this: 'Napoleon' is a *singular* term because it refers to the *single* object Napoleon, and 'horse' is a *general* term because it refers to the *several* objects (that are) horses. Ultimately, I think that this simple answer is the correct one, but there are two problems with it that I need to address. One is that many would take the reference of a general term to be (not directly the objects that are instances of it, as I just did, but) a property or a concept or some such – an object, maybe, but an 'abstract' one, or 'unsaturated', or something of the sort. So they would say that 'horse' is a general term because it refers to the property of horseness, and 'xywz' is not a general term because there is no property of xywzness for it to refer to. However, this is not going to go very far, for what makes something a property? There seem to be only three possible answers to this question. Either (a) anything whatsoever is a property (including xywzness), and then anything whatsoever (including 'xywz') is a general term. Or (b) that something be a property is defined in terms of its having instances, and then we are back to the simple answer above and properties only add a useless epicycle to this whole machinery. Or else (c) something is a property if the relevant term is part of some established vocabulary, and then we have lost our realism along the way, and we are characterizing terms by their linguistic roles *and objects by the linguistic roles of the corresponding terms*.

Since (b) seems the only plausible realist course, I can disregard properties altogether. There may very well be some such entities: mine is not an ontological stand. But, whether or not there are any, referring to them is not going to help the realist with the *logical* issue of telling general terms from nonsense any more than referring to the objects *having the properties* would. So turn now to the second problem with my simple answer. Some would say that the answer is too much of an *actualist* one, and that it consequently begs our current question. Of course, if a necessary conceptual condition for something to be a general term is that it refers to (several) objects, there cannot be general terms without instances, and the traditional policy concerning the square of opposition is a conceptual necessity. But something (these people would argue) is a general term even if it refers to nothing *existing*, so long as it *could* refer to something. 'Unicorn', after all, is a general term though there are no unicorns, and it is one because it is *possible* that there be some – whereas it is impossible that there be any xywz's.

There are two steps to understanding what is wrong with this response, and both of them can be illustrated by reference to Leibniz's critique of Descartes's ontological argument. First, what Leibniz objected to Descartes is that he had only proved that God existed *if the idea of God was consistent*. Unless this additional premiss was established, Descartes had no evidence that by combining words the way he did he was not crossing the bounds of sense and ending up with total gibberish (cfr. the combination that gives me terms like 'thoughtful windowsill', or 'red sorrow', or 'unequal equality' – the stuff poetry, and sometimes politics, but not sensible discourse, is made of).

The second step is based on Kant's objection to Leibniz's attempt at 'fixing' Descartes's proof (as reconstructed by Bencivenga, 1987). Leibniz argued that, since the idea of God is obtained by joining perfections, and perfections are positive, and hence involve no negation, there can arise no contradiction in the process (he did not consider the sort of category mistake that occurs in some of the examples above – and we too can leave those aside). And Kant objected that Leibniz could not say that there was no negation (and hence no contradiction) involved. All he could say is that, *as far as he knew*, there was no negation involved. By continuing the analysis of those 'positives' further, one might discover previously hidden negations, and maybe even establish that the idea of God was an absurd one after all. *Real* possibility (that is, something that is more than an *appearance* of possibility) can never be proved at the purely conceptual level – where only the temporary, revisable notion of *logical* possibility (possibility as far as we can tell) is accessible. Which means that, by mobilizing concepts (or any other purely logical tools) one never knows that an expression *really can* refer to objects.

How, then, could one know that? In a letter to Hilbert (also quoted by Bencivenga, 1987, p. 224), Frege insists that the only way of establishing the consistency, and hence the possible instantiation, of a set of conceptual specifications, is by giving an *actual* model of them. Which brings us back to the simple answer above. The detour through possibility proved delusive and the actualist construal is in fact the only one available to the realist:[4] the only expressions of which he can say that they are general terms are those that *do* refer to objects.[5] Whence the traditional 'strategy' of limiting the square of opposition to general terms with existential import.

Two remarks are in order, before closing. First, the realist's logic is awkward and cumbersome.[6] Both with singular and with general terms, empirical conditions must be satisfied before inferential patterns can be applied, and the only alternative (one that has been pursued regularly with singular terms but apparently not so often with general ones – though see later) is that of eliminating such expressions altogether and reducing oneself to using other expressions (say, individual variables) that cannot fail to hook up with the world. The idealist's logic, on the other hand, is much smoother. For him, general terms, too, can be defined by their grammatical roles, and even if this does not exclude thoughtful windowsills and the like, nor does it create any problem: it will just be a matter, here as with singular terms, of writing the ruling out of such monstrosities into (some of) the inferential patterns themselves. But all of this was to be expected. Language and the world are distinct, and hence there is no reason to think that the structure of the one should fit the structure of the other. If the structure of the world is what comes first and matters most, as is the case for the realist, it is no wonder that the structure of language will have to suffer, and possibly go through some regimentation: after all, one is adapting something to something else. The idealist will encounter his problems in another context: when out of language he tries to reconstruct the world.

Second, if the traditional 'existentially loaded' treatment of general terms is the most adequate to transcendental realism, why does classical quantification theory proceed otherwise? Doesn't this logic issue, according to my analysis, in an uneasy compromise between realism at the level of singular terms and idealism at the level of general terms? One could argue that this is indeed the case: that classical quantification theory is the outcome of a half-baked conceptual revolution that is only completed by moving to free logic. It is also possible to claim, however, that classical quantification theory has simply given up on general terms, and replaced them with predicates – that is, *syn*categorematic expressions with no independent meaning.[7] A consequence of this line of thought – and one that receives some confirmation by the work of Bressan, Gupta, and others - is that, if general terms are to get back into the 'classical' picture, it will have to be with heavy existential commitments.

Department of Philosophy
University of California, Irvine

NOTES

[1] This note has profited greatly from discussions with Karel Lambert.

[2] As couched in the now fashionable linguistic jargon.

[3] Among the tactics deployed to dispel this appearance, Russell's theory of descriptions is by far the most popular.

[4] I find it suggestive to quote here the parting shot by Almog (1991), one of the most consistent supporters of (what I would call) transcendental realism in the philosophy of logic: "With possibilities, less is more" (622).

[5] It is irrelevant to our present purposes that there be *several* objects involved here. If a term has at least one instance, the worst that could happen to it is that it does not have more. But that would not make it less of a term: it would only make it, possibly to our surprise, a *singular* one. In any case, there would be a clear distinction between it and nonsense.

[6] To avoid unnecessary and misguided objections, let me point out that by 'the realist's logic' I understand whatever complex tools he uses to analyze and evaluate ordinary language statements and arguments. Such complex tools *include*, but *are not limited to*, some formal systems, and of course the formal systems themselves are often simple, elegant structures. It is quite a different story when it comes to applying them – and justifying the claim that they are formal systems *of logic*.

[7] The two reactions are not unrelated, since syncategorematic expressions are defined by their roles in language. An *extreme* realist position would probably find an objectual counterpart for *all* linguistic expressions, and categorize expressions on such grounds. For an example of this extreme position, see Montague (1974).

REFERENCES

Almog, J.: 1991, 'The Plenitude of Structures and Scarcity of Possibilities', *Journal of Philosophy*, **88**, 620–622.

Bencivenga, E.: 1987, *Kant's Copernican Revolution.* New York, Oxford University Press.

Bencivenga, E.: 1990, 'Free From What?', *Erkenntnis*, **33**, 9–21.

Lambert, K.: 1967, 'Free Logic and the Concept of Existence', *Notre Dame Journal of Formal Logic*, **8**, 133–144.

Montague, R.: 1974, *Formal Philosophy*, edited by R.H. Thomason. New Haven, Yale University Press.

W.E. HERFEL AND C.A. HOOKER

FROM FORMAL MACHINE TO SOCIAL COLONY: TOWARD A COMPLEX DYNAMICAL PHILOSOPHY OF SCIENCE

1. INTRODUCTION

We are naturalists aiming to ground philosophical notions in real dynamical processes. From our perspective science is better modeled as a dynamical system than as the formal logical (inductive and deductive) machine found in the work of both the rationalists and empiricists (Hooker 1991,1995). We seek a model of science whereby accepted theory, practice and phenomena develop in mutual dynamical interaction; these, in turn, interacting with institutional organisations as well as our normative models of them. Only a dynamical conception of norms, we claim, is able to properly encompass the fundamentally social and historical nature of science while acknowledging the role played by the psychological capacities of individuals, all of which are sundered from it in the usual static formal models. In pursuit of this conception we have elsewhere examined a range of traditional positions to reveal their tacit dynamical implications or implicit underlying models (Herfel and Hooker 1996). The most promising is Kuhn's (1962) explicitly dynamical account, which is analogous to a re-organisational model of science with constant global ordering rules (normal science) interrupted by sharp, self-organised and disruptive phase transitions in which the ordering rules dissolve and reconstitute themselves in a different global pattern (revolutions). Though useful, the 'phase transition' model of scientific revolution is crude, because: (1) Historical changes in science are much more complex (Hooker 1995; Herfel 1990), (2) it is difficult to know how to adapt the physical model to the sociological situation (especially since the physical situation itself is so poorly understood), and (3) phase transition is just one type of phenomenon available within nonlinear dynamical systems. Already in such innocuous inorganic models as the Bénard system we see rich dynamical features and one of us has argued (Herfel 1996) that such nonlinear dynamics are also typical of science, while the other has argued (Hooker 1995) that the science-technology system is a non-linear dynamic system sharing many of the distinctive characteristics of living systems, which centrally display these same features. Before showing how the unique properties of nonlinear dynamic systems prove pertinent to dynamical accounts of science (Section 3), we explore just one such model (for want of space).

M.L. Dalla Chiara et al. (eds.), Language, Quantum, Music, 7–18.

2. INVESTIGATING MODELS FOR SCIENCE: SOME LESSONS FROM THE SOCIAL AMOEBAE

Living creatures are highly interactive non-equilibrium systems stabilised by flows of energy through them, and so therefore are communities of them. A well documented organic process in which an energetic constraint enables novel behaviour is aggregation and differentiation of the cellular slime mould, *Dictyostelium discoideum*. Cellular slime moulds can exist both as uni-cellular, and as multi-cellular, organisms. When there is plenty of food (they live off bacteria), they exist as single celled amoebae. When starved, however, individual cells do not die; rather, they respond to this constraint (Nicolis 1989, 326) – a reduction in energy throughput – by organizing in circular wave patterns around, and streaming toward, a single organization centre. They then form a compact aggregate which grows upward, secretes a slime sheath, tumbles over and becomes a slug-like creature which crawls away, following environmental cues, toward a more favourable location where it again transforms itself, growing a fruiting body supported by a stem to drop spore cells producing new amoebae. This latter phase is marked by cytological, not only functional, cell differentiation.

Here we consider just the onset of the aggregation process. Aggregation requires a coordination of movement, in turn requiring a communication medium, now generally agreed to be cyclic adenosine monophosphate (c-AMP), which is known to increase in cells under the stress of glucose deficiency (Harold 1986, 478). The constraint to starvation enables the formation of aggregation by stimulating production of an excess of available c-AMP. The pulsatile secretion of the substance establishes a signal which the cells follow in order to meet at the point of aggregation. From an initially homogeneous amoebic distribution an organising centre forms, and several models now exist which attempt to understand the dynamics of this process (Bazin and Saunders 1977; Garfinkel 1987; Gerisch 1986; Goldbeter and Martiel 1987; Keller 1985).

Characteristically, structure emerges from the imposition of a symmetry breaking constraint on a system of non-linearly interacting components (here amoebae), which lose some freedom of motion in the process. In a nutrient-rich environment, each individual amoeba is an organized molecular system – hence one to which many complex constraints already apply – stabilized far from equilibrium by its food intake. In this condition the system is in a dynamical stability with its environment, and symmetric with respect to inter-amoebic interactions. Starvation represents a constrained departure from that dynamic stability, one which disables independence of individual movement but enables collective behaviour which would otherwise be inaccessible and which may ultimately by-pass the starvation constraint (Herfel 1997). This represents a higher degree of organization in the following sense: it allows the system to more effectively seek out the energy gradients that maintain its existence, a feature unique to biological organization (Wicken 1987), and a deep example of what Pattee (1976) calls coordination of constraints. By increasing its global constraints on individuals it has freed up their capacity to cooperate and so provided the dynamical basis for a capacity either to feed independently or to aggregate, depending on environmental conditions. This simple adaptability enhances their capacity to act fittingly in a wider range of environments. (Here adaptability is expressed at population level, but often it is expressed at individual level through behavioural adaptability.)

Alan Garfinkel (1987) discusses various models of the dynamics of *D. discoideum* aggregation, and proposes a framework for adapting the results of these models to understanding human socio-dynamics. He identifies two types of models, the field model of Keller and Segel (1970) and various individual cell models, e.g. Goldbeter and Segel (1977). He then goes on to apply the lessons learned from modelling the social amoeba to modelling the social dynamics of human societies. We will briefly explore this account. Two problems with the Keller/Segel field model, both artifacts of linear stability analysis, are identified by Garfinkel (1987, 187): (1) The model cannot say what will happen after the onset of instability, (2) it cannot accurately predict the stabilities of real systems subject to finite perturbations (because it is confined to vanishingly small perturbations). Later, Garfinkel (1987, 203) points out that these difficulties can be overcome by employing nonlinear models of the aggregation process. (Goldbeter and Martiel 1987 provide one such model.) This strategy is the one adopted in the complex models of adaptive self-organisation that inspire our analysis. Even so, being continuous, field models omit the discreteness of the amoebae system, especially its local signal timing interrelations.

Garfinkel (1987, 190–200) then goes on to identify several aspects of aggregation left out of the individual cell model. These include: accurate details of wave dynamics, explanation of the emergence of ('autonomous') aggregation "centres", entrainment in the aggregation field and the various macro morphologies exhibited under varying aggregation conditions. In theory, one could construct more powerful models based on the dynamics of individual cells; in practice, however there are good reasons to believe that an accurate representation of these phenomena will remain elusive to the individualistic approach. After sketching the general features that such an individualistic super-model would possess, Garfinkel (1987, 201) remarks of it: "The model would be, mathematically, completely intractable, and nothing can be said about it. The complexity involved is at least that of the general problem of the entrainment of N oscillators, on top of the problem of pattern in reaction-diffusion systems added to the N-body motion problem. Any one of these problems is insuperable in general form; simultaneously they render the analytic situation hopeless." After showing that employing some of the standard simplifying assumptions largely used in overcoming such difficulties would amount to throwing the baby out with the bathwater (Garfinkel 1987, 201), Garfinkel (1987, 203) concludes that "... we are forced to adopt a field-theoretic, rather than a particle-theoretic, model of the self-organization process."

We note, however, that there are now many seminal models of non-linear dynamics based on the generation of complex global dynamics from iterated interactions among elements obeying just a few global constraints and local interaction rules. (See work by those in the Santa Fe Institute, e.g. Arthur 1994; Forrest 1991, and references therein.) We do not know whether Garfinkel considers such models 'individualistic' (since he doesn't discuss them in his work with which we are familiar), but if so, we part company with his dismissal of individual-based models. We suspect that a grand unified theory of non-linear dynamics is not going to be easily forthcoming – possibly not at all – for exactly the kinds of reasons Garfinkel gives for the insolubility of the N-oscillator problem. But whether an individual-based approach or a global field model will be most useful will most likely depend upon context. Garfinkel should agree with this observation, given his remarks concerning 'explanatory relativity' in his (1981). Garfinkel outlines a framework for applying his conclusions from the modelling of slime mould aggregation to social structure: explaining cooperation in the prisoner's dilemma and emergence in economics. We cannot explore these specific issues here, but move to the key themes that he considers to have emerged from the self-organization paradigm, namely holism and symmetry-breaking, which assert the essential dynamical roles of global system variables, constraint and coherence conditions and attractors, and catastrophe and bifurcation, which "give a richer scientific content to the notion of emergence and make possible models of physical, biological, and social processes in which periods of continuous change are punctuated by episodes of emergence" (Garfinkel 1987, 210). And, following Garfinkel here, we would add the concepts that seem essential in discussing aggregation in the slime mould, especially the notions of oscillation and entrainment. Entrainment, in the form of interactive coherence among actions, seems to be a crucial concept in understanding the evolution of social, as well as biological, organisation; aggregation and cooperation. "In the future the N-torus may be as familiar a model for biological or social space as Euclidean space is now, with pattern formation expressed in terms of various forms of oscillatory entrainment or coherence" (Garfinkel 1987, 210).

3. A FRAMEWORK FOR MODELING SCIENCE AS A SYSTEM OF NON-LINEAR INTERACTIONS

We use the themes suggested by Garfinkel's work, and that of others (e.g. those in the Santa Fe Institute, see above) to develop a cluster of concepts apt to applying nonlinear dynamics to social systems, with an eye toward treating the socio-cognitive structure of science. For each, we first present the general dynamical property and then briefly discuss its application to science.

1. Local non-linear interactions + non-equilibrium constraints yield global complexity. In the slime mould case complex and highly organised global dynamics are generated by non-linear interactions among local elements (amoebae), combined with the application of local and global constraints. The interactions are expressed in terms of one or a few well defined local interaction rules, here rules for generation of c-AMP and streaming, which gives rise to the necessary macroscopic property of co-ordination. The constraints are both local and global, and this is necessary as far as we can see, and while many of them may contribute to establishing the initial stable state, at least one must constrain the system away from that equilibrium. By contrast, the erstwhile dominant approach to modelling science has been to reduce interactions to a few simple logical rules, Scientific Method, the same for everyone, which would in principle treat all scientists identically as but local ciphers for a single global state of science. This analytic ideal is now increasingly rejected, and here we see that it is also deeply inadequate to the dynamical character of science, resulting in a false set of questions and problems. Instead we can quite properly expect the interaction rules to vary from context to context, within and without epistemic institutions, and over time as scientists learn about improving methods as they improve data and theories. (See Herfel 1996; 1997; Hooker 1991; 1995 for some elaboration.) Science actually comprises a system of individual scientists interacting non-linearly (e.g. changing each others beliefs and behaviours), subject to a set of institutional role constraints (e.g. laboratory procedure, journal reporting and refereeing conventions) and supported by, i.e. constrained away from equilibrium by, a flow of resources (energy, goods, services, money) through it. In this it is like any other social activity, a (far) more complex version of our social amoebae. But to the degree they apply, its particular institutional constraints differentiate it from other social activities and are what give it the design of an epistemic institution (e.g. Hooker 1995, Latour 1986, and references). It is these interactions which, in combination with institutional roles, generate macro-scale properties like research tradition, disciplinary grouping, cultures of criticalness and the like which are essential for science to proceed.

2. *Amplification.* Nonlinear dynamic systems can exhibit 'initial conditions sensitivity': the effect of very small events (variations, fluctuations) can become rapidly amplified to more global scales and fixed there by the energy flows in the system, so that their ordered structures thereafter play essential roles in the system's dynamics. In our example, small local timing coordinations in c-AMP emission were amplified to macro scale. Oscillatory entrainment is one of the prominent ways in which small correlations may become amplified to more global scale and fixed there. Science too shows 'initial conditions sensitivity': the details of the macrostructure of science can be sensitively dependent on activity changes by individual scientists (micro-variations) and thereafter play an essential role in the scientific system's dynamics. Watson and Crick's discovery of DNA structure (Watson 1968), e.g. applied a developing experimental technique (X-ray crystallography) to a new domain and its results were

amplified (using the resource flows) to widespread (macro-scale) acceptance producing a new coordination (entrainment?) of subsequent activity. Just this phenomenon of micro to macro-amplification lies behind all scientist's hopes of making a recognised discovery, and the 'damping out' of one's new invention because it does not stand scrutiny (or worse, because it is drowned by the sheer volume of 'noise' in the system) is at the root of their fears.

3. Self-organisation, symmetry breaking and resource flows. Our example system shows the spontaneous emergence of structure through symmetry breaking as it is constrained away from an equilibrium by alteration of energy flow through it. A dynamical symmetry in some property *P* means that the system dynamics is invariant over variations in *P*, here amoebic interaction, random movements and metabolic state; hence the system cannot constrain *P* changes – they are spontaneous for the system – and they dissipate to equilibrium with the system-environment. The self-organisation process is distinguished by the production of new macroscopic order from that in some initial small *P* variation through amplification and so represents a symmetry breaking (Collier 1996, Stewart and Golubitsky 1992). One pervasive form of this process is that of variation, selection and retention (*VSR*), where the variation is some spontaneous (i.e. system uncontrolled) symmetry-breaking production of variants amplified to macro-scale, the selection discriminates one of them and retention summarises the result of its amplification, e.g. in evolution. Similarly, science has had increasingly large resource flows directed into, and through, it over the past 400 years, and corresponding to this has been the emergence of ever wider and more differentiated institutional and cognitive structures. Each emergence of a new discipline, sub-discipline, tradition or research grouping creates a macro scale distinction where there was none originally and so breaks that previous socio-epistemic homogeneity (symmetry). Again, one pervasive form of this self-organisation is *VSR*, developed as evolutionary epistemology (Hahlweg and Hooker 1989).

4. Bifurcation. Global transformation of dynamical form resulting from broken symmetries is also characteristic of self-organisational capacities. In the self-organisation of new global order the system does not simply utilise its existing macroscopic dynamical laws to reorganise itself but transforms them through the introduction of new kinds of internal constraints. Our amoebae lose some individual freedom to move independently while the system gains organisational freedoms, to collectively move about and even fruit, it did not formerly possess. The symmetry-breaking amplification to macro-acceptance of a new scientific concept or procedure, bringing new macro-coordination constraints, establishes a new dynamics to science, a bifurcation, in which scientific experiments, methods, criteria, and communication and/or alliance patterns change somewhat thereafter in ways that are not just shallow logical re-organisation to accommodate new but compatible information. The very large scale and organisationally deep changes we call scientific revolutions, but these can only be distinguished by their dynamical effects from other complex but shallower changes occurring all the time (Hooker 1995).

5. Varieties of constraints and adaptiveness. Tighter constraint does not necessarily impoverish behaviour but, as we have just illustrated, may instead enable the emergence of new forms of previously inaccessible behaviour. A pedestrian example is provided by organisms: adopting phosphate as an energy source puts severe constraints on what can be eaten, but it is central to enabling all the distinctive attributes of living systems. Constraints can be local or global, deterministic or stochastic, static or time-varying and, most importantly, enabling as well as disabling. (Herfel 1996 argues that nonlinear dynamics are best thought of as the result of progressive application of constraints with full blown determinism being recovered by constraint to a unique allowed path.) It is through the coordination of constraints to form enabling constraints (see Herfel 1997 for an extended discussion of enabling constraints) that living systems have developed both specific adaptations and adaptability. In science it is precisely the institutionalised procedural constraints (e.g. experimental technique or mathematical modelling) that enable scientists to explore the natural world so successfully. Though they restrict individual degrees of freedom, it is these institutional constraints which are essential in enabling scientists collectively to pursue science without it falling into an incoherent babble of differing, or corrupt, voices, despite the fact that individual scientists know in any detail only a tiny fraction of their own discipline, let alone science as a whole. Science is a highly adaptive system because of its institutionalised sensitivity to environmental information (results, data) and commitment to incorporating that information into its future activity modes (Hahlweg and Hooker 1989). It is through the co-ordination of its institutionalised constraints to form enabling constraints that it expresses its adaptiveness.

6. Non-additivity and emergence. Solutions to mathematical models of nonlinear dynamical systems are non-additive, and for this reason their dynamics is non-decomposable (i.e. not derivable from local models of the individual constituents themselves), leading to genuine, because dynamically grounded, emergence. While two-body gravitational systems can be shown analytically to be stable and periodic, even three-body systems are capable of dynamical chaos, this latter possibility emerging by simply adding one more body to the system, with no change in fundamental physical law. We expect that all examples of emergent phenomena, for instance life and intelligence, can be dynamically explained in specific similar ways. Science dynamics too are non-decomposable, showing the genuine emergence of new macro-scale features not derivable solely from local models of individual scientists – see e.g. the results of the interaction of the hitherto separated bio-chemical and X-ray crystallography fields in the DNA case above. A similar situation obtained between scientific and practical technological investigation of flight, the latter making deep differences to the former, and reminding us that we are really discussing the science-technology system throughout.

7. Irreversibility, path-dependence and historicity. A non-linear dynamic system is characteristically macroscopically irreversible. Although the structure of its dynamics is determined by micro-level events, information about the exact nature of these events is lost at the macro-scale as it evolves through self-organisation and is not recoverable by running a macroscopic model of it in reverse. Such systems become 'locked-in' to the particular path that was initiated by a succession of amplified small events; its future behaviour thus path-dependent and constrained by its historical development. In such cases we can only understand why the system has come to be in the state or attractor that it is in by studying it historically. (See Arthur 1994 and Stanford colleagues for an application of this insight in economics.) Although science's dynamics is determined by local actions of individual scientists (micro-level events), information about the exact nature of most of these actions is lost at the macro-scale (e.g. in journal papers) and is not recoverable by running a macroscopic model of it in reverse. Science is also irreversible because its epistemic character leads it to being both accumulative and revolutionary, again the details are typically sensitively dependent on activity changes by individual scientists often lost at the public macro-scale. Once macroscopic structure emerges the system is typically 'locked-in' to using it thereafter as a base for new work (cf. Wimsatt and Shank 1988 and, informally, Latour 1987). In consequence, the science (-technology) system is path-dependent and thus typically constrained by its historical development and it must be studied historically to understand it.

8. Constraint duality and super-system formation. The dynamical structures that emerge in non-linear macroscopic bifurcation express constraint duality: emergent global structure constrains the behaviours of individual elements as genuinely as their interactive dynamics constrained its formation and continues to sustain it. Top-down constraint occurs simultaneously with bottom-up constraint. While a population is 'nothing but' a collection of organisms, an organism may develop in a variety of ways, depending on the physical and social environmental constraints, even while it in turn influences other organisms and alters the physical environment. Each dynamically grounded top-down constraint gives added force to the formation of a genuine super-system from the collection of interacting system elements; the amoebae, e.g. increasingly clearly form a super-system as they first aggregate and move coherently and then differentiate. The structures that emerge in scientific macroscopic bifurcation similarly represent genuine global constraints on the local behaviours of scientists; whose roles thereby lose one or more of their degrees of freedom (Herfel 1997). This expresses scientific constraint duality: the various more global structures of science then constrain the behaviours of individual scientists as genuinely as their interactive dynamics constrained its formation and continues to sustain it. Again, top-down constraint occurs simultaneously with bottom-up constraint. Each dynamically grounded top-down constraint gives added force to the formation of a genuine science super-system from the collection of interacting scientists.

To these we add two further features which concern shifts in the fundamental ways we are able to understand these systems. Hitherto we have derived our paradigms for understanding from simple systems, e.g. a single Newtonian particle, for which a complete universal global model exists. Thus completeness (all features simultaneously, analytically represented) and universality (applies in all circumstances, e.g. across all constraints, within the theory's domain of application, ideally everywhere) have formed our ideal criteria for understanding. But these ideals seem unachievable in principle for non-linear dynamic systems.

9. Model specificity/model plurality. Given its foregoing features, we do not expect a unified theory of nonlinear dynamics to emerge. The characteristic behaviours of nonlinear dynamical systems all derive from systems configuration, parameter values, and boundary conditions, not solely from the operation of its interaction laws. Hence these models must all be specific to the conditions of the phenomena being modelled, and, given dynamical bifurcation between conditions, a plurality of differing, though variously dynamically interrelated, models will generally be required to understand all the complex phenomena exhibited by systems. Consequently, we cannot expect there to be a grand unified theory of science of the simple deductivist kind idealised by formal logical models. The characteristic behaviours of science derive, not from psychological peculiarities of scientists as humans or of their kinds of interactions, all of which apply throughout social life, but from institutional role designs and their supporting norms. Hence these models must all be specific to the material and social conditions of the phenomena being modelled, and a plurality of context-sensitive models will generally be required to understand all the complex phenomena exhibited by science – though the complex global coherency requirements peculiar to science will generate a corresponding set of inferrelationships among these models. (Hooker 1999 employs ecological-genome interrelations as a model for this sort of unity.)

10. Model centredness. Typically, non-linear dynamical systems have no analytic solutions, or even analytically constructible descriptions of their pertinent dynamical behaviours, or at least none we currently know how to construct. Given the kinds of problems to overcoming this Garfinkel cites (see his last quote above), this circumstance is likely unavoidable in principle. Wherever analytic solutions or characterisations fail we have no recourse but to iterative numerical modelling to explore the dynamics. This represents a striking new feature to theoretical science, the ineliminability of concrete, specific modelling to even specify, let alone understand, the phenomena (Herfel 1995). We equally have no analytic models of the dynamics of science; certainly not over time because future knowledge is in its nature unpredictable, nor across science because the macro consequences of local interaction are not derivable from them alone. We must study science as we have always done, historically, iterating by following sequences of interactions as they spread non-linearly through the system, generalising where possible. Every scientific transition is to some extent unique and must be understood on its own terms – but from this nothing follows about the abandonment of rationality, only of simple logical models of rationality (Hooker 1991).

4. CONCLUSION: A MANIFESTO FOR AN INTEGRATION OF NORMATIVE AND DYNAMICAL MODELS

Traditionally, one way philosophical accounts have been distinguished is by the fact that they are aimed at uncovering the normative bases for human activities. In philosophy of science this has amounted to prescribing a methodology of scientific practice. Since our account is so steeped in modelling the dynamics of science one may be led to the conclusion that ours is a purely descriptive account. Certainly we believe that nonlinear dynamical models will assist in the creation of a more accurate and precise characterisation of the scientific enterprise. Nevertheless, ours is not merely a sociological exercise in the descriptive sense. First we should point out that the very notion of 'pure description' is dubious (cf. Kuhn 1962 on theory-ladenness). From our perspective, understanding, philosophic as well as scientific, is a modelling process. Modelling requires assumption. Assumption requires value judgment. Value judgment requires normativity. But our point cuts deeper than this. We are committed to the view that our norms emerge from practice and, conversely, regulate practice; so our understanding of norms must be grounded in a careful analysis of practice. Nonlinear dynamical models make intelligible processes in which norms are dynamically relevant, and show where, and how. Methods have constantly evolved in nonlinear interaction with theory and data, and just this is central to science (Hooker 1987, 1995). Methodological norms, e.g. laboratory procedures, are, and ought to be, based on empirical models of what is dynamically relevant and reliable. Hence there is neither a dichotomy between normativity and dynamics nor a reduction of one to the other. (Furthermore, we endorse reflexivity while maintaining the possibility of pluralism: one would expect that we can only learn about science through a careful modelling of the practice; however the modelling process itself must be informed by what is learned from these models.)

In this spirit one of us has proposed to replace the traditional formalist logical conception of reason with that of efficacious re- and self-organisation, applying to both individual scientists and to science (Hooker 1995). It is argued that this is a naturalist, dynamically grounded conception adequate to both the required normative role of reason and the dynamics of science. It can be fruitfully, if ultimately partially, modelled in terms of a decision theoretic approach to epistemology in which acceptances are practical actions made under a variety of utilities whose relative priorities properly differ across institutional contexts. Whether this will turn out to be the best approach is an open question.

We can imagine a range of possibilities for the normative bases upon which science has evolved, from crass economic gain at one extreme to the disinterested pursuit of truth at the other. Identifying the possibilities and attaching them empirically to portions of science, is only part of the task. In the long term we seek to evaluate the scientific process, with an eye toward improving it (to the extent that this is practicable). The norms that ought to be adopted are the ones that lead to intelligent and successful practice. So far there is no way to judge which norms will so lead without a clear understanding of their dynamical consequences. Within our framework at least this program is intelligible.

Department of Philosophy
University of Newcastle

REFERENCES

Arthur, W.B.: 1994, *Increasing Returns and Path Dependence in the Economy.* Ann Arbor, Michigan University Press.

Bazin, M.J. and P.T. Saunders.: 1977, 'An Application of Catastrophy Theory to the Study of a Switch in Dictyostelium Discoideum', in Thomas, R., ed., *Kinetic Logic: Lecture Notes in Biomathematics, no.29*. Berlin, Springer-Verlag.

Collier, J.D.: 1996, 'Information Originates in Symmetry-Breaking", *Symmetry: Culture and Science*, **7**, 247–56.

Forrest, S., ed.: 1991, *Emergent Computation.* Cambridge, MIT Press.

Garfinkel, A.: 1987, 'The Slime Mold Dictyostelium as a Model of Self-Organization in Social Systems', in Yates, F.E., ed., *Self-Organizing Systems.* New York, Plenum.

Garfinkel, A.: 1981, *Forms of Explanation.* New Haven, Yale University Press.

Gerisch, G.: 1986, 'Dictyostelium Discoideum', in Fougereau, M. and Stora, R., eds. *Aspects Cellulaires et Moleculaires de la Biologie du Developement.* Amsterdam, North Holland.

Goldbeter, A. and L. Segel.: 1977, 'Unified Mechanism for Relay and Oscillation of Cylic AMP in Dictyostelium Discoideum', *Proceedings of the National Academy of Science USA*, **74**, 1543.

Goldbeter, A. and J.L. Martiel.: 1987, 'Periodic Behaviour and Chaos in the Mechanism of Intercellular Communication Governing Aggregation of Dictyostelium Amoebae', in Degn, H., Holden, A.V. and Olsen, L.F., eds, *Chaos in Biological Systems.* New York, Plenum.

Hahlweg, K. and C.A. Hooker.: 1989, 'Evolutionary Epistemology and Philosophy of Science', in Hahlweg, K. and Hooker, C.A., eds, *Issues in Evolutionary Epistemology.* Albany, SUNY Press.

Harold, F.M.: 1986, *The Vital Force: A Study of Bioenergetics.* New York, Freeman.

Herfel, W.E.: 1990, *Coming Attractions: Chaos and Complexity in Scientific Models.* Doctoral Dissertation, Temple University.

Herfel, W.E.: 1995, 'Nonlinear Dynamical Models as Concrete Construction', in Herfel, W., Niiniluoto, I., Krajewski, W. and Wojcicki, R., eds., *Theories and Models in Scientific Processes.* Amsterdam, Editions Rodopi. Herfel, W.E.: 1996, 'On Cognitive and Social Dimensions of Science: Constructivism and Nonlinear Dynamics'. Forthcoming in Einstein meets Magritte Conference Proceedings, Orange Book.

Herfel, W.E.: 1997, 'How Social Constraints Enable Scientific Practice', in preparation.

Herfel W.E. and Hooker, C.A.: 1996, 'Cognitive Dynamics and the Development of Science', in Ginev, D. and Cohen, R.S., eds, *Issues and Images in the Philosophy of Science.* Dordrecht, Kluwer.

Hooker, C.A.: 1987, *A Realistic Theory of Science.* Albany, SUNY Press.

Hooker, C.A.: 1991, 'Between Formalism and Anarchism: A Reasonable Middle Way', in Munevar, G., ed., *Beyond Reason: Essays on the Philosophy of Paul Feyerabend.* Boston, Kluwer.

Hooker, C.A.: 1995, *Reason, Regulation and Realism.* Albany, SUNY Press.

Hooker, C.A.: 1999, 'Unity of Science', in Newton-Smith, W.H., ed., *A Companion to the Philosophy of Science.* Oxford, Blackwell.

Keller, E.F.: 1985, 'The Force of the Pacemaker Concept in the Theories of Aggregation in Cellular Slime Mold', in *Reflections on Gender and Science.* New Haven, Yale University Press.

Keller, E.F. and L.Segel.: 1970, 'Initiation of Slime Mold Aggregation Viewed as an Instability', *Journal of Theoretical Biology*, **26**, 399.

Kuhn, T.: 1962, *The Structure of Scientific Revolutions.* Chicago, University Press.

Latour, B.: 1987, *Science in Action.* Cambridge, Harvard University Press. Nicolis, G.: 1989, 'Physics of Far-From-Equilibrium Systems and Self-Organisation', in Davies, P., ed., *The New Physics.* Cambridge, UK, University Press.

Pattee, H.H.: 1976, 'Physical Theories of Biological Co-ordination', in Grene, M. and Mendelsohn, E., eds, *Topics in the Philosophy of Biology.* Dordrecht, Reidel.

Stewart I. and Golubitsky, M.: 1992, *Fearful Symmetry.* Oxford, Blackwell.

Wimsatt, W.C. and Schank, J.C.: 1988, 'Two Constraints on the Evolution of Complex Adaptations and the Means for their Avoidance', in Nitecki, M.H., ed., *Evolutionary Progress.* Chicago, University Press.

Watson, J.D.: 1968, *The Double Helix.* New York, Mentor.

Wicken, J.S.: 1987, *Evolution, Thermodynamics and Information.* Oxford: University Press.

OBSERVATION AND RELIABLE DETECTION[1]

Empiricists worry about the division between the observable and the unobservable because they want epistemic security. Direct perceptual knowledge of the observable might be defeasible, they reason, but it does not have the shaky status of phlogiston, the superego, noxious effluvia, and all the rest of those entities that turned out to be unobservable because they did not exist. I am an empirical realist, so I worry about the security of my data too. Hence our first step ought to be to take the security issue as primary and the observable versus unobservable distinction as merely one attempt to resolve that issue. That is, it is not the intrinsic status of something as 'observable' or 'unobservable' that should concern us, but how good are the detectors we use at producing secure knowledge. In that respect, the traditional human sensory modalities are neither particularly reliable nor all that versatile. We all know that eyewitness reports of who was present at a crime scene are unreliable, and that evidence in the form of 'unobservables', such as *DNA* profiles and blood types, is far more secure than the unaided observations of humans. I know that on a dark Warsaw street, my eyes are not reliable indicators of the street names on a Polish map, and although I can confidently tell the aural difference between a piano and a harpsichord, I cannot reliably discern a trumpet from a cornet. The superiority of instruments over human senses is, of course, widely recognized. Photo-finishes in horse races, electronic timing in Olympic swimming and sprints, touch sensors in fencing, automatic pilots, light meters for camera exposures, and so on are all reflections of this superiority. Readers of a certain age may remember the use of a 'personal equation' as a corrective in some scientific experiments such as scintillation counts before the era of automatic detectors. As a last example, many of the optical illusions that fool the human visual apparatus do not affect computerized image analyzers.[2]

Nor is it the size of something that determines whether it is observable or not. Male chicks are middle sized objects, and should be typical observables, but I am much more secure in my identification that this is a fly's eye under a microscope than I am in my unassisted identification that this is a male chick. Magnetic resonance imaging (*MRI*) provides one case where appeal to 'unobservables' can provide results that are superior to observation of medium sized objects with the unaided senses. Some years ago a brain tumour was detected in a patient with the aid of both *CAT* scans and *MRI*. Nevertheless, when the patient's skull was opened, no tumour was visible by optical

M.L. Dalla Chiara et al. (eds.), Language, Quantum, Music, 19–24.

inspection. A biopsy was then used to confirm the presence of the tumour and it was removed.[3]

So one has here a case of a macroscopic object that was unobservable with the naked eye yet observed with quite sophisticated instrumentation applied to 'unobservables'. Instruments are also capable of detecting aspects of unobservable entities that we do not antecedently believe they have, or even could have.We desperately want to believe that we do not have lung cancer, but there it is, the X-ray shows it. It is often claimed that the theory you believe can make you see something that is not there, or that it can prevent you from seeing something that is there. No doubt, but the power of this should not be exaggerated, and the examples that are often cited are not entirely convincing. For example, it is sometimes said that prior to William Harvey's discovery of the circulation of the blood, anatomists 'saw' holes in the septum dividing the heart because their Galenic textbooks told them they were there. But some cadavers *do* have holes in the septum – the famous holes-in-the-heart babies, for example – and it is not unreasonable to think that some students were presented with just that kind of cadaver.[4] We all know, although too infrequently remember, that the opposition between observable and theoretical entities is a category mistake, and that the proper division is between observable and unobservable entities on the one hand, and theoretical versus non-theoretical terms on the other. But what are these entities? Often talk is of observable and unobservable objects, but this is too casual. For what is observed or detected are instances of properties. Different parts of the electromagnetic spectrum give us knowledge about different properties possessed by galaxies, and different staining techniques in microscopy allow us to observe different aspects of a given specimen. One good example here comes from astrophysics. The galaxy *NGC 309* is usually classified as a classic spiral galaxy on the basis of its image in visible light, thus putting it in the same category as the Milky Way. But when viewed in the near infra-red at a wavelength of 2.1 μm, the disk only has two arms instead of three and the central disk is close to an ellipsoid. This would put it into the class of barred spirals, which are in some ways quite different from the classic spirals.[5] As another example, the galaxy *M83* when viewed by means of polarized radio continuum emission at a wavelength of 20 cm looks nothing like the spiral shape it has within the visible realm.[6]

If we accept that what is observed are properties (or their instances) then we can see that *any detection device, including the human sensory apparatus is a physical abstraction device that detects particular properties*[7], for it selects some feature(s) of the world for examination, and omits (most) others.

The kind of realism that is most easily defensible is what I call *property cluster realism* – some kind of entity is detected through instrumentation that detects a particular property, and then more and more properties of that entity are detected through different instrumental techniques. Using an analogy that I made some years ago, realism is rather like geographical discovery – first some country is discovered and next to nothing is known about it beyond its mere existence and local terrain; then more and more details are filled in as further exploration takes place. We can be wrong about some of those properties (just look at early maps of America, for example), but we are rarely wrong about the existence of the detected entity itself. (It is worth noting that the Aristotelian crystalline spheres were never detected in any way, they were merely postulated.)

Some instruments, such as the reflecting telescope, fit a simple pattern - some property of the target object initiates a causal process that is successively transformed by the instrument and the process eventually impinges on our sensory apparatus. However, these simple instruments, which as examples have tended to shape the philosophical debates, are unrepresentative of many contemporary detection devices. Many instruments now transform and analyze the data computationally before it reaches the observer, and this affects a claim made by Ian Hacking in his fascinating and important study of the role of microscopes in observation (Hacking [1981]) where he asserted that you do not need to know how microscopes work in order to use them.

The issue of whether we need to understand how microscopes or any other instruments work is part of the more general issue of whether the processes that result in observation have to be understood or whether we can take them at 'face value' as more or less veridical. Naturalists often adhere to the former, anti-naturalists to the latter. The issue has arisen before in the case of whether one needs to know how human perceptual mechanisms work in order to evaluate the veracity of the data that they give us, as opposed to simply taking the observational data as a 'given', and in the issue of whether one needs to know how telescopes work in order to, say, rebut the objections that were made to Galileo's telescopic observations.[8] Hacking here falls on the naturalist side: "To understand whether she was seeing, or whether one sees through the microscope, one needs to know quite a lot about the tools" (*op.cit.* p.307). The point here, then, is that the more we know about how some instrument works, the more justification we have about the degree of faithfulness with which the data represent the object observed. But contrast this with Hacking's immediately following remark, which is crucial for his overall conclusion that one can circumvent the dependence of observation on theory: "You do not need theory to use one" (*ibid* p.309). For the central argument of Hacking's article is designed to show that a direct comparison with known structure-preserving devices can circumvent the need to know how the instrument works. One could view this as a version of a reliabilist account, where a justification is not required, and instead direct evidence that the instrument works is appealed to. Thus, in epistemology, we have a parallel with the realist disputes of old. The instrumentalist does not feel the need to have any further justification for why his instruments work, whereas the realist does.

An important point here is that many instruments are deliberately designed to produce images that are available to the human eye. One tends to think of

Galileo's telescope as simply being made and then directly looked through, but merely getting the focal length right is important and has to be done deliberately. As with many instruments, one has to calibrate them on some standard to make sure that they are correctly representing that standard. (Recall the Hubble telescope fiasco, where the calibration instrument was itself maladjusted.) To take another simple example, when installing an internal/external thermometer, one has to know a little bit about how the thermometer works in order to obtain a reliable reading. Installing it on a north wall out of the sun (for accurate air temperatures) asking whether it will work horizontally as well as vertically (it will), realizing that the uninsulated mud room will still be warmer than the outside air, and so on are all important factors in setting up an accurate instrument.

In assessing the claim that "One needs theory to make a microscope. You do not need theory to use one." (Hacking [1981] p.309), one should be aware of how limited the scope of this claim is. What limits the applicability of this argument is its emphasis on bench microscopes, that is, instruments that have been refined to a degree that their veridicality is no longer an issue, and so the correctness of their output is moot. Hacking's claim is no doubt true for these off-the-shelf instruments. But the claim is false for many instruments that are in the process of development, and for which the issue of incorrect or misleading output is crucial. Moreover, such stock instruments are usually, although not exclusively, used to discover new instances of known properties (or, if you prefer, entities) rather than instances of previously undiscovered properties, which is where the focus of disputes about realism ought to be.

To make these points concrete, consider our earlier example of magnetic resonance imaging (*MRI*). An external magnetic field lines up the spins of the protons in hydrogen nuclei. A brief radio signal is used to distort that uniformity, and as the hydrogen atoms return to the uniform lineup, they emit characteristic frequencies of their own. Computers process this information to form an image, which represents various densities of hydrogen and its interaction with surrounding tissue. One important feature of *MRI*s is that rather than detecting water by means of its surface features, as we do in everyday life, the instrument uses part of the inner structure of water to detect its presence. This is a clear instance of detecting property instances, rather than the object itself. In fact, this enables one to avoid imaging certain entities even though they are in the 'direct line of vision' between the *MRI* scanner and the target.

An amusing story illustrates why you need to know how *MRI* devices work in order to use them effectively. The first *MRI* scanner that was put into use at Cornell Medical Center consistently produced seriously distorted images. After a number of attempts to explain what was going wrong, it was finally discovered that the Vatican Radio was broadcasting on the same frequency that the scanner was using and interfering with the operation of the device.[9] It is clear that, here, knowing how *MRI* devices work was important to using them effectively. Indeed knowing how instruments work is important mostly when they go wrong, or give unexpected outputs. (This is also true of machines, human bodies, and computer programs.) Consider the case of the discovery of the 3 degrees *K* background radiation by Penzias and Wilson in 1965, where the astronomers had to destroy all the other explanations they could think of that might have accounted for the unexpected observations. To do that takes a great deal of theoretical rather than practical knowledge.

The technique of quantitative electron microprobe compositional mapping provides another good illustration of these points. With these instruments, a spectrometer (either a wavelength or energy spectrometer) measures the count rate for each element within an element of a grid, compares it to the count rate for a pure sample of the element, and a matrix of intensities is constructed for the sample. There are two principal instrumental errors involved with compositional mapping – defocusing of the wavelength spectrometer and decollination of the energy spectrometer, and variation in the background radiation that depends on the composition of the material. Both of these correction techniques require significant understanding of how the spectrometer works. The important philosophical feature here is that we have explicit control over the parts of the observation process that are governed by theory. We know why wavelengths that are off the optical axis produce distortions in the intensity distribution and thus how to correct for it. Because that knowledge is explicit, it poses no threat to the security of the evidence that the output gives us – in fact it strengthens it by assuring us that one source of distortion has been removed.

Moreover, manipulation of the images provides additional information. If one takes compositional maps of the elements in a material, and represents them with primary colours, then superimposing those maps will immediately give a secondary colour image that represents the distribution of the joint presence of the elements. A process of perceptual abstraction can also be performed with algorithmic transformations of the data. The so-called blob splitting algorithm selects contiguous groups of pixels that are undifferentiated with respect to intensity. Then, regions that lack internal structure are picked out, highlighted, and the remaining areas suppressed.

We can conclude with a rather abstract suggestion. To be (directly) observable in the traditional sense is for some entity *E* to stand in a relation *R* to another entity (standardly a human) possessing a specified set of detection equipment *D* under some standard set of conditions *C*. Much effort was expended in this century, and it was by no means wasted effort, to determine what were the appropriate sets *D* and *C*, and what was the proper relation *R* that achieved the optimal reliability for *D*. Within the picture presented here, the issue is how to find, for a given type of entity or property *E*, a set of detection equipment D' standing in a relation R' to *E* such that under a specified set of circumstances C', the best available reliability is achieved. One will still, of course, be forced to take *D*, *C*, and *R* as the standard set in terms of which the output of D' is observed, but that is a much different issue and simply reflects the point already made that the output of D' is specifically tailored to be a traditional direct observable, such as a digital read-out, or a visual image, items for which the human detection equipment D is at least as reliable as any other.

Corcoran Department of Philosophy
University of Virginia

NOTES

[1] This is a much revised version of an informal talk given at the LMPS96 conference. The talk was illustrated with colour images that supplemented the oral argumentation in essential ways. Unfortunately, a combination of cost and copyright constraints prevented me from including those images in this paper. That is a pity, for part of the background to this paper, although it is not explicitly dwelt upon here, is that graphical output from experiments and simulations is now an essential component of much scientific activity.
[2] See Russ [1990], p.12.
[3] See Sochurek [1988], p.20 for this and other cases of the superiority of imaging over observation.
[4] This would explain, amongst other things, the claim made in Hall [1983], p.157, that: "Some anatomists, however, still maintained that the passages [in the septum] were easy to find in very young hearts, though concealed in the adult body".
[5] See Wolff and Yaeger [1993], p.20, figs 1.10c,d.
[6] *ibid*, p.21, fig. 1.10e.
[7] By which I mean 'instances of properties' but this is tiresome to repeat.
[8] See, for example, Galileo's polemical responses to Lothario Sarsi in The Assayer (Galileo [1623]).
[9] See Sochurek [1988], p.10.

REFERENCES

Galilei, G.: 1623, *The Assayer* in *Discoveries and Opinions of Galileo*. Translation, introduction, and notes by Stiliman Drake. Garden City, NY, Masterworks Program.

Hacking, I.: 1981, 'Do We See Through a Microscope?', *Pacific Philosophical Quarterly*, **62**, 305–322.

Hall, A.R.: 1983, *The Revolution in Science: 1500–1750*. London, Longinans.

Russ, J. C.: 1990, *Computer Assisted Microscopy*. New York, Plenum Press.

Sochurek, H.: 1988, *Medicine's New Vision*. Eastam (Pa), Mack Publishing Company.

Wolff, R. and Yaeger, L.: 1993, *Visualization of Natural Phenomena*. New York, Telos, Springer-Verlag.

ALAN MUSGRAVE

CONCEPTUAL IDEALISM AND STOVE'S GEM

Let me begin with a true story. I once gave a talk at the Sociological Institute at Amsterdam University. A Dutch sociologist, Gerard de Vries, was deputed to comment on my talk. He said my picture of different theories about just one world was utterly naive: different theories, different worlds. He gave an example: "When the concept 'person with an IQ two standard deviations above the mean' was invented *new entities* came into being". So, I replied, there are two ways of making babies, the way we all know and love, namely love, and this new way, psychological theorising. I was then told that the new entities were not babies, were not even persons with an IQ two standard deviations above the mean. I could form no clear idea of what kind of thing they were. My ex-friend Gerard de Vries was, of course, a conceptual idealist.

This paper is a tribute to David Stove, who died last year. In 'Idealism: a Victorian Horror-story' (contained in his 1991), Stove shows that since the seventeenth century most important Western philosophers have been idealists of one kind or another. Stove speculates that idealist philosophy is a kind of religion substitute:

"... stemming from the very same emotion as religion stems from, there is a kind of belief which is more irrational, in fact very much more irrational, than ordinary religious beliefs are. I mean what in philosophy is called idealism ... *All* idealists are engaged, above everything else, in satisfying the religious demand for the universe to be reassuring or consoling, or at the very least, *kindred* [or mind-like]" [p. 87]

Stove shows that down the ages one argument or cluster of arguments has been the chief intellectual prop of philosophic idealisms. Stove calls the argument 'The Gem'. The first Gem is to be found in Berkeley, who argues that we cannot even think or conceive of objects which exist "without [or outside] the mind":

"To make this out, it is necessary that you conceive them existing unconceived or unthought of; which is a manifest repugnancy. When we do our utmost to conceive the existence of external bodies, we are all the while only contemplating our own ideas. But the mind, taking no notice of itself, is deluded to think it can and does conceive of bodies existing unthought of, or without the mind, though at the same time they are apprehended by, or exist in, itself. A little attention will discover to anyone the truth and evidence of what is here said, and make it unnecessary to insist on any other proofs against the existence of *material substance*." [*Principles of Human Knowledge*, paragraph 23.]

What is going on here? Berkeley's conclusion is that we cannot "conceive of

M.L. Dalla Chiara et al. (eds.), Language, Quantum, Music, 25–35.

bodies existing unthought of, or without the mind" – for example, we cannot have trees-without-the-mind in mind. The only premise he produces to demonstrate this is that when we try to conceive of external objects "at the same time they are apprehended by [the mind]" – for example, we cannot have trees-without-the-mind in mind, without having them in mind. So, the argument proceeds from the premise that you cannot conceive of external objects without conceiving of them, to the conclusion that you cannot conceive of external objects. Or in Stove's homely version:

"You cannot have trees-without-the-mind in mind, without having them in mind. Therefore, you cannot have trees-without-the-mind in mind." [1991: 139–40]

The premise of this argument is a tautology. The conclusion is not a tautology – far from it! The argument is therefore invalid: you cannot validly obtain a non-tautological conclusion from a tautological premise. Yet Stove documents the fact that subsequent idealist views depended solely upon versions of The Gem!

For example, Kant rested his philosophy entirely upon a Gem. Observing that we cannot think of things without bringing them under the categories of our thought, Kant concluded that we cannot think of things-as-they-are-in-themselves. What, then, can we think of? Why, of things-as-thought-of-by-us, of course. So Kant distinguished things-in-themselves (*Ding-an-sich*) from 'phenomenal things', things-as-perceived-by-humans or things-as-conceived-by-humans or things-as-thought-about-by-humans. And he distinguished the noumenal world, which contains the things-in-themselves, from the phenomenal world, which is the only one we know about.

Now as the case of Kant illustrates, to accept the conclusion of The Gem is not yet to be an idealist. It is merely to accept the epistemic thesis that we cannot think about, let alone know, things-as-they-are-in-themselves. Kant insisted that he was not an idealist and even produced several 'refutations of idealism', as he hopefully called them. However, once you accept the conclusion of The Gem pressure mounts to go for fully-fledged idealism, for two reasons.

First, The Gem leaves us labouring under an inescapable cognitive disadvantage. The things we can think about or know about are not the real things, the things-as-they-are-in-themselves. Our thinking or knowing about a thing somehow contaminates or pollutes it, and turns it into something else, a thing-as-thought-about-by-us. It is as if we are cognitive sinners: having eaten of the fruits of the Tree of Knowledge, we cannot really know things-as-they-are-in-themselves. Obviously, doing away with the noumenal world and going for idealism escapes all this. Idealism removes any suggestion that we labour under an inescapable cognitive disadvantage. And it makes the world and the things in it kindred to us, mind-like to the extent that they are partly constituted or created by our concepts.

The second reason why idealism attracts, once you have accepted The Gem, is that noumenal things, things-in-themselves, are very *odd* things. For a start, their postulation is, or verges upon, inconsistency: we can know nothing of them and we can know that they exist. Again, Kant's things-in-themselves possess none of the properties of phenomenal things: no colours, smells, tastes, shapes or sizes, motions or weights. Kant, like Berkeley before him, obliterates the distinction between primary and secondary properties and makes all apparent properties secondary and subjective. But noumenal things are queerer still. They do not exist in space and time, which are 'forms of sensibility' which house only phenomenal things. Nor do noumenal things help cause our experiences, causality being a 'category of the understanding' which applies only to phenomenal things. Kant says noumenal things 'ground' our experiences. But this 'grounding' is wholly mysterious. So things-in-themselves are nowhere, at no time, and do nothing. Besides, as Schopenhauer was to protest, Kant should not have assumed that mind-independent reality consists of different *things*-in-themselves. And Kantian talk of the Moon-in-itself, for example, is even worse since it already imposes our Moon-concept. No, for all we know the noumenal world may just be a single, undifferentiated blob of pure being. Now again, all of these difficulties disappear if we abridge the Kantian metaphysics, do away with the noumenal world and its things-in-themselves, and opt for idealism.

So far I have considered just two historically important Gems. But there is an enormous number of different Gems. Their general form is:

> You cannot *X* things unless *C*, a necessary condition for *X*-ing things, is met. Therefore, you cannot *X* things-as-they-are-in-themselves.

You get a Gem by substituting for *X* and *C*. Substitutes for *X* include 'perceive', 'have in mind', 'think of', 'conceive of', 'talk of', 'refer to', 'theorise about', 'interact with', and so on. Once the substitute for *X* is known, the substitute for *C* usually becomes obvious. For example, if the substitute for *X* is 'talk of' or 'refer to', the substitute for *C* is 'using words'.

There is an infallible sign or symptom that a Gem and a form of conceptual idealism is nigh – what I call a '*hyphenated entity*'. Kant's hyphenated entities were things-as-they-are-in-themselves as opposed to things-as-conceived-by-humans. Here is an incomplete list of some others:

things-as-perceived-by-us
things-as-thought-about-by-us
things-as-known-by-us
things -as-talked-about-by-us
things-as-referred-to-by-us (or referred-to-things)
things-as-related-to-by-us (or related-to-things)
things-as-interacted-with-by-us (or interacted-with-things)

Usually, once you have the hyphenated entity you can reconstruct the form of conceptual idealism being proposed and the version of The Gem which lies behind it. For example, consider interacted-with-things. The Gem here is "We cannot know things without interacting with them. Therefore, we cannot know things-as-they-are-in-themselves". Once we are persuaded by this argument, we have to say what we *can* know. And the answer is obvious – interacted-with-things, of course. Hyphenated entities are what conceptual idealists need to express their philosophy. As before, as with Kant, we have not yet reached conceptual idealism – things not interacted with by us might still exist. But as before, there is irresistible pressure to abridge our metaphysics by doing away with them.

Most of what I have said so far is due, essentially, to David Stove. But before I go any further I must record a slight disagreement with Stove. He says that philosophic idealism died out around 1940, because by then there were not enough philosophers looking for a religion-substitute (1991: 96). He also says that in 1940 the Gem "was poised to enter upon the most brilliant phase of its career" (1991: 165). But this is to miss the close connection between the Gem and a version of idealism. As I said, once you accept a Gem, the pressure to adopt a form of conceptual idealism is almost irresistible. And in fact, as we shall see, conceptual idealism fuelled by Gems, implicit and explicit, is a dominant metaphysics of our age.

Berkeley and Kant assumed that there was only one 'phenomenal world'. Contemporary philosophic wisdom has outgrown that assumption. What we have now is idealism relativised to conceptual or linguistic scheme, or to theory, or whatever. There is no unique world-as-conceived-of-by-humans. The world-as-conceived-by-the-Aristotelian differs radically from the-world-as-conceived-by-the-Newtonian. The world-of-the-Eskimo is not the same as the world-of-the-Kalahari-bushman. This gets really exciting if we drop human chauvinism and bring in non-human animals too. The world-of-the-chimpanzee is different from the world-of-Albert-Einstein, and both are worlds apart from the world-of-the-honeybee.

Of course, this idealist talk need not be taken seriously. We can see it just as a fancy way of drawing attention to the diversity of experience or thought or talk *about the world.* On this view, all hyphenated entities are *Ersatz* entities. The Moon-as-experienced-by humans is just the Moon – and the same goes for all other hyphenated entities, including the Kantian Moon-in-itself. And a profundity such as "The Moon-as-conceived-of-by-Aristotelians was perfectly spherical" is just philosopher's gobbledy-gook for "Aristotelians thought that the Moon is perfectly spherical".

But conceptual idealists do take their talk of different worlds seriously. They think it utterly naive to suppose that all experience, thought and talk is of one world which is (largely) independent of experience, thought and talk. Arthur Fine writes:

The difficulty is that whatever we observe, or, more generously, whatever we causally interact with, is certainly not independent of us. This is the problem of *reciprocity*. Moreover, whatever information we retrieve from such interaction is . . . information about interacted-with-things. This is the problem of *contamination*. How then, faced with reciprocity and contamination, can one get entities both independent and objective? Clearly, the realist has no direct access to his World. [1986:151.]

Notice the hyphenated entity: interacted-with-things. Notice the implicit Gem: "We cannot interact with independent things without interacting with them. Therefore, we cannot interact with independent things". Implicit in all this is a silly account of independence which no realist should accept. The only independent entities are things we cannot interact with. The only independent entities are platonic entities, which do not exist in space and time and cannot interact with beings like us who do. But realists need not be platonists. When a realist says that the Moon is independent of us, what she means is that it is non-mental, existed long before we did, continues to exist when we are not perceiving it, and so on.

As for Fine's problem of contamination, that was supposed to show that if we look up and ascertain that there is a full Moon we gain information about an interacted-with-Moon rather than the Moon-in-itself. Where is this interacted-with-Moon? Is it inside our heads, like a Moonish sense-datum? I do not know.

I do know that Fine adheres to Hilary Putnam's so-called '*internal realism*'. Objects do not exist independently of conceptual schemes. It is the deployment of a conceptual scheme which carves up the world into objects. The Moon which we interact with and talk about is an object that we help create by interacting with it and talking of it. This is said, believe it or not, to be part of a *natural* ontological attitude to the teachings of science, which takes those teachings seriously. Well, some pretty mundane teachings of science imply that the Moon did not pop into existence when somebody first deployed the Moon-concept. No, the Moon existed long before we or any other concept-mongers did. It is not the sort of thing that can be created by thought or talk. And so on. As for carving up the world into objects with our words, that is – well, that is just a hopeless metaphor.

Yet this hopeless metaphor encapsulates the dominant constructivist metaphysics of the age. An early constructivist was Benjamin Lee Whorf, who thought people with different languages live in different worlds. Thomas Kuhn said that scientists with different paradigms live in different worlds. There was all that stuff about 'the social construction of reality'. Nelson Goodman thinks that "We can have words without a world, but no world without words or other symbols" (1978: 00). Think on it: "In the beginning was the Word, and the word was with God, and the Word was God" (John, 1: 1). Goodman insists that he is not talking in metaphors: "When I say that worlds are made, I mean it literally" (1980: 213). He says that Kant's noumenal world is "not worth fighting for" (1978: 20). Rorty thinks it "well lost" (1972). Notice how

Goodman and Rorty accept Kant's distinction between the noumenal and phenomenal worlds and then abridge it. Realists do not fight for the noumenal world either. But that is because realists do not accept Kant's distinction in the first place. For realists the noumenal and phenomenal worlds are the same – simply the world. As Michael Devitt has pointed out (1991: 235), obscure constructivist views are endemic in 'structuralism', 'post-structuralism' and French thought in general. Before he died of AIDS, Foucault said that AIDS was a mere construct! These doctrines are sweeping through language and literature departments throughout the western world. Fish, a literary theorist, thinks that countries can be produced by talking: "The France you are talking about will always be the product of the talk about it . . ." (1980: 199).

There is a kind of litmus-paper test for conceptual idealism. Michael Devitt ascribes to Lloyd Reinhardt a story involving Abraham Lincoln (Devitt 1991: 241–2). (I recall first seeing this in Lewis Carroll, but I cannot find it now.) Lincoln asks somebody "If tails as well as legs were called 'legs', how many legs would a donkey have?". "Five, Mr Lincoln", comes the reply. And if any of you are tempted by that reply, then you are tainted with conceptual idealism. Lincoln had it right: "No, sir, for you cannot make a tail into a leg by calling it one." It is basically a use/mention confusion. Lincoln's question is being confused with a different one: "If tails as well as legs were called 'legs', how many things that a donkey has would be called 'legs'?". 'Five' is the right answer to *that* question.

I do not know how to convey to you how pervasive this conceptual idealism is, except to pile up examples. So here are a few more. John Mackie proposed an 'error theory' of moral judgements, according to which we 'project' moral predicates onto a world which lacks moral properties and always speak falsely. Simon Blackburn liked the idea of projection but did not like the idea that it involves us in error. So he decided to square the circle and say that we can project things onto a world which lacks them and make no mistake. He called this 'quasi-realism'. Jennings extends this to science to yield 'scientific quasi-realism'. The theoretical entities and properties of science are "projected onto the world, they are the offspring of our theories" (1989: 243). ". . . the theoretical things we talk about in our talk are the *products of our talk*" (1989: 238). So *talking* produces electrons, black holes, genes, or whatever. From which it follows that none of these things existed before people started talking. Which conflicts, of course, with everything that science teaches about such entities.

If talking produces theoretical entities, it becomes hard to see how we can speak falsehoods. Once upon a time some people talked of witches and subscribed to a body of theory about them. For a while witch-theorists were quite successful: suspected witches were identified, tested for witch-hood in accepted ways, and when they tested positive (as they usually did) disposed of in accepted ways, too. Plain folk think that, despite these successes, witch-theory was false and there were no witches. Quasi-realists think this metaphysically naive. Given witch-theory, there is a 'projected world' of which that theory is true. There used to be witches after all, not noumenal witches in some inaccessible noumenal world, but projected witches in a projected world. Alan Chalmers once said the same, in a popular text endorsed by Britain's Open University: "In medieval Europe, witches really inhabited the common-sense world while in modern times they do not ..." (1976: 131). In fairness to Chalmers, let me add that he dropped this from the second edition of his book.

I doubt that a sixteenth-century witch-hunter, prone to bang the table and insist that there really are witches, would thank Jennings for telling him that he was right, but only about his 'projected world'. The witch-hunter's project was to find real witches in the real world, not to 'project' witches into a world of his own making. Jennings says that he accepts the ordinary notion of truth. Obviously, however, that notion is in for a drubbing. Suppose a witch-hunter armed with a consistent witch-theory confronts an opponent armed with a consistent no-witch theory. Is there a truth of the matter, as we would ordinarily suppose? Jennings thinks not (1989: 239):

"... theories with which we no longer agree ... are seen as having projected different properties and entities onto the world, and based on their projections, they spoke truly. We now, of course, project differently ..."

So there are projected witches in the witch-hunter's world, and no projected witches in our world. And that is all we can say. To say that there are no witches in the real world is utterly naive.

An afterthought. I said earlier, when discussing Fine, that when a realist says the Moon is objective and mind-independent what she means is that "it is non-mental, existed long before we did, continues to exist when we are not perceiving it, and so on". Was this not naive? Has not quantum mechanics shown that Berkeley was basically right? N. D. Mermin, appealing to quantum mechanics, writes (1981: 405): "We know that the Moon is demonstrably not there when nobody looks". Several things need to be said here. First, it is not Berkeley's view that the Moon is there when somebody looks and is not there when nobody looks. Berkeley's view is not that physical objects pop in and out of existence according to whether a human being pays them any attention. To think this is to make Tolstoy's Mistake. As a young man, Tolstoy read Berkeley and came to fear for his sanity. He tells (in his *Autobiography*) how he would turn his back on tables and chairs, and then turn around quickly to see if he could catch them fading out of existence like the Cheshire Cat! Young Tolstoy was doubly foolish. He misunderstood Berkeley. Then he misunderstood his misunderstanding. If *esse est percepi* (to cite Berkeley's own slogan), then you cannot perceive an object half existing or fading out of existence. No, Berkeley's view was different. Berkeley's view was that physical objects do not exist at all. For Berkeley, all that exists are 'finite spirits' (such as you and me), the ideas we experience, and an 'infinite spirit' (God) who puts the ideas into our minds.

Incidentally, my assertion that Berkeley denied the existence of external objects was recently described as an 'egregious error' by Jonathan Dancy (1994: 215). But the egregious error is Dancy's, not mine. As Dancy's 1987 book on Berkeley makes clear, he is just one of the many philosophers anxious to *make excuses* for Berkeley.

Returning to Mermin, he thinks quantum mechanics has shown that Tolstoy's misunderstanding of Berkeley is actually the way things are. The Moon pops into existence whenever a human being pays it attention. Stove is typically dismissive (1991: 99):

> Now it might perhaps be true that Professor Mermin depends for his ease of mind on being an object of attention ... But that he depends for his very existence on being an object of attention, is entirely out of the question: it is much more likely (to say the least) that one or more of his scientific theories is wrong.

I would have said "wrong or wrongly interpreted". For Mermin's view is actually just one *interpretation* of quantum mechanics, and it is not the only one. Recently I was at a conference in Beijing where a Chinese philosopher of science was advocating views similar to Mermin's. Allan Franklin, who is a physicist, leaned over to me and whispered "Thank God for the oceans, they keep tabs on the Moon when nobody is looking!". That was a very good remark. Can the oceans not be viewed as a measuring instrument, constantly monitoring the Moon's continued existence by registering its effects on them? Can the same not be said of any two physical systems which constantly interact with one another? I am not a physicist and I do not know the answers to these questions. I do know that to say, with Mermin, that the Moon is only there as long as we humans monitor its existence is to buy into idealism.

Close allies of conceptual idealists are what I call '*epistemic truth theorists*'. These seek to define truth as some epistemic feature of a statement. The favourite is the ideal limit theory of truth advocated by Hilary Putnam and Brian Ellis, and defended by Carl Matheson. According to this view, "the truth is to be equated with the limit of ideal scientific practice: what is true is just perfect science pursued to its conclusion" (1989: 253). Matheson makes it clear that this is not an empirical claim. It is not the claim that science is our best epistemic engine, so that all truths will as a matter of fact be thrown up by 'perfect science'. That is a claim which a mad-dog scientific realist convinced of the omnipotence of scientific methods might make. (It is also, by the way, a claim that has been logically refuted. It has been proved that not all truths are knowable. The universe is not even congenial to us to that minimal extent!) No, Matheson's claim is a conceptual claim: "It is a *conceptual* truth, that all truths would be part of our ideal science" (1989: 253, footnote). Matheson does not tell us what he thinks 'perfect science' or 'ideal scientific practice' is. Nor does he argue for his presumption that it will reach some sort of 'limit'. (In fact, good philosopher that he is, he is going to dispute this later.) One thing is clear, though. However 'perfect science' or 'ideal scientific practice' is to be defined, it cannot be wholly or partly in terms of truth. To be told that ideal scientific practice is that which reaches truth, and then to be told that truth is what ideal scientific practice reaches, is to be told nothing. No, ideal scientific practice must be defined in truth-free terms. Then we will be told something non-trivial when we are told: "This is what ideal scientific practice is – and truth is, by definition, what it reaches in the limit".

This opens Matheson up to Moore's 'Open Question Test'. However 'perfect science' is to be defined, it remains an open question whether what it reaches in the limit is true. The sceptic does not contradict herself when she says that it is possible that even perfect science pursued to its limit might issue in falsehoods.

It is primarily to dispose of extreme sceptical scenarios like this that epistemic truth theories are proposed at all. But their anti-sceptical virtues are entirely spurious. The sceptic says that perfect science might issue in untruth in one sense of the term 'true' – the commonsense sense. The epistemic truth-theorist says this cannot happen in another sense of the term 'true' – the epistemic sense. You cannot dispose of scepticism by changing the subject. Nor, for that matter, do epistemic truth theories make the truth any more accessible to us – which is their other proclaimed virtue. To recognise the truth I am supposed to find out what perfect science is and what it will throw up when pursued to its limit. How do I do that? I suppose that perfect science is that practised by God (if he exists). But what God will be believing at the end of time is completely inaccessible to you and me.

By the way, and digressing, Stove said that idealism was fuelled by the search for a religion-substitute, a philosophy that would make the world congenial to us. No doubt there is much truth in that. But I think idealism is fuelled just as much by a *horror of scepticism*. Philosophers are, still, and often without realising it, certainty-freaks. To be certain that something is true we must have eliminated all relevant alternative possibilities. Along comes the sceptic. We might be disembodied spirits fooled by Descartes' Evil Genius. We might be Putnam's disembodied brains-in-a-vat, fooled by a super computer. (Actually, Putnam tries to prove that we can't be disembodied brains-in-a-vat fooled by a super computer. But he does not succeed.) Everything might have been created by God five minutes ago, folk with memories, fossils in the rocks, and so on. Philosophers exert themselves to show that these things are not possible. They waste their time. All these things are possible. No contradiction is involved in any of these sceptical suppositions. Of course, that is not saying much. Most things are *possible*. Sceptics give no shred of evidence that any of these things is *actual*, nor could they. Nor, to be fair to the sceptics, need they produce any evidence. Their target is the certainty-freak. We should move beyond this tired debate. (If I may be permitted a plug, this is a central theme of my book *Common Sense, Science and Scepticism*.)

Finally, consider Matheson's assertion that "It is a *conceptual* truth, that all truths would be part of our ideal science". In particular, consider the occurrence of the word 'truth', as opposed to the plural 'truths', in this assertion. Applying Matheson's professed theory to the term 'truth' turns his assertion into "It is a conceptual part of ideal science pursued to its limit, that all truths would be part of our ideal science". In other words, Matheson thinks that the ideal limit theory of truth will itself be thrown up by ideal science. I defy him or anyone else to find this theory stated in the writings of any scientist. And I defy him or anyone else to give a reason why this theory is bound to be stated by scientists of the future. No, the ideal limit theory of truth is not science but philosophy. And bad philosophy to boot.

Let us see just how bad. I mentioned earlier that Matheson, good philosopher that he is, would dispute the assumption that ideal scientific practice is bound to converge on a single limit. And so he does (1989: 253-4):

> There is no reason to believe that different species would arrive at the same theories as us, even in the ideal limit. Nor is there reason to believe that a moderately successful field physics and a moderately successful particle physics would eventually converge ... Thus the ideal limit theory might better be called the ideal *limits* theory. According to it, your community's limit might not be my community's limit; what is true for you might not be true for me. In allowing for the possibility of relativism, the ideal limit theory contradicts global realism.

So an ideal field physics which says there are no particles might be true. And an ideal particle physics which says there are no fields might also be true. This looks like a contradiction. But nowadays philosophers leap over contradictions effortlessly: "(To remove the contradiction, just say that Wave theory is true for my community and Particle theory is true for yours.)" (1989: 254). Here truth-relativism is invoked as an answer to a *reductio ad absurdum* argument. The answer is technically correct. The law of contradiction assumes that truth is not relative to theory, community, conceptual scheme, or anything else. Those who think truth is relative to one of these things will not succumb to a *reductio*. They will "remove the contradiction" by invoking their relativism.

They pay a price, of course. Matheson confidently asserts that his ideal limit theory *contradicts* global realism (1989: 254). But why does he not invoke his relativism to leap over this contradiction, too? Why does he not say that his ideal limit theory of truth is true for his idealist philosophical community and that the global realist theory of truth is true for the global realist community?

As a member of the latter community, I stick with the law of contradiction and assume that *reductio* arguments have some teeth. Indeed, I wonder how philosophy can be pursued at all in communities other than my own. Philosophers have only philosophical argument to appeal to. We are not scientists who can appeal to observation and experiment to settle our disputes. The most potent philosophical argument is the *reductio*: show that a philosophical position leads to a contradiction, and you know that it is false. If *reductio* does not work, nothing will. Nothing *does* work, in the community to which Matheson belongs.

The foregoing was naive, I know. It assumes that *reductio* has teeth because it takes it for granted that a contradiction cannot be true. But nowadays so-called 'paraconsistent logic' is a small growth industry. There has recently been published a 718-page collection of articles on the matter (Priest et al. 1990).

In paraconsistent logic some contradictions are true, as Hegel and Engels said long ago. And because some contradictions are true, some *reductios* do not work. Or as Graham Priest told me "*Reductio* is not always valid". I always thought an argument-form was either valid, full stop, or invalid, full stop. But not any more. I asked Graham *when reductio* was valid and when not. He said it was a good question. I do not recall him answering it.

Department of Philosophy
Otago University

REFERENCES

Dancy, J.: 1987, *Berkeley, An Introduction.* Oxford and New York, Blackwell.

Dancy, J.: 1994, 'Review of Common Sense, Science and Scepticism by Alan Musgrave', *Mind*, **103**, 214–216.

Devitt, M.: 1991, *Realism and Truth*, 2nd edn. Oxford and Cambridge, Blackwell.

Fine, A.: 1986, 'Unnatural Attitudes: Realist and Instrumentalist Attachments to Science', *Mind*, **95**, 149–177.

Fish, S.: 1980, *Is There a Text in this Class?.* Cambridge, MA, Harvard University Press.

Goodman, N.: 1978, *Ways of Worldmaking*. Indianapolis, Hackett Publishing Company.

Goodman, N.: 1980, 'On Starmaking', *Synthese*, **45**, 211–215. Reprinted in Goodman 1984: 39–44.

Goodman, N.: 1984, *Of Mind and Other Matters.* Cambridge, MA, Harvard University Press.

Jennings, R.: 1989, 'Scientific Quasi-Realism', *Mind*, **98**, 225–245.

Matheson, C.: 1989, 'Is the Naturalist Really Naturally a Realist?', *Mind*, **98**, 247–258.

Mermin, N.D.: 1981, 'Quantum Mysteries for Everyone', *Journal of Philosophy*, **78**, 397–408.

Priest, J. et al.: 1990, *Paraconsistent Logic: Essays on the Inconsistent*, Philosophia Verlag.

Rorty, R.: 1972, 'The World Well Lost', *Journal of Philosophy*, **69**, 649-665.

Stove, D.: 1991, *The Plato Cult and Other Philosophical Follies.* Oxford, Basil Blackwell.

CARLO PENCO

HOLISM IN ARTIFICIAL INTELLIGENCE?[1]

Before Fodor and Lepore's criticism, *holism* was a very widely accepted position both in philosophy and in A.I. Fodor and Lepore claim that A.I. is "almost everywhere holistic".[2] I disagree with such a sweeping contention and I will give some evidence for such disagreement. I will not consider connectionist systems, but I will confine my remarks to symbolic A.I.[3] Contrary to what Fodor and Lepore claim, I will suggest that mainstream symbolic A.I. is an attempt at implementing molecularist theories of meaning (theories where the meaning of an expression is determined not by the overall system of language to which the expression belongs, but by subparts of it). I will consider three classical cases of A.I. research: (1) procedural semantics as implemented in toy worlds, (2) semantic networks and frames, (3) contextual reasoning. I will try to show that in all these cases we find evidence against an holistic view of meaning, and hints towards a molecularist view that does not degenerate into holism.

1. MEANING AS PROCEDURE – THE CASE OF TOY-WORLDS

The first theory of meaning devised by artificial intelligence was the idea of meaning as procedure. The idea itself has been variously developed, both in psychological terms (e.g. Johnson-Laird) and as a general theory which should integrate model theoretical semantics (Woods)[4]. Nevertheless, the original formulation's core remains untouched. Such a core is the idea that a representation of the meaning of an expression is given by the procedure attached to that expression. In Winograd 1972, this idea appears very similar to Frege's idea of sense,[5] supplemented with Austin's theory of speech acts. Most important, procedures are compositional, as Fregean senses are intended to be.[6] These early results are not easy to interpret in the light of our present concern, particularly because there is an apparent tension within Winograd's views on meaning given in his early papers: on the one hand, he explicitly assumes an holistic vision of meaning; on the other hand his technical apparatus devised to represent meaning seems to be a molecularistic one (if not strictly atomistic). It is possible to accept the technical devices, while rejecting his metaphysical assumption.

Winograd's overall picture of meaning appears to be explicitly holistic: the meaning of any concept depends on its interconnection with all the other

M.L. Dalla Chiara et al. (eds.), Language, Quantum, Music, 37–48.

concepts in the model (1973, p.167). Given that the task of defining meaning as such is impossible, Winograd suggests that, rather than trying to prove things about meaning, we can design procedures which can operate with the model and simulate the processes involved in human use of meaning. Therefore, procedures are meant as an alternative to meaning understood in vaguely holistic terms; they should represent the use of concepts. All language use can be thought of as a way of activating procedures. An apparent result of this strategy should be a shift from a theory of meaning to a model of understanding.

Nevertheless, on the other hand, Winograd goes on in speaking of meanings; he suggests that an expression as 'pick up' has different meanings depending on whether it refers to a single object or several (p.174) (in the latter case it means put away); some other times he speaks directly of meaning as a set of procedures (e.g. "the different possibilities for the meaning of 'the' are procedures which check various facts about the context, then prescribe actions such as 'Look for a unique object in the data base which fits the description' ..."(p.175).

We may therefore say that in Winograd's work meanings are represented by procedures plus their context of use: for instance the same procedure serves to express the meaning of two different words, 'pick up' and 'put away'; the only difference is the context in which the procedure is used, in one case the context of the action of picking up a single block, in the second the context of picking up all the blocks. What shall we conclude? We might say that the procedure represent a basic core meaning and, depending on the context and way of application, it represents two different meanings or two different aspects of meaning. Winograd is not completely clear on this matter. But it is seems clear, anyhow, that procedures are constant and do not depend on the whole system. They are written as definite LISP programs and always perform the same steps when activated. Procedures represent a sort of core meaning, which gives the basic strategy for using a word. This core may be articulated, depending on the context, expressing what can be considered different aspects of meaning, or, as Winograd says, different meanings attached to the same procedure.

There is something definitely anti-holistic in this representation of meaning: meaning-as-use represented as a specific procedure attached to a word. Procedures run in the same way in all different contexts, and we cannot say that they depends on the overall system. We might say that, notwithstanding Winograd's adherence to holism from a philosophical point of view, his meanings-as-procedures are more easily identifiable with an atomistic or molecolaristic stance: we cannot make the meaning dependent on the entire system. The meaning is always relative to a procedure + the specific context in which it runs. The result is some kind of molecolaristic interpretation: meanings depend on a procedural core, which remains fixed, and on different applications of this core.

Winograd 1981 has analysed the value and the limitations of his early theory.

Among the limitations we find the inadequacy of the definition of meaning of the words and the difficulty of dealing with commonsense knowledge. These may be considered two main problems which lie behind this project, casting doubts on such a representation of meaning:

(i) The relative flexibility of the system is based on rigidly defined procedures. But most of our concepts are often vague, and they have often many different applications. A single rigid procedure attached to a predicate cannot represent the vagueness of most of our concepts (even if some step in this direction has been done with the idea of meaning = procedure plus context).

(ii) The analysis of language is restricted to simple artificial situations (toy-worlds). But we need to analyse language in real situations. A knowledge base that would be adequate to real situations would be too big and complex to be used by a program without running the risk of combinatorial explosion.

The A.I. milieu reacted to these limitations with ingenuity. And the response to these limitations has also thrown, or may throw some light on the problem of a definition of meaning. The following sections deal with limitations (i) and (ii) respectively.

2. MEANING AS PROTOTYPE – THE CASE OF FRAME SYSTEMS

Minsky 1975 deserves credit for introducing the notion of frame. Minsky's idea of frames as conceptual structures with default values was accepted and made both more precise and logically expressible in different formalisms: Winograd himself used similar ideas in KRL (Bobrow-Winograd 1977), KL-ONE family of semantic networks developed a theory of default values, Hayes 1979 translated the idea of frame into logical formalism; after these early results the idea of frame and frame-nets has become a common tool in A.I. and has been described in most textbooks of LISP or of introduction to A.I. I assume here a general knowledge of frames.

Frames are part of a frame system, where each value of a slot may be [connected with] another frame. Hereditary relations will license the assertion that, if x is a tiger, then x is an animal (because it is a mammal). The default representation provided by the frame can be understood as a fixed meaning given to the lexical item representing the concept; this meaning – the stereotype – is supposed to be the one that best represents the average use of the expression in the linguistic community. But other uses and other meanings may be attached to the same expression: other properties, instead of the default properties, may be the values of the slots of the frame. We may enclose them in a set of possible values (e.g. the stereotypical table has four legs, but it may have also three legs: in the "legs" slot there will be a default value "4", but different values are allowed). The information of the context will help to decide which values to choose.

This does not amount to say that default values and the other possible values embedded in a frame representation are what is common to all speakers: probably there is no set of information which is the intersection of the information available to all speakers belonging to a linguistic community. But certainly meaning – average use – is determined by the majority, as influenced by experts and their definitions. In case of conflict people defer to experts. Therefore, a certain stability in the definition of meaning reflects the practice and the needs of a linguistic community. Frames with default values are stable structures; the availability of alternative values for each slot of a frame is what gives new life and flexibility to the idea of meaning.

The stereotypical definition of a concept may be intended as representing the meaning of the predicate used to express the concept. But, even if we accept that a stable, stereotypical definition of meaning is given by the interplay of average use and deference to experts, it is still unclear to what extent individual frames depend on the overall system. We find the idea of global connections at the origin of semantic networks: in Quillian 1967 differences in meaning between two lexical items were individuated with the differences concerning *all* the connections activated with other nodes of the network, starting from the set of nodes immediately defining the items. From Quillian's point of view, in order to understand the meaning of an item, and compare it with another, we need to run the entire data base. But this sounds very uncomfortable. Although this idea (spreading activation) seems fascinating, there is something awkward as interpretation of what meaning is: intuitively we do not need to understand the entire system of lexicon in order to understand a sentence like "Peter killed the tiger": we need to know just a few things about tigers and killing. This intuitive attitude is reflected in the restrictions that are imposed on artificial systems: if we are building a frame system, we must take care that individual steps in its computations do not involve the whole system, on pain of combinatorial explosion. A technical requirement deriving from the finite nature of our machines leads to restrictions on our representation of meaning. We have at least two strategies for dealing with this problem:

(a) One strategy consists in making a difference between conceptual definitions and factual information. This kind of research is somehow an attempt at recovering a viable distinction between sentences (or inferences) that are necessary to define meaning and sentences that are not; it is an attempt at performing the role of the distinction between analytic and synthetic sentences, which had been so effectively criticized by Quine 1953. One example is the distinction between 'assertional' and 'definitional' developed in KL-ONE systems; this distinction has been criticized for not being well grounded (see e.g. Marconi 1997, pp.38–41); certainly a main problem is to find a *principled* distinction which could take the place of the old analytic-synthetic distinction; but what do we count as 'principled'? Principles which define the distinction between two kinds of inferences might derive from the needs of implementation. If machines are limited artefacts (as we are limited agents), we need, at least for computational economy or computational necessity, to distinguish basic semantic information from factual information, or basic information about the literal meanings of words from information about different applications. If we abandon the principles behind the analytic-synthetic distinction, we may still find some pragmatic principles which justify the distinction between two sorts of inferences, the one defining basic uses of words, the other defining occasional applications of them. We may refer here to the pragmatic stance taken by Brandom 1994, where he distinguishes "the properties governing correct use in which the concepts grasped by individuals consist", on the one hand, from "the disposition to apply concepts, make inferences, and perform speech acts, in which an individual's grasping of a concept consists", on the other (p.636). A philosophical attitude of this kind is well followed in the practice of A.I.: any 'viable' (= compatible with an A.I. system) representation of meaning as inferential role is bound *not* to include *all* (or *most*) possible inferences. Many A.I. programs may be considered as attempts to respect this restriction.[7] The same point had been already considered by McCarthy in his early work. In his 'Advise taker' project (1968) he defined the concept of immediate inference: in order to understand a situation we do not need to make explicit all the inferences from the relevant premises, but only their immediate consequences, beginning with the inferences which require just one step in the deductive process. The idea is barely sketched; however, it does point to the necessity of controlling the risk of combinatorial explosion of inferences derivable from a single premise.

(b) Another strategy consists in defining, with some degree of arbitrariness, a set of relevant contexts: if we are speaking of tigers and refer to somebody shooting a tiger, we need some idea of what happens in big-game hunting, but we do not need to know anything about skiing or going to a restaurant. This strategy is reflected in the practice of partitioning a semantic network into an organised, hierarchical structure, where super-concepts control sets of concepts under their nodes. Hereditary aspects, widely discussed in the literature (see Frixione 1994), do not entail that in order to understand the meaning of a lexical item you need to go through the entire net: the path of hereditary relations is followed inside a specific semantic field, and what happens in other semantic fields is not at all relevant to understanding or defining the meaning of the items in the field under consideration. Information about birds (e.g. that penguins don't fly) is not relevant to understanding what a tiger is. Shank's scripts are another classic example of this kind of strategy. The idea of a script (think of the famous script "restaurant") is just the idea of encoding a certain amount of information in a single unit, somehow autonomous and not depending on other information. High level representation such as scripts are treated in formalisms which use, to a certain extent, a top-down representation of knowledge, where procedures (demons) are used to activate the script or scenario relevant to a given situation (an early example is BORIS, by Lenhert and others 1983).

In both cases (a) and (b), frame systems can be seen as representing the average use of words in a language. They might include links to experts' knowledge: part of commonsense knowledge about the world is the acknowledgement that experts exist for most of our fields of interest. Admitting a link to scientific definitions of each term is a representation of such an ability to defer. This representation must presuppose individual sites for scientific definitions, and the availability of such sites to people looking for more precise information and definitions. In both cases, what counts as semantic competence is relatively fixed. We must assume some 'idealisation' of the speaker, which does not amount to claiming that speakers share *exactly* the stereotypical meaning, the set of properties stereotypically defined. If the representation of meaning is the representation of the practice of a language in a community, the stereotype is the set of features which are the most constant in this practice.

Indeed, some problems seem untouched by the strategy of the frame system: understanding a language does not consist only in the mastery of different concepts; on the contrary, sometimes we have to understand a speech which requires an amount of information that seems to be difficult to extract from a set of frames or from a script. Such limitations of frame analysis have been pointed out by Haugeland (1979), who says that "common sense can be organised only partially according to subject matter. Much of what we recognise as 'making sense' is not 'about' some topic for which we have a word or idiom, but rather about some (possibly unique) circumstance or episode". The strategy of contextual reasoning – the last topic in our brief survey of not-holistic views in symbolic A.I. – is a possible way to face these limitations.

3. MEANING AS USE IN A CONTEXT – THE CASE OF CONTEXTUAL REASONING

The idea behind the strategy of toy-worlds was that language is strictly intertwined with action, so that we need to know a lot about the world in order to use language. Therefore, programs like SHRDLU had vast knowledge of a limited world (a toy-world) so that we could analyse the workings of language as in a thought experiment, without all the complications of interaction in a real situation. But we can also think of the strategy of toy worlds as a paradigm of how language works. We may imagine the general workings of language as split among different toy worlds, each with its own language and basic knowledge.

The main limitations of the idea of toy worlds was the impossibility to pass from a limited representation of a single small domain to a general representation of the complexity of the world. But with such a re-interpretation, we may see that these old fashioned A.I. programs still give us suggestions on the development of A.I. The main idea could be expressed in the following way: instead of relying only on a network of frames, in order to understand the meaning of a sentence we must consider the sentence within something like a toy world. The representation of our knowledge and our language cannot be given once and for all in a single system, but must be given in different systems connected each other. We find today many general frameworks which can be considered an implicit development of this suggestion (for a comparison see Bouquet 1997), e.g.:

(i) projects like CYC, developed initially by Lenat, which propose the construction of a very big knowledge base organized in micro-theories, which represent defined knowledge on an aspect of the world. A logical foundation of this kind of approach is found in Guha 1991.

(ii) Mental spaces, a theory originally proposed by Fauconnier and developed by Dinsmore 1991 in a formalism based on the idea that knowledge is partitioned in logical spaces.

(iii) Multi-context theories by Giunchiglia 1993, which rely on intuitions early given by McCarthy1987,1993. Multi-context theories (M-C theories from now on) are based on the implementation of first order logic in systems such as FOL , on the line of Weyhrauch's 1980. In M-C theories a context is a complete description of a particular set of objects and actions formally given as a triple $\langle L, A, R\rangle$, where L is a language (with a vocabulary of words used in the context), A is a set of axioms (a body of specific information), and R is a set of inference rules. Each context is therefore represented as a single complete formal system.

I will try to show some anti-holistic suggestions given in these kinds of approach. I will refer hereafter to the third line of thought given above. First of all, we have to consider the relation of M-C theories with the earlier frame theories. In principle, there is no conflict between frame analysis of the lexicon and M-C theories devised for dealing with commonsense reasoning. Because of their original interest in reasoning and action, M-C theories lack a general concern for lexical semantics and the representation of lexical meaning. However, M-C theories could be enriched with the idea of concepts as frames.

Actually, contexts might play also a role similar to Schank's Scripts. The main difference in this case consist in the formalization: first-order logic allows greater simplification and generality of the system.

All this looks highly programmatic. Even if the discussion on lexical meaning is not yet fully developed, however, some features of multi-context theories are worth noting for the purpose of our discussion; they actually give some conceptual tools which could help the building of a molecolaristic stance in A.I.

(i) *locality* – With M-C theories, we abandon the idea of a single language to which all sentences belong; there are *languages*, and the same sentence (the same string of characters) may belong to different languages: languages are embedded in contexts, each of which deals with a special part of the knowledge as is normally organized in a linguistic community. The deeply anti-holistic intuition behind this move was clearly expressed by Giunchiglia (1993): reasoning is usually performed on a subset of the global knowledge base; we never consider all we know but only a small subset. Any system which does not capture this point suffers from what Giunchiglia calls the problem of (lack of) locality [Giunchiglia 1993 is mirroring an original point made by McCarthy 1987 on the problem of (lack of) generality].

(ii) *travelling through contexts* – All contexts are on the same level (there is no super-context), so that relations among contexts can be dealt with easily. Simple algorithms implement rules for (a) entering and exiting a context, (b) taking some elements from a context (or a whole context) into another; (c) sharing inference rules among contexts.

(a) entering and exiting a context is the first operation given in McCarthy 1987 to show that a context is always referred to in any assertion; while making a derivation inside a context, we may enter the context, make the derivation and assert the conclusion as valid relative to the context. Exiting the context, the conclusion itself must be referred to only with an index for the context in which it has been derived.

(b) can be treated with forms of *lifting*, expressible as

$$Lift(C1, C2) \leftrightarrow (\forall x(true(x, C1) \rightarrow true(x, C2))).$$

This means that you may take what is true in a context $C1$ as true inside another context $C2$. This step (which could be done also with subparts of contexts) is fundamental for (iii) below.

(c) was developed by Giunchiglia 1993 with the technique of bridge rules: they are rules which permit to pass from a premiss in a context, through some steps, to a conclusion which is valid in another context.

(iii) *dealing with individual situations* – How can this simple conceptual machinery face the problem of commonsense reasoning in individual, not necessarily typical situations? A possible answer is the idea of working context. The rationale for this idea is that, when we face a situation, we pick up information from different contexts; formally, we construct a working context by 'lifting' axioms and rules from the relevant contexts. Thus we can deal with individual situations without having to take the whole data base into account. Facing an individual situation, we pick up from different contexts exactly what is necessary to understand and solve the problem in question. If unexpected obstacles were to arise, other contexts could be called inside the working context. This idea is a development of the idea of 'default' worked out in semantic networks; facing a problem we find a default solution (e.g. in order to fly from London to Moscow we only need a small amount of knowledge concerning flying); we look for other solutions in front of new data (if we discover that we have lost our ticket, or our luggage, or our clothes, more knowledge must be taken into account, knowledge which we would normally disregard in drawing our normal inferences).

The discussion of lexical semantics through contexts has still to be defined; but the general strategies given above seems open towards some kind of molecularistic stance.

4. CONCLUSION

The alternatives to holism are atomism and molecularism. According to the former, meaning is defined atomistically for each single word. For the second, meaning is defined by subparts of language, not by the whole language. Molecularism gives therefore the background for a non-holistic definition of understanding.

We have to distinguish carefully between a representation of meaning and a representation of understanding. A representation of meaning might follow the lines of a stereotypical representation with default values as we have seen in the discussion of frames: here we have a representation of something socially shared in a language as a social product. But understanding is related to individual ability to use the language; and we cannot make the assumption that every speaker grasp exactly the same stereotype defined in the language as a social product. Even if we accept that there is 'a common store of thoughts' and concepts which are transmitted from one generation to the next (Frege 1892), we must recognise individual failures to take note of such thoughts and concepts. We need a weaker definition of *understanding*: a person understands a concept if he/she understands *some* of its (plausible, relevant, and easy) inferential relations.

Molecularism has to face the challenge posed by Fodor and Lepore (the challenge being: either molecularism collapse into holism or it is obliged to adhere to a rigid distinction between analytic and synthetic propositions – which is normally refused after Quine's criticism). An answer to the challenge was made and formulated in a very precise way by Perry 1994 and Marconi 1997 in their discussion of Fodor and Lepore's definition of anatomism.[8] Grossly speaking the answer is formulated as follows: in order, for two people, to understand a sentence *P*, it is not necessary that exist some set of sentences which two people have to share, but it is necessary that two people share some set of sentences. This solution gives a molecularistic theory of understanding (semantic competence) which seems to escape the criticism given by Fodor and Lepore, and nicely fits our requirements of a weak definition of understanding.

Perhaps we do not strictly *share* the same meanings, as given in idealised stereotypical definitions; in communication we just *converge* towards some set of common inferences. But, even if we do not properly share meanings or stereotypes, there is, however, something we do share in communication: some basic strategies by which we converge towards common inferences. Such strategies could be exemplified by the ability to switch from a context to another, to import information from a context, to build up new contexts from a given one, to refer to experts' contexts in cases of uncertainty. While looking for a representation of meaning as something shared by all speakers of a language, we reach the tentative conclusion that what is shared are not exactly *meanings*, but *strategies* to find ourselves at home in any context.

Department of Philosophy
University of Genova

NOTES

[1] I am grateful to Paolo Bouquet, Ernest Lepore, and Diego Marconi for comments on earlier versions of this paper. Part of the work has been done at the University of Rochester, with the kind help given by the Staff of the Department of philosophy in letting me working there in 1994. An early version of the paper has been given at a Conference on Holism organized by Rosaria Egidi in Rome, 1994.

[2] See Fodor and Lepore 1992 and subsequent essays. The negative consequences of holism had already been clearly detected by Dummett 1973, to whom Fodor and Lepore refer. Fodor and Lepore probably refer to a general feeling in the environment of A.I. in the eighties, which might be well represented by Haugeland 1979: here four forms of holism are described and it is assumed that A.I. somehow demonstrated at least commonsense holism. Commonsense holism is the claim that *the whole of commonsense knowledge is relevant at each step of the interpretation of a sentence.* Because of that, structures like frames and scripts are like an encyclopaedia entry, with links to larger structures and cross references to other concepts, suggesting an holistic definition of the concepts which are supposed to be represented by these techniques. I will try to give a different interpretation of these kind of structures.

[3] Here I endorse, for the sake of the argument, the classical view that connectionism is the study of subsymbolic mechanisms and processes of the mind, while symbolic A.I. is the study of symbolic mechanisms of our culture or the study of cognitive systems which supervenes the processing of individual minds (Smolensky 1988; Clark 1991). An argument for holism in dealing with psychological descriptions of mental processing is given in Block 1995. Block's result, if correct, is a strong evidence in favour of an holistic approach to the working of the mind; this approach seems to be devoted to studying the working of individual mental contents (narrow contents). I will not deal with such an approach, which could be perhaps accounted for in a connectionistic framework; I will discuss instead researches in A.I. whose aim is an idealized representation of cognitive contents as knowledge shared by limited agents.
[4] I tried a general reinterpretation of the idea of meaning-as-procedure as a way to represent the idea on meaning-as-use given in later Wittgenstein in Penco 1992.
[5] See Marconi 1997.
[6] There are strong suggestions for an argument from compositionality principle against holism. A case in that direction is given by Fodor & Lepore in their criticism to Block's Advertisement for a semantics for psychology. See F&L pp.181–182. They apply their argument against the possibility of a viable Conceptual Role semantics which professes to be holistic, but inevitably falls in some kind of analytic-synthetic distinction, which blocks holism. Also Dummett 1991 argues on the conflict between holism and compositionality (ch.10, §5).
[7] An example is the work of Norvig 1989, who designs an algorithm computing a limited set of *proper* inferences quickly, without computing *all types* of inferences. Proper inferences are defined as *plausible*, *relevant*, and *easy*. Quick computation of a small set of proper inferences yields a partial interpretation, which can be used as input for further processing.
[8] A general point on the problem had already been given by George Rey (1983, p. 259). For Perry 1994, strong anatomism is expressed by: $\forall P \exists Q \Box$ (*P is shared* $\rightarrow$ *Q is shared*); while molecularism requires: $\forall P \Box$(*P is shared* $\rightarrow$ $(\exists Q) Q$ *is shared*). Marconi 1997 gives a different formulation. A discussion of the point where Fodor and Lepore suggest a misunderstanding of molecularism is made by Penco 1999.

REFERENCES

Block, N.: 1995, 'An Argument for Holism', *Proceedings of the Aristotelian Society,* **XCV**, 151–169.

Bouquet, P.: 1997, *Contexts*, PhD dissertation, University of Genova.

Brandom, R.: 1994, *Making it Explicit*. Cambridge, Mass., Harvard University Press.

Clark, A.: 1991, *Microcognition*. Cambridge, Mass., MIT Press.

Dummett, M.: 1973, *Frege. Philosophy of Language*. London, Duckworth.

Dummett, M.: 1991, *The Logical Basis of Metaphysics*, (especially ch.10: 'Holism'), London, Duckworth.

Fodor, J. and Lepore, E.: 1992, *Holism. A Shopper's Guide*. Oxford, Blackwell.

Frixione, M.: 1994, *Logica, Significato e Intelligenza Artificiale*. Milano, Angeli.

Giunchiglia, F.: 1991, 'Multilanguage Systems', *Proceedings of AAAI Spring Symposium on Logical Formalization of Commonsense Reasoning*.

Giunchiglia, F.: 1993, 'Contextual Reasoning', in Dalla Pozza C and Penco C, eds, *Epistemologia, Special Issue: I linguaggi e le macchine*, **XVI**, 345–364.

Haugeland, J.: 1979, 'Understanding Natural Language', *Journal of Philosophy, LXXVI*, 619–632.

Hayes, P.: 1979, 'The Logic of Frames' in Metzig, D., ed., *Frame Conceptions and Text Understanding*, De Gruyter and in Brachman, R.J. and Levesque, H.J., eds., *Readings in Knowledge Representation*. Morgan Kaufman, 1985.

McCarthy, J.: 1968, 'Programs with Common Sense' in Minsky, M., ed., *Semantic Information Processing*. Cambridge, Mass., MIT Press.

McCarthy, J.: 1987, 'Generality in A.I.', reprinted in McCarthy *Formalizing Commonsense* (edited by V. Lifschitz). Norwood, NJ, Ablex, 1990.

McCarthy, J.: 1993, 'Notes on Formalizing Contexts' in *IJCAI Proceedings,* Morgan Kaufman, 555–560.

Marconi, D.: 1997, *Lexical Competence*. Cambridge, Mass, MIT Press.

Norvig, P.: 1989, 'Marker Passing as a Weak Method for Text Inferencing', *Cognitive Science*, 469–620.
Penco, C.: 1992, 'Significato, Uso e Procedure', *Lingua e stile,* **XXVII**, 251–263.
Penco, C.: 1999, 'Olismo e Molecolarismo', in Dell'Utri, M., ed., *Prospettive dell'olismo*, forthcoming.
Perry, J.: 1994, 'Fodor and Lepore on Holism', *Philosophical Studies*, **LXIII**, 123–138.
Rey, G.: 1983, 'Concepts and Stereotypes', *Cognition*, **XV**, 238–261.
Smolensky, P.: 1988, 'On the Proper Treatment of Connectionism', *Behavioural and Brain Sciences,* **XI**, 1–74.
Weyhrauch, R.W.: 1980, 'Prolegomena to a Theory of Mechanized Formal Reasoning', *Artificial Intelligence,* **XIII**, 133–176.
Winograd, T.: 1972, *Understanding Natural Language*. New York, Academic Press.
Winograd, T.: 1973, 'A Procedural Model of Language Understanding' in Schank-Coldby, eds., *Computer Models of Thought and Language*. San Francisco, W.H. Freeman Company.
Woods.: 1981, 'Procedural Semantics as a Theory of Meaning' in Joshi, A.K. and Webber, B.L., eds., *Elements of Discourse Understanding*. Cambridge, Cambridge University Press.

BENOIT DESJARDINS

MODEL SELECTION FOR CAUSAL THEORIES

1. INTRODUCTION

One of the most central problems in scientific research is the search for explanations to some aspect of nature. This often involves a cycle of data gathering, theorizing, and experimentation. In many scientific fields, including medicine, data comes in the form of statistical distribution information, representing the value of different features for a sample in a population. One of the tasks in research is to discover some structure in that data. In particular, one is interested in finding something about the causal processes explaining the statistical data, in the form of a theory or a model of the aspect of nature under study. Such causal model can then be used as a basis for explanation and experimentation.

Discovering causal processes generating a body of statistical data is a difficult endeavor. Statistical tools have been designed and used for many years for that purpose. But one of the principal problems with these traditional statistical tools has been their inadequacy at finding the causal structure in the data. These tools are more appropriate at finding correlations, and this is different from causation [8, 14].

In the last decade, people in the philosophical and artificial intelligence community have started to work on novel approaches at using algorithms to automatically discover the causal structure in a body of data, given a certain number of hypotheses [7, 9, 15, 16, 18]. Examples of such algorithms are: WL [21], PC [18], IG [20], and FCI [18].

The most successful approach so far uses the independence relations present in a body of statistical data to infer the causal structure present in the data. Initially sketched by Pearl's group at UCLA [9], it has been principally developed by Glymour's group [18] at CMU and the University of Pittsburgh. Their formal approach is graph theoretical [22], and efforts are made at establishing strong connections between graph theoretic properties and statistical aspects of causal influences. Using fundamental assumptions about the data (the Markov condition and the faithfulness condition), their latest algorithms construct a maximally informative structure called a POIPG (see later) which represents all the independence facts true in a body of statistical data, and is used to infer all causal relations present in the data.

Such POIPGs can be viewed as causal theories, which are satisfied by several

M.L. Dalla Chiara et al. (eds.), Language, Quantum, Music, 49–59.

causal models describing the exact causal relations in the data. In fact, a causal theory can be satisfied by infinitely many different causal models because of the presence of latent variables in the models. One can always add infinitely many latent variables to a model without changing the pattern of causal influences between the measured variables. If we assume causal sufficiency, then the theory will only be satisfied by a handful of causal models [3]. Glymour's approach [18] is particularly effective when dealing with causally sufficient statistical information, but yet is also able to detect the presence of latent common causes between pairs of measured variables. It however doesn't go much beyond that in the exploration of the structure of the unobservable (except in the linear case, by using information about vanishing tetrad differences) [17].

In this short paper, we will not assume causal sufficiency, and we will explore the problem of generating and testing models with latent variables satisfying some arbitrary causal theory [2]. Since infinitely many such models can be generated, part of the efforts will involve the determination of a finite subset of interesting models satisfying the causal theory.

2. FORMAL PRELIMINARIES

The causal models we investigate in this paper contain causal graphs. So we now introduce some important notions from graph theory [18, 22].

By **variable**, we mean a random variable representing some quantity. A variable is called **observable** (or measured) if statistical information is available on it. Otherwise, it is called unobserved (or **latent**).

A **graph** is a pair $(\mathbf{V}_G, \mathbf{E}_G)$ such that $\mathbf{V}_G$ is a non empty finite set of elements called **vertices** (or variables), and $\mathbf{E}_G$ is a finite family of unordered pairs of elements of $\mathbf{V}_G$ called **edges**. The subscript G will be omitted most of the time when talking about a particular graph. A **digraph** is a graph in which the pairs in **E** are ordered pairs. If $(V_1, V_2) \in \mathbf{E}$, then we say there is an edge from V_1 into V_2, represented as $V_1 \rightarrow V_2$.

An **undirected path** in a graph is a finite subset $\{(V_i, V_{i+1}) : i = 1 \text{ to } n\}$ of **E**. We say that the path is from V_1 to V_{n+1}. The path is **directed** if the graph is a digraph. A **simple directed open path** in a digraph is a directed path such that for each $i \neq j$, $V_i \neq V_j$. The path is **closed** if $V_{n+1} = V_1$. A **collider** in a graph G is a pair of directed edges $< (V_1, V_2), (V_3, V_2) > \in \mathbf{E}$. A pair of directed edges that is not a collider is called a **definite non-collider**. A variable V_1 is an **ancestor** of a variable V_2 (or V_2 a **descendant** of V_1) if $V_1 = V_2$ (by convention) or there is a simple directed path from V_1 into V_2. A variable V_1 is a **parent** of a variable V_2 (or V_2 a **child** of V_1) if there is a directed edge from V_1 into V_2.

A **simple digraph** (or directed acyclic graph, DAG) is a digraph in which **E** is a set, and in which there are no simple directed closed paths (no loops). Figure 2 shows two examples of such DAGs. By convention, latent variables are represented by circles, and observables by squares. A **extended simple digraph** is a n+1 tuple $(\mathbf{V}, \mathbf{E}_1, ..., \mathbf{E}_n)$ such that for each i, $(\mathbf{V}, \mathbf{E}_i)$ is a simple digraph. In other words, this is a DAG with more than one kind of edges. A **causal graph** is a directed acyclic graph in which for each $(V_1, V_2) \in \mathbf{E}$, V_1 is a direct cause of V_2. A **causal model** is a structure $M =< G, P >$, where G is a causal graph over a set of variables $\mathbf{V}$, and P is a probability distribution over $\mathbf{V}$. For improved clarity, we will often identify causal models with their causal graphs in situations not involving P specifically.

The expression $I(V_1, V_2, \{W_1, ..., W_N\})$ represents the conditional independence of variables V_1 and V_2 given the set of variables $\{W_1, ..., W_N\}$. A causal model $< G, P >$ satisfies the **Markov condition** if every variable in G is conditionally independent of its non-parents and non-descendants given its parents. $< G, P >$ satisfies the **faithfulness condition** if all and only the conditional independence relations true in P are entailed by the Markov condition applied to $< G, P >$.

In a causal graph, two different variables V_1 and V_2 are **d-separated** given a set of (other) variables $\mathbf{W}$ if and only if there is no undirected path U between V_1 and V_2 such that every collider on U has a descendent in $\mathbf{W}$ and no other variable on U is in $\mathbf{W}$. They are **d-connected** if and only if they are not d-separated. An important theorem is: In a causal model M, variables V_1 and V_2 are d-separated given a set of variables $\{W_1, ..., W_N\}$ if and only if $I(V_1, V_2, \{W_1, ..., W_N\})$. An **inducing path** (relative to a set $\mathbf{S}$ of variables) from variable V_1 to variable V_2 in a causal graph is a path U from V_1 into V_2 $(V_1 \neq V_2)$ such that every variable in $\mathbf{S} \setminus \{V_1, V_2\}$ is a collider on U, and every collider on U is an ancestor of either V_1 or V_2. There is an inducing path from variable V_1 to variable V_2 iff V_1 and V_2 are not d-separated given any subset of $\mathbf{S} \setminus \{V_1, V_2\}$. The inducing paths make the fundamental connection between statistical independence relations and causal relations.

An **inducing path graph** (IPG) relative to a set $\mathbf{S}$ of variables is an extended simple digraph $G = (\mathbf{V}, \mathbf{E}_1, \mathbf{E}_2)$ such that if $(V_1, V_2) \in \mathbf{E}_1$, then there is an inducing path relative to $\mathbf{S}$ from V_1 into V_2, but no inducing path from V_2 into V_1. If there is also an inducing path from V_2 into V_1, then $(V_1, V_2) \in \mathbf{E}_2$. An IPG is essentially a DAG with two kinds of edges: an edge in $\mathbf{E}_1$ will be represented as $V_1 \rightarrow V_2$, and an edge in $\mathbf{E}_2$ as $V_1 \leftrightarrow V_2$, as in Figure 1. From each causal graph, it is possible to compute the corresponding IPG, by determining the inducing paths in the causal graph.

3. CAUSAL THEORIES AND MINIMAL MODELS

Given a set of assumptions, one can determine from independence relations present in a set of statistical data, a general causal theory T characterizing all causal relations between the different variables. A causal theory can be viewed as the deductive closure of the set of I relations produced by the Markov condition, using specific derivation rules. This causal theory is called a 'partially oriented inducing path graph' (POIPG), which is essentially an inducing path graph (IPG) with partially specified edges. All the edges are in the theory, but some of the orientation information is missing. To avoid unnecessary complexity in this paper, we will be using IPGs instead of POIPGs to represent our theories, although the problem of extending a POIPG into IPGs is not without complexities. An IPG simply represents a stronger theory than a POIPG. This theory will be satisfied by several T-equivalent causal models (originally called faithful entailment equivalent models [18]). We aim to construct T-equivalent causal models that will satisfy some arbitrary theory, and to define general methods to try to falsify (or distinguish between) some of these T-equivalent causal models.

In this paper, we will be using a simple representative example (Figure 1) to illustrate our general methods. But first, given that a causal theory is satisfied by infinitely many T-equivalent causal models, it is necessary to restrict the models under study to a finite class of the simplest possible models. We therefore now introduce the notion of a minimal model.

Let $\mathbf{S}_1$ be the set of all causal models. Let $\mathbf{S}_2$ be the set of all IPGs. Let M be a causal model in $\mathbf{S}_1$ with its causal graph G. Let e be in $\mathbf{E}$, the set of all edges of G. Let v_1, v_2 be in $\mathbf{V}$, the set of all vertices of G. The operator $\mathbf{IPG}() : \mathbf{S}_1 \to \mathbf{S}_2$ is defined such that $\mathbf{IPG}(M)$ is the inducing path graph of M. The operator $\mathbf{RE}_e() : \mathbf{S}_1 \to \mathbf{S}_1$ is defined such that $\mathbf{RE}_e(M)$ has the same causal graph as M, but with edge e removed. The operator $\mathbf{MV}_{v_1,v_2}() : \mathbf{S}_1 \to \mathbf{S}_1$ is defined such that $\mathbf{MV}_{v_1,v_2}(M)$ has the same causal graph as M, but with variable v_2 substituted for v_1 in all edges of G. Thus v_1 and v_2 become merged in the new causal graph.

Let M be a model for a causal theory T, and let G be its causal graph. Let $\mathbf{O}_1$ be the set of all possible operators $\mathbf{RE}_e()$, where e is an edge in G. Let $\mathbf{O}_2$ be the set of all possible operators $\mathbf{MV}_{v_1,v_2}()$, where v_1 and v_2 are distinct variables in G. A **reducing transformation** $\mathbf{R}$ is a finite composition of operators in $\mathbf{O}_1 \cup \mathbf{O}_2$. If there is a reducing transformation $\mathbf{R}$ such that $\mathbf{IPG}(\mathbf{R}(M)) = \mathbf{IPG}(M)$, then the model $\mathbf{R}(M)$ is a reduction of M. The model is **minimal** (or irreducible) if for every reduction $\mathbf{R}(M)$ of M, we have $\mathbf{R}(M) = M$.

In other words, a minimal model is a model in which it is not possible to remove an edge or to merge two variables in its graph without always ending up with a new model which has a different inducing path graph. Thus minimal models are the simplest models that will satisfy a causal theory. In the example, Figure 1 shows a very simple causal theory T, represented by an IPG, and Figure 2 shows the causal graphs of the only two minimal models that satisfy theory T. The difference between the two minimal models in this case is very small. With more complex theories, the minimal models can present significant differences.

Before studying the question of testing minimal models satisfying a causal theory, it is necessary to find a systematic approach to generating such minimal models given some arbitrary causal theory. Both problems are complex, and have recently been investigated by the author. Although all the details are included in [3–5], we will now provide a simple overview of our strategies.

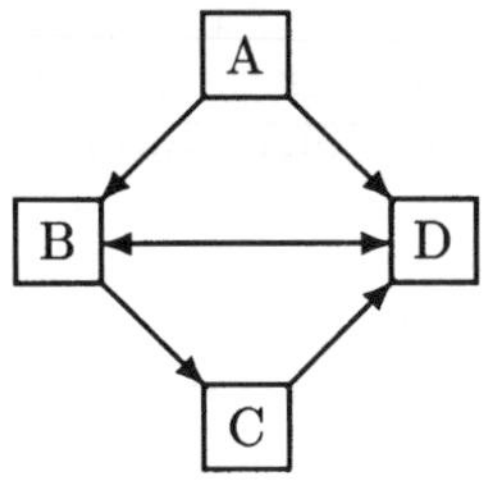

Figure 1. Some causal theory T

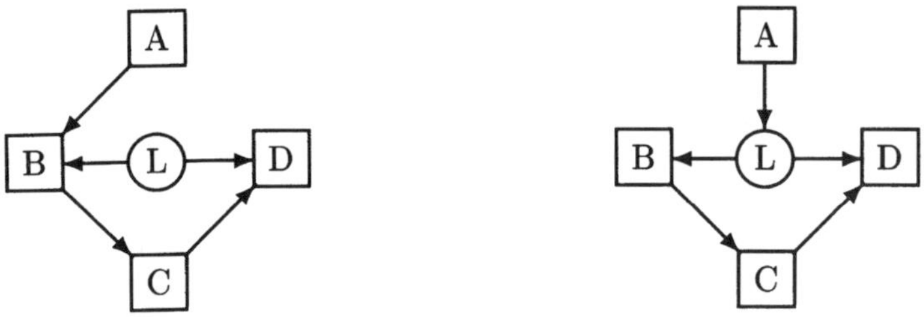

Figure 2. The causal graphs of the only two minimal models satisfying T: M_1 (left), M_2 (right)

4. GENERATING MINIMAL MODELS

To fully understand that question, we need to clarify the relation between IPGs and causal models. Take a causal model M with causal graph G over the set $\mathbf{V}$ of variables $\{v_1, \ldots, v_n\}$. The IPG of G over the subset $\mathbf{V}'$ of the observable variables in $\mathbf{V}$ is a subgraph G' of the complete graph on $\mathbf{V}'$. In G', if there is a directed edge $A \rightarrow B$, then either A is a direct cause of B, or there is an indirect inducing path between A and B. As a reminder, an inducing path between A and B relative to $\mathbf{V}'$ is a path U from A into B involving a combination of variables in $\mathbf{V}'$ (observables) and variables in $\mathbf{V}'' = \mathbf{V} \setminus \mathbf{V}'$ (latents). Variables in $\mathbf{V}'$ on U must all be colliders and ancestors of A or B. Variables in $\mathbf{V}''$ on U can be pretty much anything, but if they are colliders, they must be ancestors of A or B. So the causal relations between observable variables in the causal model is strongly determined by the edges in the IPG, while the causal relations involving latent variables have a very high degree of freedom. In G', if there is a bidirected edge $A \leftrightarrow B$, then by acyclicity, there must be a latent common cause L between A and B. This might be a direct common cause, or a common cause of some other pair of variables (that could include A or B), which spreads to A and B through the structure of the inducing paths.

In our initial approach to the generation of causal models, we used an important constraint on minimal models to reduce the computational complexity of the task. That constraint is that a minimal model must not contain any direct causal edge between two latent variables. Overall, this constraint reduces the computational complexity of the problem, without affecting the final results in most cases. The computational advantage comes from the simplified structure of inducing paths in graphs satisfying the assumption. This enables us to specify well defined bounds to the structure of any minimal model satisfying a theory (IPG) T.

With the assumption, what is the structure of an inducing path between two observable variables in a minimal model? We know by the assumption that there cannot be two adjacent latents in an inducing path in the causal graph of a minimal model. Furthermore, except for the endpoints of the path, there cannot be two adjacent observables in an inducing path. This is because of the definition of the inducing path. Assume without loss of generality that there is an inducing path between variables A and B containing and edge $C \rightarrow D$, where C and D are observables different from the endpoints of the path. An inducing path between A and B relative to a set of observable variables containing at least A, B, C and D would force C to be a collider in the path, which it cannot be because of the edge $C \rightarrow D$. Thus any inducing path between A and B must involve an alternation of observables and latents (except for the endpoints), in which every observable is a collider, and every latent is a common parent. This special structure of the inducing paths enables us to efficiently generate minimal models directly from a slight modification of the IPG of theory T.

Let T be a causal theory, represented by an IPG. We will apply a simple transformation to T, which will make it the graph of a maximal model T′ In T, simply replace every edge of the form $A \leftrightarrow B$ by a group of edges of the form $B \leftarrow A \leftarrow L_{AB} \rightarrow B \rightarrow A$, where L_{AB} is a new latent variable. Note that **IPG**(T′) = T always (assuming we allow the temporary cyclicity introduced by the transformation). We also define the following additional operator, to take care of embedded latent variables: the operator $\mathbf{EL}_{v_1,v_2,v_3,l_1}() : \mathbf{S}_1 \rightarrow \mathbf{S}_1$ is defined such that $\mathbf{EL}_{v_1,v_2,v_3,l_1}(M)$ has the same causal graph as M, but if there is a pattern $[v_1 \rightarrow v_2, v_1 \rightarrow v_3, l_1 \rightarrow v_2, l_1 \rightarrow v_3]$ in the model where l_1 is a latent variable and v_1, v_2 and v_3 are observable variables, it will be replaced by the pattern $[v_1 \rightarrow l_1, l_1 \rightarrow v_2, l_1 \rightarrow v_3]$. An **extended reducing transformation** is a reducing transformation that also include operators of the **EL**() type. An **extended reduction** is defined in a similar fashion.

We have the following main theorem: If $M = < G, P >$ is a minimal model satisfying T, and G does not contain direct causal edges between any two latent variables, then M is an extended reduction of T′. The theorem is proven in [3, 4]. A simplified version of the initial transformation of T, in which edges of the form $A \leftrightarrow B$ are instead replaced by edges of the form $A \leftarrow L_{AB} \rightarrow B$ would have been much more elegant, and does enable us to generate most minimal models, but not all of them. A complete analysis is included in [3].

Thus every minimal model satisfying a causal theory can be obtained by finding an appropriate extended reducing transformation of T′. In practice, it is not necessary to test large numbers of transformations. We can indeed produce generation algorithms which are quite computationally efficient. Relaxing the assumption previously described increases the computational complexity of the model generation task [3].

Besides providing a useful tool for generating models, our results can be used to answer many specific statistically interesting questions. For example given some distribution information postulating the existence of more than one latent common cause, it helps us determine when two common causes can be the same, and when they must be different.

5. SELECTING MINIMAL MODELS

Once a set of causal models is generated, all of which satisfy a theory T (and therefore the set of *I* relations present in the statistical data), it is possible to find general methods that will allow a researcher to test (or distinguish between) these different models [11–13, 19]. For example, in Figure 2, both minimal models satisfy the same causal theory of Figure 1. Since both minimal models share the same *I* relations on the observable variables, it is necessary to go beyond these *I* relations to be able to distinguish them. A simple solution is to use an experimental approach, which involves manipulations on the piece of nature having generated the original empirical information, the acquisition of additional data and the production of a new theory.

Glymour [18] describes a **controlled experiment** as a procedure involving a new variable (a manipulation variable *M*), which is applied to one of the observable variables *V* in the data. A variable *V* is experimentally manipulated if all causal influences to it are eliminated by the manipulation, and the manipulation affects other variables in the data only through the manipulated variable *V*. Graph theoretically, this corresponds to the following procedure: in the original model of the non-experimental population, all the edges into *V* are broken, and there is a new edge from *M* to *V* inserted. Also, there must be no path between *M* and any other variables in the model that does not contain the edge from *M* to *V*. Figures 3 and 4 represent the new theories after manipulations of variables *B* and *C* of the previous minimal models. The new theories of the manipulated phenomenon will not in general be satisfied by the same causal models. This will enable us to distinguish the two original models based on the new theory produced by the manipulation variable *M*.

The only caveat to the experimental approach is that it requires the acquisition of additional empirical data on the piece of nature under study, and the experimental control of one or more specific variables. In practice, this is not always possible. Can non-experimental approaches be used to distinguish minimal models? We will now offer two such approaches.

The first approach is based on the principle of instrumental variables, a concept often used in economics [1]. The use of instrumental variables for

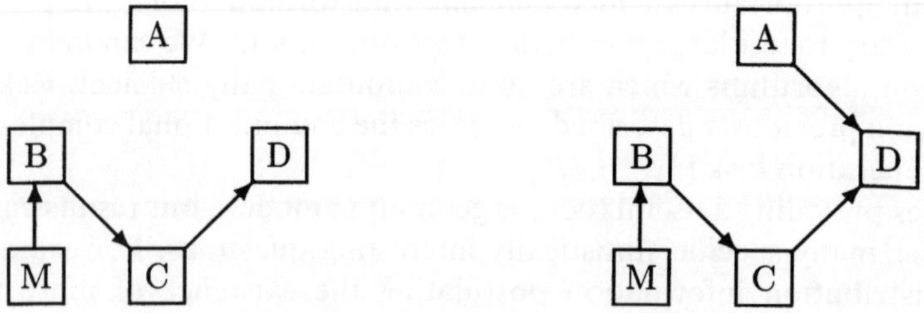

Figure 3. The minimal models with a manipulation on B

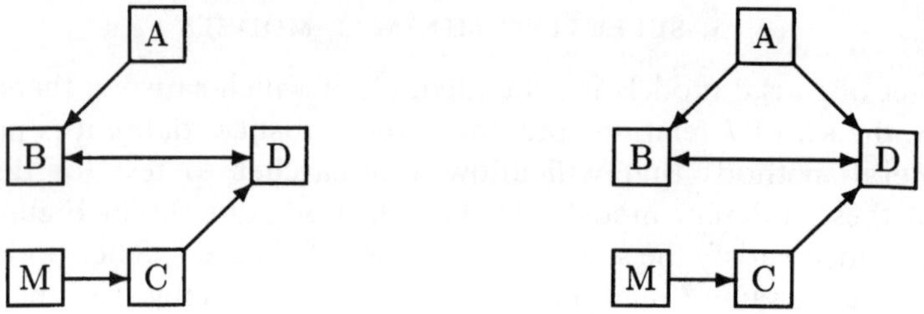

Figure 4. The minimal models with a manipulation on C

nonparametric models with latents has recently been sketched by Pearl [13]. It is necessary to first translate the two models we are trying to distinguish into structural equations [1]: for M_1: B = h(A,L), C = g(B), D = f(C,L), and for M_2: B = h'(L), C = g'(B), D = f'(C,L), L = e(A). A variable Z is defined as an **instrumental variable** relative to an ordered pair (X, Y) of variables, if X and Y are generated by the processes $x = g(z,u)$ and $y = h(x,u)$, where g and h are arbitrary deterministic functions, and U is a random latent variable independent of Z. But in M_1, variable A is such an instrumental variable, since $b = h(a,l)$, $d = f(g(b),l)$ and A is independent of latent L. However A is not an instrumental variable in M_2, since among other things, A is not independent of L.

Pearl proves [13] that a necessary condition for A to be an instrumental variable in this case is that the conditional distribution $P(d,b|a)$ in M satisfies $max_b \sum[max_a P(d_i, b|a)] \leq 1$, where the sum is over all the values of variable D. This constraint can be directly evaluated by looking at the statistical data at hand, and does not require the acquisition of additional data. Unfortunately this is only a necessary constraint. Thus if the data does not satisfy the constraint, model M_1 is falsified and we will infer that model M_2 is the correct minimal model truly representing the causal influences in the data at hand. But if the data does satisfy the constraint, then this does not even provide supportive evidence for model M_1 or against M_2, as the constraint is not sufficient to demonstrate the instrumentality of variable A. What we would prefer to find are necessary and sufficient criteria for falsifying a model. This will be provided by the second approach.

For the second observational approach, it is worth noting that to create the initial theory T, only independence relations between observables have been used. However the presence of latents in M_1 and M_2 creates additional I relations involving latents. The Markov condition on each of the minimal models of Figure 2 entails the following I relations (grouping symmetric ones together): For M_1: $I(a, L, \{\})$, $I(L, a, \{\})$, $I(c, a, \{b\})$, $I(c, L, \{b\})$, $I(d, a, \{L, c\})$, $I(d, b, \{L, c\})$. For M_2: $I(b, a, \{L\})$, $I(c, a, \{b\})$, $I(d, a, \{L, c\})$, $I(d, b, \{L, c\})$, $I(c, L, \{b\})$.

Thus by considering all the variables in each model, there is a difference in the resulting set of I relations. It is therefore possible in each case to extend the initial theory into two stronger but different theories. As previously indicated, a theory is the deductive closure of the set of I relations produced by the Markov condition, using specific derivation rules [6, 10]. By neglecting the symmetry rules, we produce the following larger sets of I relations. The I relations common to both models are $I(c, a, \{b\})$, $I(c, L, \{b\})$, $I(d, a, \{c, L\})$, $I(d, b, \{c, L\})$, $I(c, a, \{b, L\})$, $I(d, a, \{b, c, L\})$, $I(d, b, \{a, c, L\})$, $I(a, c, \{b, d, L\})$, $I(a, d, \{b, L\})$, $I(c, L, \{a, b\})$. The I relations unique to M_1 are $I(a, L, \{\})$. And the I relations unique to M_2 are $I(b, a, \{L\})$, $I(a, c, \{L\})$, $I(a, d, \{L\})$, $I(a, c, \{L, d\})$, $I(a, b, \{L, d\})$, $I(a, b, \{L, c\})$, $I(a, b, \{L, c, d\})$.

It is easy to determine the truth of I (independence) relations involving only observable variables by looking at the available statistical information on these observables. But no data is available on the latent variables, so it is not directly possible to determine the truth of I relations involving at least one latent variable. Clever indirect methods are needed to determine their truth or falsehood. We now propose such a method, based on distributions on the original data.

Consider the expression: $\sum_i P(b_i|a) \cdot P(d|a, b_i, c)$, where the sum is over all possible values of b in the original data. This equation was first proposed by Verma [20]. We will show that for minimal model M_1 of Figure 2, this expression is not a function of a, whereas for minimal model M_2, it is.

In M_1, one infers that (*) $\sum_i P(b_i|a) \cdot P(d|a, b_i, c)$ is a function $g(c, d)$, since

$$\begin{aligned}\sum_i P(b_i|a) \cdot P(d|a, b_i, c) &= \sum_j \sum_i P(b_i|a) \cdot P(d|a, b_i, c, L_j) \cdot P(L_j|b_i, c, a)\\ &= \sum_j P(d|c, L_j) \cdot \sum_i P(b_i|a) \cdot P(L_j|b_i, a) = \sum_j P(d|c, L_j) \cdot P(L_j|a)\\ &= \sum_j P(d|c, L_j) \cdot P(L_j) = g(c, d)\end{aligned}$$

This relation, non-inferable from the I relations on the observables alone, was inferred using new I relations with the latent variable L: $I(d, a, \{c, L\})$, $I(d, b, \{c, L\})$, $I(c, L, \{a, b\})$, $I(a, L, \{\})$ all hold in M_1, but $I(a, L, \{\})$ does not hold in M_2. Since (*) does not involve specific values of latent variables, it is possible to evaluate it using only the original statistical data on the measured variables. There is no need to gather additional empirical data on the variables. If the data does not satisfy constraint (*), model M_1 will be falsified, and if it does satisfy it, model M_2 will be falsified (after some minor additional verifications). A general method at producing constraints like (*) given two causal models, as well as an analysis of the problem can be found in [3, 5].

6. CONCLUSION

This short paper gave a brief overview of our work on the generation and selection of models for causal theories. We have briefly described an approach at generating a set of simple models given a causal theory T. This task is computationally intensive, and major efforts are spent at creating computationally efficient algorithms for model generation. Given the assumption that no two latent variables can be directly causally related, we have been able to clearly bound the space of all possible models satisfying a causal theory T. The relaxation of this main assumption offers more flexibility, but at the price of increased computational complexity.

We have also described different approaches at model selection for causal theories. An experimental approach requires the acquisition of new data, and is not always possible for the problem at hand. But we have shown that in some cases, it is nevertheless possible to distinguish between two models satisfying the same causal theory by using information that goes beyond simple independence relations between observable variables.

University of Pittsburgh

REFERENCES

1. Bowden, R.J., Turkington, D.A.: 1984, *Instrumental Variables.* Cambridge University Press, Cambridge, MA.
2. Desjardins, B.: 1995, *Equivalence of Causal Theories.* Preprint.
3. Desjardins, B.: *Formal Discovery of Causal Structure.* PhD dissertation (in progress).
4. Desjardins, B.: *Generating Equivalent Causal Models, A Computational Perspective* (to be published).
5. Desjardins, B.: *On the Non-Isomorphism of Equivalent Causal Models* (to be published).
6. Geiger, D., Pearl, J.: 1989, 'Logical and Algorithmic Properties of Conditional Independence and Qualitative Independence', *Royal Statistical Soc B.*
7. Glymour, C., Scheines, R., Spirtes, P., Kelly, K.: 1987, *Discovering Causal Structure.* Academic Press, New York.
8. Mosteller, F., Tukey, J.: 1977, *Data Analysis and Regression, A Second Course in Regression.* Addison-Wesley, Massachusetts.

9. Pearl, J.: 1988, *Probabilistic Reasoning in Intelligent Systems.* Morgan Kauffman, San Mateo CA.
10. Pearl, J., Geiger, D., Verma, T.S.: 1990, 'The Logic of Influence Diagrams', in Oliver, R.M. and Smith, J.Q., eds., *Influence Diagrams, Belief Networks and Decision Analysis.* John Wiley and Sons Ltd, Sussex, England, 67–87.
11. Pearl, J.: 1995, 'Causal Diagrams for Empirical Research', *Biometrika*, **82(4)**, 669–709.
12. Pearl, J.: 1995, *On the Identification of Nonparametric Structural Models.* Technical Report, UCLA Cognitive Sciences Laboratory, Nov 1995. To appear in *Latent Variable Modeling with Application to Causality.* Springer Verlag, Lecture Notes Series.
13. Pearl, J.: 1995, 'On the Testability of Causal Models with Latent and Instrumental Variables', in Besnard, P., Hanks, S. eds., *Uncertainty in AI 11.* Morgan Kaufmann, San Francisco, CA, 435–443.
14. Rawlings, J.: 1988, *Applied Regression Analysis.* Wadsworth, Belmont, CA.
15. Schacter, R.: 1986, 'Evaluating Influence Diagrams', *Operations Research*, **34(6)**.
16. Spiegelhalter, D., Dawid, A., Lauritzen, S., Cowell, R.: 1993, 'Bayesian Analysis in Expert Systems', *Statistical Science*, **8(3)**.
17. Spirtes, P.: 1991, 'Building Causal Graphs From Statistical Data in the Presence of Latent Variables', in Skyrms, B., ed., *Proceedings of the IX International Congress on Logic, Methodology and Philosophy of Science, Uppsala, Sweden*
18. Spirtes, P., Glymour, C., Scheines, R.: 1993, *Causation, Prediction, and Search.* Springer-Verlag, New York.
19. Spirtes, P., Verma, T.: 1994, *Equivalence of Causal Models with Latent Variables.* Technical Report CMU-Phil-33, 1994.
20. Verma, T., Pearl, J.: 1991, 'Equivalence and Synthesis of Causal Models', *Proceedings of the Sixth Conference on Uncertainty in AI, Mountain View, CA*, pp 220–227.
21. Wermuth, N., Lauritzen, S.: 1983, 'Graphical and Recursive Models for Contingency Tables', *Biometrika*, **72**.
22. Wilson, R.J.: 1985, *Introduction to Graph Theory.* Longman, Essex, England.

SERGIO FAJARDO

NONSTANDARD ANALYSIS AND A CLASSIFICATION OF PROBABILITY SPACES

1. INTRODUCTION.

Since 1960 and with the admittedly unappealing name of Nonstandard Analysis, infinitesimal and infinite quantities have made a comeback into the mathematical playground. The return, after Bishop Berkely's fierce attack in the early years of the past century, hasn't been easy. The following paragraph of the well known Probabilist, Joseph L. Doob ([D]) is very explicit and touches upon several issues related to this paper.

"The development of science is not a simple progression from one advance to the next. Judged by hindsight, the development is slow, proceeds in zigzag course, with many wrong turns and blind alleys, and frequently moves in directions condemned by leading scientists. In the 1930s Banach Spaces were sneered at as absurdly abstract, later it was the turn of locally convex spaces, and now it is the turn of nonstandard analysis. Mathematicians are no more eager than other humans to embrace new ideas, and full acceptance of mathematical probability was not realized until the second half of the century. In particular, many statisticians and probabilists resented the mathematization of probability by measure theory, and some still place mathematical probability outside analysis . . .".

This paper is about new ideas in mathematics. It deals with developments emerging from the interaction of three areas: Mathematical Logic, in particular Model theory, Nonstandard Analysis and Probability Theory. Our purpose is simple: to introduce the main ideas and results through a special, quickly understandable and yet interesting case.

The contents are as follows. Section 2 gives some basic definition needed for better readability and further reference. Section 3 presents material directly inspired by the impressive results obtained by Keisler, [K1], using nonstandard methods to solve Stochastic Differential Equations. In order to explain why he obtained those results, he came up with a new approach, that can be called: Model Theory of Stochastic Processes. Section 4 brings into the picture a completely new theory, the Theory of Neometric Spaces. Once again Keisler is behind the main ideas, the inspiration for this theory came from a problem that has always been around the Nonstandard mathematical community: explain in standard terms what is done using nonstandard analysis. This was achieved in the series of papers [FK1–4]. Putting together these two sections we came to the title of the paper: we have at hand a Classification of Probability Spaces.

M.L. Dalla Chiara et al. (eds.), Language, Quantum, Music, 61–71.

Warning

An observation has to be made: we have written the paper so that a reader with no knowledge of nonstandard analysis can still go through it and enjoy and grasp the main ideas. Nonstandard analysis only appears in definitions of some concepts but here don't explictly use it. So if you come across a nonstandard point that you don't understand, read on that there are plenty of good ideas around!

2. BASIC DEFINITIONS

Necessarily we have to assume familiarity with basic notions of probability theory. Nonetheless we include some of them here in order to fix notation and for the reader that may one to check them out. Good references are [B], [EK].

Definition 2.1 Let $\underline{\Omega} = (\Omega, \mathcal{F}, P)$ and $\underline{\Gamma} = (\Gamma, \mathcal{G}, Q)$ be probability spaces.

(a) A real valued *Random Variable* X on $\underline{\Omega}$ is an $\mathcal{F}$-measurable function $X : \Omega \to \mathbb{R}$. If instead of taking values in $\mathbb{R}$ we allow the function to take values in a complete, separable metric spaces M, we say that it is an M-valued random variable.

(b) Given random variables X and Y defined on spaces $\underline{\Omega}$ and $\underline{\Gamma}$ respectively, we say that X and Y have *the same Distribution* if for each Borel subset A of $\mathbb{R}$, $P(X^{-1}(A)) = Q(Y^{-1}(A))$. We denote this relation by $X \equiv_0 Y$. Notice that this definition also makes sense for the case of M-valued random variables, simply taking Borel subsets of M in place of the Borel subsets of $\mathbb{R}$.

A useful characterization of the relation $\equiv_0$ for M-valued random variables, possibly living on diferent probability spaces, is illustrative:

$X \equiv_0 Y$ if and only if for each $f : M \to \mathbb{R}$, continuous and bounded $E(f(X)) = E(f(Y))$, which means $\int_\Omega f(X)dP = \int_\Gamma f(Y)dQ$.

As indicated in the introduction, we are going to classify probability spaces according to their 'Richness', here is a definition of a well known concept that is the minimum we require. See [HK] section 4.

Definition 2.2

(a) A Probability Space $\underline{\Omega} = (\Omega, \mathcal{F}, P)$ is *Atomless* (or *Non-Atomic*) if for each set $A \in \mathcal{F}$ with $P(A) > 0$, there is a $B \in \mathcal{F}$, with $B \subseteq A$, such that $0 < P(B) < P(A)$.

(b) Let $\mathcal{G}$ and $\mathcal{D}$ be sub-σ-algebras of $\mathcal{F}$, with $\mathcal{G} \subseteq \mathcal{D}$. $\mathcal{D}$ is Atomless over $\mathcal{G}$, if for each set A$\in \mathcal{D}$ with $P(A) > 0$, there is a $B \in \mathcal{D}$, with $B \subseteq A$, such that on a set of positive measure $0< P(B|\mathcal{G})< P(A|\mathcal{G})$.

(c) $\Omega = (\Omega, \mathcal{F}, \mathcal{P})$, is said to be $\aleph_1$-*Atomless* (or $\aleph_1$-Non-Atomic) if for each countably generated σ-algebra $\mathcal{C}$ contained in $\mathcal{F}$, $\mathcal{F}$ is $\aleph_1$-atomless over $\mathcal{C}$.

Finally we have to introduce a well known family of Probability spaces built using nonstandard analysis. For a good introduction to nonstandard analysis and probability we recommend [AFHL] and the references there in. The collections [C1,2] contain several expositoty papers. In spanish, you can read [F1].

Definition 2.3 A *Hyperfinite Loeb* space is a probability space of the form $(\Omega, L(\Omega), L(\mu))$ where Ω is a hyperfinite set, $L(\Omega)$ is the Loeb σ-algebra over the set of internal subsets of Ω, μ is the internal counting measure over Ω and $L(\mu)$ is its associated Loeb measure over L(Ω). The hyperfinite size (i.e the internal cardinality) of a hyperfinite set A is denoted by $|A|$.

3. UNIVERSALITY, SATURATION AND HOMOGENEITY

In this section we introduce, in the basic probabilistic setting, the different ways of comparing probability spaces that have been studied from a Model Theoretic point of view of Probability Theory. The first concept is very well known in Probability but it is disguised, as we will quickly see, with a familiar name in Logic.

Definition 3.1 A Probability space $\underline{\Omega}$ is said to be *Universal* if for every random variable X on an arbitrary space $\underline{\Gamma}$ there is a random variable Y on $\underline{\Omega}$ such that $X \equiv_0 Y$. Note: this property holds for all M-valued variables, with M metric, complete and separable.

Are there Universal Spaces?. Of course, a very familiar space is Universal: the unit interval with the Borel σ-algebra and Lebesgue measure is the typical example. In fact, there are plenty of Universal spaces around, here is the reason (the proof is left as an exercise).

Proposition 3.1 *A space Ω is Universal if and only if it is Atomless.*

Naturally our Hyperfinite Spaces are Universal: using this proposition it follws as an immediate corollary, since it is trivial to see that Hyperfinite spaces are Non-atomic. Moreover, with a little extra work it is equally easy to verify that Hyperfinite spaces are $\aleph_1$-atomless. On the other hand, a direct proof of Universality is also illustrative ans easy (see for example, [K2,3] or [F1]).

Building on the definition of Universality we can introduce our next notion; first, observe the following:

Suppose X is random variable on $\underline{\Lambda}$ and Y is a random variable on a Universal space $\underline{\Omega}$ such that $X \equiv_0 Y$, if we are given a new random variable X' on $\underline{\Omega}$ and consider now the pair (X, X'), by universality of $\underline{\Lambda}$ we can get on it a pair (Z, Z') such that $(X, X') \equiv_0 (Z, Z')$, but let's remark something simple and important: in principle we can't guarantee that the variable Z is the variable X, they may be quite different but of course, they must have the same distribution. This observation leads to a new concept, stronger than Universality, also inspired by Model Theory, see [HK] and [K2].

Definition 3.2 A Probability space $\underline{\Omega}$ is said to be *Saturated* if it has the following property: given a random variable X on an arbitrary space $\underline{\Gamma}$ and Y on $\underline{\Omega}$ such that $X \equiv_0 Y$, if X' is another variable on $\underline{\Gamma}$ there exists Y' on $\underline{\Omega}$ such that $(X, X') \equiv_0 (Y, Y')$.

The natural question is now: are there saturated spaces?. One quick non-example: the Unit interval is not saturated! What went wrong?.

In order to answer this question, let's introduce some variations on the definitions previously given: if in the definitions of Universality and Saturation we restrict the attention to simple random variables (i.e. taking finitely many values), we obtain, in principle, weaker notions respectively called *Simple Universality* and *Simple Saturation.* The following proposition explains what goes on.

Proposition 3.2

(a) *Universality = Simple Universality.*

(b) *Universality = Simple Saturation.*

(c) *Saturation is stronger than Simple Saturation (i.e. Universality)*

Proof: (a) and (b) are left as an exercise, (c) was explained before.

We need to exhibit Saturated Spaces, there are plenty of them.

Theorem 3.1 *Hyperfinite Spaces are Saturated.*

A nice proof of this theorem will be given below. In fact, there are many more saturated Spaces: Hoover and Keisler in [HK], section 4, deal with this issue and state that using the results from Maharam [M1,2] it is easy to show the existence of Saturated Spaces; for example, uncountable powers of the unit interval are saturated.

There is an interesting follow up to the saturation case of Proposition 3.1 above, where we characterized universal spaces as Nonatomic.

Proposition 3.3 *A space $\underline{\Omega}$ is Saturated if and only if it is $\aleph_1$-atomless.*

We postpone the proof of the left to right half of this proposition in order to introduce later on another method of proof.

It is now time to introduce a property rarely found in the 'standard' literature on Probability Spaces: Homogeneity. It has nothing to do with the two previous ones, but we will see what happens when combined with Universality. (see [K3]).

Definition 3.3

(a) A bijection $h : \Omega \to \Omega$ is called an *Automorphism* of $\underline{\Omega}$ if it preserves measures and $h(\mathcal{F}) = \mathcal{F}$.

(b) The space $\underline{\Omega}$ is said to be *Homogeneous* if for every pair of random variables X and Y on $\underline{\Omega}$ with $X \equiv_0 Y$ there is an Automorphism h of $\underline{\Omega}$ such that $X(\omega) = Y(h(\omega))$ a.s.

Observe the following simple fact: if there is an automorphism h of Ω such that $X(\omega) = Y(h(\omega))$ a.s. then $X \equiv_0 Y$.

Once again we ask the obvious question: Are there Homogeneous Spaces? At this stage the answer should be no surprise.

Theorem 3.2 *Hyperfinite Spaces $\underline{\Omega}$ are Homogeneous.*

The proof of this result is a natural nonstandard argument and its details can be worked out by the interested reader or found in [K3] or [F1]. Let's remark that the Unit space is not homogeneous: this fact can be checked directly or is an immediate consequence of the following one together with the previous observation that it is not Saturated.

Theorem 3.3 *If $\underline{\Omega}$ is Universal and Homogeneous then it is Saturated.*

Proof: Let $\underline{\Gamma}$ be an arbitrary space and X a variable on it, such that $X \equiv_0 Y$ with Y a variable on $\underline{\Omega}$. Now we are given a new variable X' on $\underline{\Gamma}$ and want to find Y' on $\underline{\Omega}$ such that $(X, X') \equiv_0 (Y, Y')$. By universality of the space $\underline{\Omega}$ there exists a pair (Z, Z') such that $(X, X') \equiv_0 (Z, Z')$.

In consequence, $Z \equiv_0 Y$ and using homogeneity in $\underline{\Omega}$, there is an automorphism of Ω such that $Y(\omega) = Z(h(\omega))$ a.s. We now use this h in order to define Y' in the obvious way: $Y'(\omega) = Z'(h(\omega))$. Then we get $(Z, Z') \equiv_0 (Y, Y')$ and consequently $(X, X') \equiv_0 (Y, Y')$ as we wanted to show. □

As before we ask: Are there Saturated Spaces which are not Homogeneous? The answer is yes and it is a good exercise. Let's add an interesting remark: the proof of Theorem 3.3 is what can be called a 'Soft' proof, very much on the line of well known Soft Model Theoretic arguments. This is no coincidence and shows the Model Theoretic inspiration behind the Classification Theory we are developing.

Here is another example of a proof following a Soft Model Theoretic argument, as promised, we proceed to sketch a proof of the following half of Proposition 3.3: Saturation implies $\aleph_1$-atomlessness.

Proof: Given a countably generated σ-algebra $\mathcal{G} = \sigma(G_n : n \in \mathbb{N})$ with $G_n \in \mathcal{F}$, we can assume without loss of generality that the G_n's form a partition of Ω. Let $A \in \mathcal{F}$ with $P(A) > 0$, we want to get $B \in \mathcal{F}$ with $B \subseteq A$ such that $0 < P(B|\mathcal{G}) < \mathcal{P}(A|\mathcal{G})$.

Now let $\underline{L}$ be a fixed Hyperfinite space. By Universality, there exists a sequence $(D_n)_{n \in \mathbb{N}}$ such that $(\underline{\Omega}, (\chi_{C_n})_{n \in \mathbb{N}}) \equiv_0 (\underline{L}, (\chi_{D_n})_{n \in \mathbb{N}})$, and by Saturation of L there exists a set D with $(\underline{\Omega}, (\chi_{C_n})_{n \in \mathbb{N}}), \chi_C) \equiv_0 (\underline{L}, (\chi_{D_n})_{n \in \mathbb{N}}), \chi_D)$. Using the fact that $\underline{L}$ is $\aleph_1$-atomless (see the remark after proposition 3.1), there exists $E \subseteq D$ with $0 < \lambda(E|\sigma(D_n)) < \lambda(D|\sigma(D_n))$.

Using Saturation of $\underline{\Omega}$, we can find B such that there exists

$$(\underline{\Omega}, (\chi_{C_n})_{n \in \mathbb{N}}), \chi_C, \chi_B) \equiv_0 (\underline{L}, (\chi_{D_n})_{n \in \mathbb{N}}), \chi_D, \chi_E)$$

To finish the proof, we leave to the reader to verify that this set B does exactly what we want. □

What is new here?. This proof is an example of the so called 'Come-Back Problem' ([F2]) which intuitively does the following: first prove a given property within Hyperfinite space where, due to its Nonstandard features, things are simpler and then use a Model Theoretic (Soft) argument prove that it is true elsewhere. Of course this general statement can't be true, but there are many instances where it works, the problem is then to identify such properties. (See [F2] for details).

4. RICHNESS

Now it is time to move in another direction. As mentioned in the introduction, Nonstandard Analysts have looked for alternative ways of explaining their results to the 'standard' mathematical community (recall Doob's statement). Work in this direction has recently led to new and promising developments: the Theory of Neometric Spaces. The first steps towards this theory appeared in the paper 'From Discrete to Continuous Time' due to Keisler [K4], which explained some of the main features of nonstandard practice via a Forcing Theorem which almost" used no nonstandard analysis. Stepping on this paper, the series of papers [FK1–4] went all the way so that the nonstandard analysis disappears of the picture and a standard theory, the Theory of Neometric Spaces, was born and with it the original purpose was achieved. The papers mentioned above and the survey paper [K5] are suggested for a thorough study of Neometric Spaces.

The main examples of the Theory of Neometric Spaces come from Probability, and in particular, Stochastic Analysis. Using the new concepts we are able to introduce a new richness property of probability spaces. The approach is completely different from the one in the previous section, but, surprise, they turned out to be closely related. We will see. First let's introduce some notation and present the new basic concepts.

We use script letters $\mathcal{M}, \mathcal{N}, \mathcal{O}$ for complete metric spaces which are not necessarily separable, and let ρ, σ, τ be their metrics. Given two metric spaces $(\mathcal{M}, \rho)$ and $(\mathcal{N}, \sigma)$, the *product metric* is the metric space $(\mathcal{M} \times \mathcal{N}, \rho \times \sigma)$ where $(\rho \times \sigma)((\mathrm{x}_1\mathrm{x}_2), (\mathrm{y}_1, \mathrm{y}_2)) = \max(\rho(\mathrm{x}_1, \mathrm{y}_1), \sigma(\mathrm{x}_2, \mathrm{y}_2))$.

Definition 4.1 Let $\mathbb{M}$ be a collection of complete metric spaces $\mathcal{M}$ which is closed under finite products, and for each $\mathcal{M} \in \mathbb{M}$ let $\mathcal{B}(\mathcal{M})$ be a collection of subsets of $\mathcal{M}$, which we call *basic sets*. By a *neocompact family* over $(\mathbb{M}, \mathcal{B})$ we mean a triple $(\mathbb{M}, \mathcal{B}, \mathcal{C})$ where for each $\mathcal{M} \in \mathbb{M}, \mathcal{C}(\mathcal{M})$ is a collection of subsets of $\mathcal{M}$ with the following properties, where $\mathcal{M}$ and $\mathcal{N}$ vary over $\mathbb{M}$.

(a) $\mathcal{B}(\mathcal{M}) \subseteq \mathcal{C}(\mathcal{M})$;

(b) $\mathcal{C}(\mathcal{M})$ is closed under finite unions; that is, if $A, B \in \mathcal{C}(\mathcal{M})$ then $A \cup B \in \mathcal{C}(\mathcal{M})$;

(c) $\mathcal{C}(\mathcal{M})$ is closed under finite and countable intersections;

(d) If $C \in \mathcal{C}(\mathcal{M})$ and $D \in \mathcal{C}(\mathcal{N})$ then $C \times D \in \mathcal{C}(\mathcal{M} \times \mathcal{N})$;

(e) If $C \in \mathcal{C}(\mathcal{M} \times \mathcal{N})$, then the set $\{x : (\exists y \in \mathcal{N}), (x, y) \in C$ belongs to $\mathcal{C}(\mathcal{M})$, and the analogous rule holds for each factor in a finite Cartesian product;

(f) If $C \in \mathcal{C}(\mathcal{M} \times \mathcal{N})$, and D is a nonempty set in $\mathcal{B}(\mathcal{N})$, then

$$\{x : (\forall y \in D)(x, y) \in C\}$$

belongs to $\mathcal{C}(\mathcal{M})$, and the analogous rule holds for each factor in a finite Cartesian product.

The sets in $\mathcal{C}(\mathcal{M})$ are called *neocompact sets*. The neocompact family $(\mathbb{M}, \mathcal{B}, \mathcal{C})$ induces a family of metric spaces with extra structure,

$$\mathcal{M} = (\mathcal{M}, \mathcal{B}(\mathcal{M}), \mathcal{C}(\mathcal{M})),$$

which we call *neometric spaces*. A neometric space thus consists of a complete metric space $\mathcal{M} \in \mathbb{M}$ and two families $\mathcal{B}(\mathcal{M})$ and $\mathcal{C}(\mathcal{M})$ of subsets of $\mathcal{M}$. We call $(\mathbb{M}, \mathcal{B}, \mathcal{C})$ the neocompact family *generated by* $(\mathbb{M}, \mathcal{B})$ if $\mathcal{C}(\mathcal{M})$ is the collection of all sets obtained by finitely many applications of rules (a)–(f).

We have at hand a very simple example: the family generated by $(\mathbb{S}, \mathcal{B})$ where $\mathbb{S}$ is the collection of all complete metric spaces, and for each $\mathcal{M} \in \mathbb{S}, \mathcal{B}(\mathcal{M})$ is equal to the set of all compact subsets of $\mathcal{M}$. It is not hard to see that the family of compact sets is closed under all of the rules (a)–(f), so the collection of neocompact sets $\mathcal{C}(\mathcal{M})$ generated by $(\mathbb{S}, \mathcal{B})$ is just $\mathcal{B}(\mathcal{M})$ itself: every neocompact set is compact.

The neocompact families we are interested in have a property, that closely resembles the behavior of the family of compact sets in Topology. Here it is.

Definition 4.2 A neocompact family $(\mathbb{M}, \mathcal{B}, \mathcal{C})$ has the *Countable Compactness Property* if for each $\mathcal{M} \in \mathbb{M}$, every decreasing chain $C_0 \supseteq C_1 \supseteq \cdots$ of nonempty sets in $\mathcal{C}(\mathcal{M})$ has a nonempty intersection $\bigcap_n C_n$.

Now we want to focus our attention in the Classification Theory that we began to develop in the previous section, so let's go directly to Probability spaces and forget some facts about general neocompact families. Again, first some basic concepts (see the book [EK]).

Let $M = (M,\rho)$, $N = (N,\sigma)$, and $O = (O,\tau)$ be complete separable metric spaces. $L^0(\Omega,M)$ is the set of all P-measurable functions from Ω into M, identifying functions which are equal P-almost surely. ρ_0 is the metric of convergence in probability on $L^0(\Omega,M)$,

$$\rho_0(x, y) = \inf\{\varepsilon : P[\rho(x(\omega), y(\omega)) \leq \varepsilon] \geq 1 - \varepsilon\}.$$

The space of Borel probability measures on M with the Prohorov metric

$$d(\mu, \nu) = \inf\{\varepsilon : \mu(K) \leq \nu(K^{\varepsilon}) + \varepsilon \text{ for all closed } K \subseteq M\}$$

is denoted by Meas(M). It is again a complete separable metric space, and convergence in Meas(M) is the same as weak convergence. There is a very useful characterization of the compact sets in this space: A set $C \subseteq \text{Meas}(M)$ is said to be *tight* if for each $\varepsilon > 0$ there is a compact set $K \subseteq M$ such that $\mu(K) \geq 1 - \varepsilon$ for all $\mu \in C$.

Each measurable function $x : \Omega \to M$ induces a measure $\text{law}(x) \in \text{Meas}(M)$, and the function $\text{law} : L^0(\Omega,M) \to \text{Meas}(M)$ is continuous, due to the fact that convergence in probability implies weak convergence.

Definition 4.3 Let $\underline{\Omega} = (\Omega, \mathcal{F}, \mathcal{P})$ be a probability space, and let $\mathbb{M}_\Omega$ be the family of all the metric spaces $\mathcal{M}_\Omega = L^0(\Omega,M)$ where M is a complete separable metric space. A subset B of $\mathcal{M}_\Omega$ is *Basic*, $B \in \mathcal{B}(\mathcal{M}_\Omega)$, if either

(1) B is compact,

(2) $B = law^{-1}(C)$ for some compact set $C \subseteq Meas(M)$, for some compact $C \subseteq Meas(M)$.

Definition 4.4 $\underline{\Omega}$ is said to be a *Rich Probability Space* if the measure P is atomless on $\mathcal{F}$ and the neocompact family generated by $(\mathbb{M}_\Omega,\mathcal{B})$ has the countable compactness property.

In order to give the reader a taste of the new theory, let's introduce two notions for neocompact families analogous to familiar notions for metric spaces (notice what happens if you leave out the prefix Neo).

Definition 4.5

(a) A set $C \subseteq \mathcal{M}$ is *neoclosed* in $\mathcal{M}$ if $C \cap D$ is neocompact in $\mathcal{M}$ for every neocompact set D in $\mathcal{M}$.

(b) Let $D \subseteq \mathcal{M}$. A function $f : D \rightarrow \mathcal{N}$ is *neocontinuous* from $\mathcal{M}$ to $\mathcal{N}$ if for every neocompact set $A \subseteq D$ in $\mathcal{M}$, the restriction $f|A = \{(x, f(x)) : x \in A\}$ of f to A is neocompact in $\mathcal{M} \times \mathcal{N}$.

Here is a list of properties that Rich Probability Spaces have. The detailed proofs appear in [FK1].

1. No 'Ordinary' Probability space is Rich. An 'Ordinary' Probability spaces is of the form (M, $\mathbb{B}$(M), μ) where M is a complete separable metric space and μ is the completion of a Borel probability measure on $\mathbb{B}$(M).

2. Every compact set is Neocompact. Every Neoclosed set is closed. Every Neocontinuous function is continuous. The composition of Neocontinuous is Neocontinuous.

3. The image of a Neocompact set by means of a Neocontinuous function is Neocompact.

We are now ready to state the main theorem about Rich Spaces, it explains why we have used the name Rich. It captures the following idea: 'if you know how to approximate, that is enough'. If you have some familiarity with the methods for proving existence theorems in Analysis, will quickly notice the importance of the theorem: the natural part of the proofs is 'to approximate', the difficult one is to use the approximations in order to guarantee the existence of the limit. So, intuitively our theorem says: if you are in a Rich Space, you just have to prove 'the easy part' of an existence proof !

This is exactly, to put it bluntly, what existence proofs using Nonstandard Analysis have shown, in nonstandard terms. Here we have shown it through the theory of Neometric spaces. The implications of this result are very interesting, and the paper 'Existence Theorems in Probability Theory' develops at length this subject. Here is the result. (A word on notation: D^ϵ denotes the set of points in the space which are at a distance $\leq \epsilon$ from D).

Theorem 4.1 **(Approximation Theorem)** Let Ω be a *Rich Probability Space*. Let A be neoclosed in $\mathcal{M}$ and f: A$\rightarrow$ $\mathcal{M}$ be neocontinuous from $\mathcal{M}$ into $\mathcal{N}$. Let B be neocompact in $\mathcal{M}$ and D neoclosed in $\mathcal{N}$ and assume that for each $\epsilon > 0$, we have

$$(\exists x \in A \bigcap B^{\epsilon}) f(x) \in D^{\epsilon}$$

Then

$$(\exists x \in A \bigcap B) f(x) \in D$$

Now that you have seen the theorem, there are many questions that may come to mind, for example: Are there Rich Spaces?. Are there any interesting and useful Neocompact sets, Neoclosed sets, Neocontinuous functions ?. What is the relationship, if any, between Richness and the concepts developed in the previous section? The answer to the first is the obvious one, but its proof is not trivial, see [FK2].

Theorem 4.2 *Hyperfinite Spaces are Rich.*

The second one is easy: yes there are plenty, but you have to read about them in the [FK1].

The third question is more difficult, but it does have an answer: Keisler in the unpublished paper 'Rich and Saturated Adapted Spaces' [K6] shows that in Probability Spaces Richness = Saturation. But, and here we want to wake up your curiosity, he shows that if we move from plain Probability Spaces and random variables, to Adapted Probability Spaces and Stochastic Processes, where there are analogous notions to all the ones presented in this paper (see [HK], [K], [FK1,2]) we will discover that the equality doesn't hold. In fact, he proves a very deep result showing that in discrete time they are the same, but in continuous time they are not. Behind his results lies an important issue in mathematics: the transition from discrete to continuous time. This seems the appropriate time and place to finish this paper: we would have achieved our objectives if the reader is curious to find out what else we have done.

ACKNOWLEDGEMENTS

This work was partially supported by El Banco de la Republica de Colombia, Colciencias and the Universidad de los Andes.

Department of Mathematics
University of Los Andes

REFERENCES

[AFHL] Albeverio, S., Fenstad J.E., Hoegh-Krohn, R. and Lindstrom, T.: 1986, *Nonstandard Methods in Stochastic Analysis and Mathematical Physics.* New York, Academic Press.

[B] Billingsley, P.: 1988, *Probability and Measure.* Wiley.

[C1] Cutland, N.: 1988, *Nonstandard Analysis and its Applications.* Editor. LMSST 10. Cambridge University Press..

[C2] Cutland, N. *et al.*: 1995, *Developments in Nonstandard Analysis.* Pitman Research Notes. Longman.

[D] Doob, J.: 1994, 'The Development of Rigor in Mathematical Probability (1900–1950)', in Pier, Jean-Paul, editor, *Developments of Mathematics 1900-1950*, Basel, Birkhauser Verlag, Basel.

[EK] Ethier, S. and Kurtz, T.: 1986, Markov Processes. Wiley.

[F1] Fajardo, S.: 1990, *Introduccion al Analisis No-estandar y sus Aplicaciones en Probabilidad.* Fondo editorial Acta Cientifica Venezolana.

[F2] Fajardo, S.: 1987, 'Intrinsic Stochastic Processes', *Revista Colombiana de Matemáticas*, **21**.

[FK1] Fajardo, S. and Keisler, H.J.: 1996, 'Existence Theorems in Probability Theory', *Advances in Mathematics*, **120**.

[FK2] Fajardo, S. and Keisler, H.J.: 'Neometric Spaces', *Advances in Mathematics*, **118**.

[FK3] Fajardo, S. and Keisler, H.J.: Long Sequences and Neocompact Sets. In [C2].

[FK4] Fajardo, S. and Keisler, H.J.: Neometric Forcing. Submitted.

[HK] Hoover, D.N. and Keisler, H.J.: 1984, 'Adapted Probability Distributions, *Trans Am Math Soc*, **286**, 159–201.

[K1] Keisler, H.J.: 1984, 'An Infinitesimal Approach to Stochastic Analysis', *Memoir Am Math Soc*, **297**.

[K2] Keisler, H.J.: 1991, Hyperfinite Models for Adapted Probability Logic, *Annals of Pure and Applied Logic*, **52**.

[K3] Keisler, H.J.: Infinitesimals in Probability Theory. In [C1].

[K4] Keisler, H.J.: 'From Discrete to Continuous Time', *Annals of Pure and Applied Logic*, **52**, 99–141.

[K5] Keisler, H.J.: A Neometric Survey. In [C2].

[K6] Keisler, H.J.: Rich and Saturated Adapted Spaces. To appear.

[M1] Maharam, D.: 1942, 'On Homogeneous Measure Algebras', *Proc Nat Acad Sci USA*, **28**.

[M] Maharam, D.: 1950, 'Decompositions of Measure Algebras and Spaces', *Trans Am Math Soc*, **69**.

MARIA CARLA GALAVOTTI

SOME REMARKS ON OBJECTIVE CHANCE (F.P. RAMSEY, K.R. POPPER AND N.R. CAMPBELL)

1. The debate on 'objective' chance embraces a wide range of conflicting positions, from the interpretation of chance in terms of (meta)physical propensities to a view of chance rooted in a subjectivistic framework. Though the propensity interpretation perhaps prevails over the subjective or personally oriented conception, both of them gather some consensus among philosophers, probabilists and scientists. Intermediate and alternative positions have also emerged, some of which can be considered variants of the above, while others differ in more substantial ways[1]. Leaving aside the variety characterizing this debate, I will concentrate on the propensity and subjective interpretations of chance.

A subjective interpretation of chance was devised by F.P. Ramsey in some writings dating back to the last period of his short life. Most relevant in this connection are the notes of 1928 'Chance' and 'Reasonable Degree of Belief', as well as 'General Propositions and Causality' of 1929, both published in *Foundations of Mathematics*[2]. Important remarks are also contained in 'Truth and Probability' (1926)[3] and in some of the notes appearing in the collection *Notes on Philosophy, Probability and Mathematics*[4]

The propensity interpretation is usually attributed to K.R. Popper, who put the theory forward around the middle of the Fifties in the paper 'The Propensity Interpretation of the Calculus of Probability and the Quantum Theory'[5], to take it up again in the *Postscript to the Logic of Scientific Discovery*[6] and in an essay called *A World of Propensities*[7].

In spite of their divergencies, the positions upheld by Ramsey and Popper seem to share some – albeit different – features of the view of chance advanced by N.R. Campbell[8]. A rather interesting figure of a physicist with strong epistemological interests, Campbell is best known for his systematic book on the foundations of science called *Physics. The Elements*[9] and for the treatise *An Account of the Principles of Measurement and Calculation*[10]. Authors like R. Carnap, C.G. Hempel and E. Nagel frequently refer to his work. Campbell's ideas had a strong influence on the conception of theories upheld by Ramsey[11], who nevertheless opposed his objectivist view of probability and chance. Campbell's notion of chance, on the other hand, is astonishingly similar to Popper's notion of propensity.

In what follows the main traits of Ramsey's and Popper's views of chance

M.L. Dalla Chiara et al. (eds.), Language, Quantum, Music, 73–82.

will be recalled and compared with Campbell's. This comparison will suggest some general considerations on this most controversial notion.

2. Ramsey's chance is framed in his subjective view of probability, according to which probability is the degree of belief of a person in the occurrence of events. Chance is defined as degree of belief of a special kind. Its peculiarity is that of being always referred to a 'system of beliefs', rather than to the beliefs of certain people in particular situations. A distinctive feature of the systems to which chance is referred is that of including laws and other statements which describe the behaviour of phenomena under consideration, like correlation statements. Such laws, in conjunction with the empirical knowledge possessed by the users of the system, entail degrees of belief representing 'chances', to which the actual degrees of belief held by users should approximate. This notion typically applies to 'chance phenomena' like games of chance, whose behaviour is described by systems that cannot be modified by the addition of deterministic laws ruling the occurrence and non-occurrence of a given phenomenon.

On the basis of this general definition of 'chance', Ramsey qualifies 'probability in physics' as chance referred to a more complex system, namely to a system making reference to scientific theories. In other words, probabilities occurring in physics are derived from physical theories. They can be taken as 'ultimate chances', to mean that within the theoretical framework in which they occur there is no way of replacing them with deterministic laws. The objective character of chances descends from the objectivity peculiarly ascribed to theories that are commonly accepted as true.

The 'truth' of theories within Ramsey's philosophy is not to be taken in a realistic sense, being rather accounted for in pragmatical terms. In this connection Ramsey holds the view, usually attributed to C.S. Peirce but also present in N.R. Campbell's work, that theories which gain 'universal assent' in the long run are accepted by the scientific community and taken as true. Along similar lines he characterizes a 'true scientific system' with reference to a system to which the opinion of everyone, grounded on experimental evidence, will eventually converge.

According to this pragmatically oriented view, chance attributions, like all general propositions belonging to theories – including causal laws – are not to be taken as propositions. They represent 'variable hypotheticals', or 'rules for judging', apt to provide a tool with which the user meets the future.

To sum up, for Ramsey chances are theoretical constructs, but they do not express realistic properties of 'physical objects', whatever meaning be attached to this expression. Chance attributions indicate a way in which beliefs in various facts belonging to science are guided by scientific theories.

As I have argued elsewhere[12], Ramsey's notion of chance can be seen as the natural complement to the view upheld by the other founder of the subjective interpretation of probability, namely B. de Finetti. The latter extended the static notion of probability in terms of coherent belief systems to a dynamical theory that includes a method of inference which allows the updating of subjective probabilities. This enabled the subjective interpretation of probability to stand on its own feet and to account for statistical inference. However, in his extreme subjectivism de Finetti could find no room for such notions as 'objective chance' and 'probability in physics'. The application of probability to natural sciences does not receive extensive discussion in his writings. He tends to regard science simply as a continuation of everyday life and to think that subjective probability is perfectly adequate for both. With his attitude de Finetti contributed to the conviction, widely upheld by scientists in various fields, that the subjective notion of probability is only suited for dealing with human behaviour and does not apply to natural sciences.

By combining de Finetti's views on probability and inference with Ramsey's notion of objective chance, however, a more articulated position is obtained, which can provide a good basis for an interpretation of probability in science in tune with a pragmatic and non-realistic epistemology.

3. Popper's theory of propensity stands on the opposite side with respect to Ramsey's notion of chance. In his (1957) Popper proposed the propensity theory as an answer to the problem of assigning an interpretation to quantum mechanical probabilities. More recently, however, he took a different position, according to which the notion of propensity is not intended only for application to quantum mechanics, but it becomes the basis for a much more general programme, meant to account for all sorts of causal tendencies operating in the world.

Popper's propensity theory is deemed 'purely objective' in view of the fact that propensities are defined 'physically real'. Furthermore, propensity theory is said to be 'metaphysical' in a twofold sense, in the first place because it refers to nonobservable properties, secondly because it provides "a coherent programme for physical research" (Popper 1957, p. 65).

Popper defines propensities as "relational properties of the experimental set-up" (*op. cit*, p. 68). They answer the need to 'take as fundamental the *probability of the result of a single experiment, with respect to its conditions*'. Such a need originates in connection with quantum mechanics, where the frequency interpretation of probability clashes with the problem of single case probability attributions. In (Popper 1957) the propensity interpretation is presented as a variant of the frequency interpretation, suited for assigning probabilities to the single case. As emphasized by Popper, the 'single case' is to be referred to the experimental arrangement (or set-up) in which an experiment takes place, not to single objects like particles or dice.

In subsequent years Popper's position underwent considerable changes. In (Popper 1982b) the propensity interpretation is presented as a variant of the classical view of probability. Propensities are defined as 'weighted possibilities' and seen as measurable expressions of the tendency of a possibility to realize itself upon repetition (Popper 1982b, p. 70). In this vein, Popper claims that propensity theory provides a straightforward solution to the main problems related both to the frequentist and classical interpretations of probability. On the one hand, propensity theory is said to solve the problem of the single case posed by frequentism, on the other hand it is granted the merit of avoiding determinism, to which the classical interpretation is strictly linked. Surprisingly enough, Popper does not even mention the subjectivist interpretation of probability worked out by Ramsey and de Finetti. He associates subjectivism with the classical interpretation and is thereupon led to the claim that subjectivism presupposes determinism, a claim which certainly does not hold for the position upheld by Ramsey and de Finetti.

According to Popper, propensity is a "new physical (or perhaps metaphysical) hypothesis" analogous to Newtonian forces (Popper 1982a, p. 360). It is the hypothesis that "every experimental arrangement (and therefore *every state of a system*) generates propensities" (*ibidem*). In some cases, namely when they are referred to mass phenomena or to repeated experiments, propensities can be measured through frequencies. In other cases they cannot be measured, they can only be estimated "speculatively" (Popper 1990, p. 17). In all cases, statements about propensities, like all probability statements, have to be testable. "To make a statement about probability – Popper says – is to propose a *hypothesis* ... In proposing this hypothesis, we can make use of all sorts of things – of past experience, or of inspiration: it does not matter *how we get it*; all that matters is *how we test it*" (Popper 1957, p. 66). In other words, the acceptability of such statements depends on their testability.

This raises the question of how probabilistic statements are tested. The answer is outlined in (Popper 1982b). First of all Popper makes a distinction between 'probability statements', seen as statements about frequencies in *virtual* sequences of experiments, and 'statistical statements', or statements of relative frequency, seen as statements about frequencies in *actual* sequences of experiments (1982b, p. 70). As already seen, according to the propensity theory probability statements are assignments of "weighted possibilities". Now, the weights attached to such possibilities are measures of "(conjectural) virtual frequencies, *to be tested by actual statistical frequencies*" (*ibidem*).

To illustrate this, Popper gives an example. If we have a die containing a piece of lead that we can place as we please near one or the other of its faces, we can say that by placing the piece of lead near the face 6 we increase the weight of the possibility that the die turns up on the opposite face, namely face 1. Here 'weight' can be taken as the measure of the propensity of obtaining a certain result upon repetition of an experiment. Such a measure is assumed to coincide with the virtual frequency of that result in a virtually infinite sequence of repetitions of the same experiment. This conjecture will be expressed by a probability statement that can be tested by performing a real sequence of experiments. This will lead to a statistical statement expressing the frequency thus obtained, to be then compared with the one that has been conjectured.

Probability statements expressing propensities are therefore tested by means of observed frequencies. Their acceptability depends on the possibility of performing a test of this kind.

4. As already said, albeit Ramsey's and Popper's views on chance show major divergencies, both of them have something in common with Campbell's perspective.

As seen above, Ramsey shares with Campbell the idea that those theories which gain "universal assent" can be called "true". This bears a strong resemblance to "Peirce's notion of truth as what everyone will believe in the end" as Ramsey puts it (Ramsey 1931, p. 253). The notion of 'universal assent' plays a crucial role within Campbell's epistemology, where it is regarded as a general criterion of the 'valuableness' of scientific judgments[13].

On the other hand, Ramsey rejects Campbell's notion of chance, which combines the idea that chance is a physical property with the conviction that probability is an empirical notion. To the first of these tenets Ramsey simply objects that "there are no such things as objective chances in the sense in which some people imagine there are, e.g. N. Campbell" (Ramsey 1931, p. 206). As to the second, namely to Campbell's claim that probability, taken in the frequentist sense, "will permit an experimental *demonstration* in certain cases that events are equally probable", Ramsey's comment is that "this is a mistake simply" (Ramsey 1991, p. 258). In the same vein, in his note 'Chance' Ramsey claims Campbell is mistaken when he takes as an "empirically established fact" a statement of the form "In n consecutive throws the number of heads lies between $n/2 \pm \epsilon/n$)"[14].

Not only it is not possible for a statement to this effect to be established empirically, because "any such law would be broken if we took enough instances of it" (*ibidem*), but the very notion of "infinite series of throws" which stands at the core of frequentism, cannot be grounded in experience. For Ramsey Campbell's notion of probability is then ill-founded and so is the strictly related notion of chance upheld by Campbell.

On the contrary, Campbell's notion of chance is strikingly similar to Popper's. Although Popper never refers to Campbell in his writings, it is hard to imagine that Campbell's work was unknown to him.

Campbell takes chance to be a physical property belonging to a system. In (Campbell 1922), while discussing the chance of a die turning up six, he defines chance as "something uniformly associated with and changing with the structure of the die just as (say) the electrical resistance" (p. 68). He goes on to say that "this uniform association of the resistance with the other characteristics of the die is what we assert when we say that the resistance is a physical property of the die, and accordingly the chance of turning up six is a physical property as much as the resistance" (*ibidem*).

It is worth noting that, according to Campbell's definition, chance is referred to systems. Chance can be attributed to the single case only in a somehow derivative way. In fact, systems typically include assertions of association between the properties of objects, which can only be ascertained on the basis of series of observations. Incidentally, the idea that chance is referred to systems including laws is also shared by Ramsey, as we saw in section 2.

While taking chance as definable only within a system, Campbell suggests a relational view of it, which is reflected by the way chance is measured. As should be expected, given the pivotal role assigned to measurement within Campbell's epistemology, the author addresses the issue of the measurement of chance. In this connection, Campbell claims that chance cannot be measured directly, like fundamental magnitudes such as length or weight. Chance is measured as a derived magnitude. The process of derived measurement applicable to chance according to Campbell "involves the two fundamental magnitudes – number of events and number of trials" (*op. cit.*, p. 67).

Let us focus on this point. According to Campbell, measurement of a fundamental property requires definitions of equality and addition, plus the specification of some numerical value to be "assigned arbitrarily to some one property, which with all others can be compared by means of these definitions" (*op. cit.*, p. 73). In the case of chance, this arbitrary requirement is made by attributing the value 1 to the chance of an event which always happens as a result of a trial. Addition can be defined in a straightforward way, saying that if we wanted to calculate the chance of an event A consisting in the happening of either x, or y, or z (x, y and z being mutually exclusive alternatives), we should put the chance of A equal to the sum of the chances of x, y, and z.

Some difficulties arise in connection with the definition of equality. In this regard – so goes Campbell's argument – it is usually held that chances are equal in virtue of the principle of indifference, or sufficient reason. However, he finds unacceptable the idea that the properties of a system can be determined by means of an *a priori* principle. To the contrary, he claims that the only ground for asserting the equality of chances could be given by observed frequencies. His clue is therefore to use frequencies; the measurement of chance is to be performed through the measurement of frequencies. In Campbell's words: "I can find no proposed definition of the equality of chances that is both applicable to experimental facts and independent of frequency; and I conclude, therefore, that there is not for chance ... a fundamental process of measurement independent of the derived" (*op. cit.*, p. 74).

As a matter of fact, when we calculate the chance of an event we sometimes behave as if it were measurable by the fundamental process. This is what we do, for example, when we calculate the chance of drawing any one card, or an ace, from a deck of 52. This looks perfectly legitimate, insofar as we know in the first place "of how many individual events the event under consideration ... is the sum", and secondly "that the chances of these individual events are equal" (*op. cit.*, p. 75). However, in most cases we do not have a complete knowledge of all of the events involved. Furthermore, it is often the case that the evidence that would be needed in order to calculate by derived measurement the equality of their chances is not available. Calculation of chances made in the absence of this information "is then purely theoretical, and the only legitimate use that can be made of it is to confirm or reject the theory by means of a comparison of the calculated chance with that determined experimentally by the derived measurement" (*ibidem*). Campbell's conclusion is that "true chance, the property of the system, is always and inevitably measured by the derived measurement and not by the fundamental" (*ibidem*).

The strict linkage between chance and frequency is repeatedly stressed by Campbell: "chances – he says – are important for physics only in so far as they represent relations between frequencies" (*op. cit.*, p. 69). Chance is a theoretical concept, which plays an important role within physical theories, but it "enters into the laws predicted by these theories only because it represents a relative frequency"[15].

5. We face here three conflicting views of chance. Within Ramsey's perspective, chance derives its objective character from theories. When probabilities are assigned to events which fall under the scope of a scientific theory, such evaluation is suggested by the theory itself. According to the subjectivist interpretation, it represents a conviction of the evaluating subject, but it is a conviction of a very special kind, since it is grounded in common knowledge shared by the scientific community. Frequencies will have an important part in this process, as they enter into the evaluation of probabilities like any other piece of relevant information that might be available to the person who makes the probability assignment. Indeed, no subjectivist would deny the essential role played by frequencies in all fields of science, nor the fact that in most cases statements of chance actually inform about frequencies. So much granted, one should keep in mind that according to the subjectivist interpretation the linkage between chance and frequency is provided by theories. The explication of chance is ultimately in terms of belief; it is not inextricably linked to that of frequency.

Things are quite the opposite with Popper's view. Here chance is explicated in terms of propensities and these represent "physical realities" (Popper 1990, p. 12). "Propensities – Popper says – like Newtonian attractive forces, are invisible and, like them, they can act: they are *actual*, they are *real*" (*op. cit.*, p. 18). Propensities are much more than objective chances. They not only inspire an objective theory of probability, but give rise to an indeterministic view of the world according to which the latter "can now be seen as a world of propensities, as an unfolding process of realizing possibilities and of unfolding new possibilities" (*ibidem*). However, propensity statements have to be testable by means of observed frequencies. The metaphysics of propensities, in order to gain any applicability, is to be subjected to the test of frequencies.

Campbell's position stands somehow half way between the two opposites. It comes close to Ramsey's with regard to the interpretation of theories and truth, but it resembles Popper's when it comes to the definition of chance. The obvious difference is that while Campbell insists on the measurement of chance, Popper emphasizes testability. Nevertheless, derived measurement plays within Campbell's perspective the same role ascribed by Popper to tests, and in both cases frequencies provide the essential tool by means of which such operations can be performed.

As a matter of fact, Campbell's ideas on the measurement of probabilities were challenged by H. Jeffreys on the basis of the claim that "the additive rule for probabilities is not a rule of measurement in the physical sense, nor is it even a postulate. It is a convention" (Jeffreys 1938, p. 147). Jeffreys adds that "the representation of probabilities by numbers is made in order to make use of our mathematical technique, but without the addition rule it could be done in an infinite number of ways" (*ibidem*).

I will not go into detail regarding Jeffreys' position and his controversy with Campbell. Instead, I shall focus on the affinities between Campbell's and Popper's notions of chance, stressing that, once the assumption has been made that chance is a physical property, one has to resort to frequencies in order to secure the applicability of such a notion of chance. This, however, opens the door to some of the objections traditionally moved against frequentism. Ramsey's rejection of Campbell's notion of chance moves precisely along these lines.

Popper's position is more complex. Quite apart from the metaphysics of propensities, whose appeal depends on anyone's philosophical taste, one should not forget that the propensity interpretation was initially put forward to solve the problems connected with the interpretation of quantum mechanical probabilities. As is well known, this is a most debated issue[16]. Doubts have been cast upon Popper's conviction that propensities can adequately represent quantum probabilities. Convincing arguments against this tenet have been given, among others, by S. Leeds[17].

On the other hand, it can be argued that, if combined with de Finetti's theory of probability and inference, Ramsey's position could point out a way of providing a suitable notion of objective chance, apt to be applied to probability assignments in physics and science in general, which is in tune with the subjective interpretation of probability. Important work in this direction has been made by various authors, including B. Skyrms and R. Jeffrey[18].

Those who have a pragmatically oriented view of scientific theories and truth will find it congenial, though it is likely to be considered unsatisfactory by those who take a realistic attitude towards scientific knowledge. The choice, however, is not merely philosophical. It depends also on the suitability of this notion of chance to adequately represent physical probabilities. As I said at the outset, the debate is open.

Department of Philosophy
University of Bologna

NOTES

[1] For a recent discussion of these theories see Howson (1995).
[2] See Ramsey (1931).
[3] Also in Ramsey (1931).
[4] See Ramsey (1991). For a more extensive account of Ramsey's notion of 'chance' see Galavotti (1995).
[5] See Popper (1957). See also Popper (1959).
[6] See especially volumes I and III, namely Popper (1982a) and (1982b).
[7] See Popper (1990).
[8] See Campbell (1920), (1921) and (1922).
[9] See Campbell (1920).
[10] See Campbell (1928).
[11] This has been argued, among others, by U. Majer. See his (1989) and (1991).
[12] See Galavotti (1991) and (1995).
[13] See Campbell (1920), pp. 20 ff. and (1921), chapters 2 and 5.
[14] Ramsey (1931), p. 206. On this point see also Galavotti (1995), pp. 334–335.
[15] *Ibidem.* Obviously, Campbell has in mind statistical physics.
[16] See Howson (1995) on this point.
[17] See Leeds (1984).
[18] See Skyrms (1984) and Jeffrey (1992) and (1996).

REFERENCES

Campbell, N.R.: 1920, *Physics. The Elements.* Cambridge: Cambridge University Press; reprinted as *Foundations of Science*, New York: Dover, 1957.

Campbell, N.R.: 1921, *What is Science?* London: Methuen; second edition New York: Dover, 1953.

Campbell, N.R.: 1922, 'The Measurement of Chance', *Philosophical Magazine*, **44**, 67–79.

Campbell, N.R.: 1928, *An Account of the Principles of Measurement and Calculation.* London–New York: Longmans.

Campbell, N.R.: 1938, 'Symposium: Measurement and its Importance for Philosophy', *Aristotelian Society Supplementary Volume XVII*, 121–142.

Galavotti, M.C.: 1991, 'The Notion of Subjective Probability in the Work of Ramsey and de Finetti', *Theoria,* **57**, 239–259.

Galavotti, M.C.: 1995, 'F.P. Ramsey and the Notion of 'Chance'', in J. Hintikka and K. Puhl, editors, *The British Tradition in 20th Century Philosophy. Proceedings of the 17th International Wittgenstein Symposium.* Wien: Hölder-Pichler-Tempsky, 330–340.

Howson, C.: 1995, 'Theories of Probability', *British Journal for the Philosophy of Science* **46**, 1–32.

Jeffrey, R.C.: 1993, 'De Finetti's Radical Probabilism', in B. de Finetti, *Probabilità e induzione*, edited by P. Monari and D. Cocchi. Bologna: CLUEB.

Jeffrey, R.C.: 1996, 'Unknown Probabilities', *Erkenntnis*, **45**, special volume on 'Probability, Dynamics and Causality' edited by D.Costantini and M.C.Galavotti, 327–335.

Jeffreys, H.: 1938, 'Symposium: Measurement and its Importance for Philosophy'', in *Aristotelian Society Supplementary Volume XVII*, 143–151.

Leeds, S.: 1984, 'Chance, Realism, Quantum Mechanics', *The Journal of Philosophy* 81, pp. 567-678.

Majer,U.: 1989, "Ramsey's Conception of Theories: An Intuitionist Approach", *History of Philosophy Quartely*, **6**, 233–258.

Majer, U.: 1991, 'Ramsey's Theory of Truth and the Truth of Theories: A Synthesis of Pragmatism and Intuitionism in Ramsey's Last Philosophy', *Theoria*, **57**, 162–195.

Popper, K.R.: 1957, 'The Propensity Interpretation of the Calculus of Probability and the Quantum Theory', in S. Korner, editor, *Observation and Interpretation.* London: Butterworth, 65–70.

Popper, K.R.: 1959, 'The Propensity Interpretation of Probability', *British Journal for the Philosophy of Science*, **10**, 25–42.

Popper, K.R.: 1982a, *Realism and the Aim of Science*, edited by W.W. Bartley III. London: Hutchinson.

Popper, K.R.: 1982b, *Quantum Theory and the Schism of Physics*, edited by W.W. Bartley III. London: Hutchinson.

Popper, K.R.: 1990, *A World of Propensities*. Bristol: Thoemmes.

Ramsey, F.P.: 1931, *The Foundations of Mathematics and Other Logical Essays*, edited by R.B. Braithwaite. London: Routledge and Kegan Paul.

Ramsey, F.P.: 1991, *Notes on Philosophy, Probability and Mathematics*, edited by M.C. Galavotti. Naples: Bibliopolis.

Skyrms, B.: 1984, *Pragmatics and Empiricism*. New Haven and London: Yale University Press.

MARIA GRAZIA SANDRINI

MISAPPREHENSIONS ABOUT SIGNIFICANCE TESTS AND BAYESIANISM

1. The opposition to Bayesian approach to inverse inference, after the abundant use made of it by Laplace, the first to introduce it into scientific method, formed and developed, as we know, in the positivistic climate of the late nineteenth century. Boole, Venn, and later R.A. Fisher, to mention only three well-known names, saw in the application of Bayes theorem to cases of inverse inference the danger of introducing arbitrary elements connected with the *a priori* probabilities, in their opinion unjustifiable in the scientific research, which should always and only pursue objectivity. Such opposition seemed to Fisher all the more justified by the fact that for him there was no need to fall back on *a priori* probabilities, given the availability of alternative non-Bayesian methods, which he had himself helped to work out, including significance tests. These can be easily and pleasantly introduced by Fisher's well-known example of 'the tea lady', i.e. the woman declaring that she can tell, on tasting a cup of tea, whether the milk was put in as first or the tea.

What about such a claim? In Fisher's view the question is to set up an experiment testing the hypothesis that the woman is simply guessing. This had to be an experiment virtually proof against Chance. Fisher proposes to present the woman, in random order, eight cups of tea and milk, four prepared one way and four the other. The woman's task is to divide the eight cups into two homogeneous groups of four. In considering the suitability of the experiment, we must consider that the woman has only one chance in seventy of making the right selection purely by chance. To Fisher this seems a degree of probability sufficiently low to refute the hypothesis that the woman's verdict, if she is successful, is a lucky guess.

The force of this experiment lies logically in the form of the disjunction: "*either* an exceptionally rare chance has occurred, *or* the theory of random distribution is not true".[1]

We can draw some considerations from this example. First and foremost we note that a significance test makes a distinction between possible experimental results, dividing them into two opposite classes: on the one hand, the results showing a significant difference from a certain hypothesis; on the other, those doing not. The line of demarcation between the two classes, or level of significance, is fixed by the discretion of the experimenter, although in practice it is often a standard one.[2]

M.L. Dalla Chiara et al. (eds.), Language, Quantum, Music, 83–93.

The element characterizing the method of levels of significance consists nevertheless of the hypothesis to be compared with experimental data. As Fisher says, it is 'characteristic of all experimentation' and is the only hypothesis in relation to which an experiment can show itself to be 'significant' because it is the only one susceptible of contradiction by experimental results:

in relation to any experiment we may speak of this hypothesis as the 'null hypothesis', and it should be noted that the null hypothesis is never proved or established, but it is possibly disproved in the course of experimentation. *Every experiment may be said to exist only in order to give the facts a chance of disproving the null hypothesis.*[3]

Roughly speaking, the null hypothesis may be defined as the assertion that, in relation to a given experimental situation, hypothesizes a purely random happening, a lack of any cause or regularity in the unfolding of the phenomena under consideration. "It might be argued – Fisher writes at the end of his discussion of the mentioned example – that, if an experiment can disprove the hypothesis that the subject possesses no sensory discrimination between two different sorts of object, it must therefore be able to prove the opposite hypothesis, that she can make some such discrimination. But this last hypothesis, however reasonable or true it may be, is ineligible as a null hypothesis to be tested by experiment, because it is inexact. If it were asserted that the subject would never be wrong in her judgements, we should again have an exact hypothesis, and it is easy to see that this hypothesis could be disproved by a single failure, but could never be proved by any finite amount of experimentation".[4]

Here Fisher emphasizes a characteristic that the null hypothesis shares with no other one: the null hypothesis can be always formulated precisely, that is it has an exact mathematical significance making possible to plan and evaluate the possible results of an experiment. "It is evident – says Fisher – that the null hypothesis must be exact, that is free from vagueness and ambiguity, because it must supply the basis of the 'problem of distribution', of which the test of significance is the solution".[5] Now, this exactness is not usually shared by any of the other opposite hypotheses. How can we translate into an exact formula, which allows us to evaluate the probabilities of the possible results, an assertion as 'Mrs. X can tell, by taste alone, whether the tea or the milk was put in first'? As Fisher observed, only a formulation in strictly universal terms would be exact; but is the universal hypothesis the only one opposite to the null hypothesis?

If we are doing research on a possible correlation between smoking and lung cancer, it is immediately obvious that the null hypothesis can be formulated as the hypothesis of the absolute independence between smoking and cancer, according to which any concurrence between the two factors would be purely accidental. That is to say, according to this hypothesis, the fact that being a heavy smoker would not increase the probability of getting lung cancer ill. But which are the non-null hypotheses? The impossibility of giving a more precise connotation to the vagueness of such hypotheses is reflected in the impossibility of determining what they exactly imply and, as a consequence, in the impossibility of planning the necessary experiments to subject to significance tests. Who can say how many probabilities are there that a smoker will get lung cancer ill under the generic hypothesis that a significant correlation exists between smoking and cancer?

These considerations enable us to progress towards a deeper explanation of the role and import of significance tests.

Firstly we must agree that, as Fisher has often reaffirmed, significance tests cannot and must not be interpreted as a method of deciding between two hypotheses, although they often are. I agree with Fisher claiming that "a test of significance contains no criterion for 'accepting' a hypothesis".[6] In particular, confuting the null hypothesis does not mean accepting the opposite one because *the latter may not exist.* Significance tests can be better understood in their essence if they are thought of as being applied to a *single* hypothesis, the null hypothesis precisely.[7] Looked at this way, significance tests and null hypothesis appear as tools for the analysis of experimental data rather than as factors of decision method. Secondly, I think that vagueness of the hypotheses set against the null hypothesis shows some difficulties in treating this type of problem by the Bayesian method. If it is true that Laplace used the Bayes theorem for problems of this type, we know that he did quite particularly: he used Bayes theorem in order to refute as highly improbable the hypothesis imputing to Chance the happening of certain constant and repeated phenomena, in favour of a generic deterministic hypothesis. For example, applying the theorem to the movements of rotation and revolution in the planets – all of them following the direction of the sun rotation – Laplace found that one could bet more than four thousand billion to one that such a phenomenon was not due to Chance. However, Laplace set against the chance hypothesis only one hypothesis, quite universal in its form, allowing an easy quantitative evaluation of the related likelihood in Bayes formula. As to *a priori* probabilities, he reckoned them equiprobable. In this way all the elements of Bayes formula could be easily quantified. But when the opposite hypothesis to the null hypothesis is not rigorously determined, the Bayesian method becomes inapplicable. The problematic aspect of the Bayes theorem, therefore, does not lie, as Fisher and objectivists believe, on the *a priori* probabilities of the hypotheses (for which the assumption of equiprobability or of any subjective probability would not, after all, be scandalous!) but on the possibility of quantifying the values of the likelihoods related to all possible hypotheses. Had Fisher focused his attention on this aspect, instead of losing himself, in the wake of Boole and Venn, in a sterile battle over principle, his anti-Bayesian attitude would have rested on a stronger foundation.

2. The tea-lady example is only an ideal one, which can be appropriately treated in the classic theory of probability. There are, however, experimental situations of a greater complexity which cannot be tackled in the same way. The significance tests applicable to them are much more complex, though based on the same logical reasoning. Numerous different kinds of significance tests

are used in handling statistics. The best known are still *Student's t* and K. Pearson's χ^2 tests, the first being suitable for the analysis of quantitative data, of which the mean values and standard deviations can be calculated, the second being useful for the analysis of qualitative data reducible to two categories (e.g. the treated ones and the ones not treated with a certain chemical).

A classic Darwin's experiment will serve to illustrate an experimental situation frequently arising in scientific studies. Darwin was trying to determine whether there was a difference in the speed of growth between self-fertilised and cross-fertilised plants, by comparing the height they had reached by a certain date. The organization of the experiment requires each self-fertilised plant to be paired with a cross-fertilised one in conditions as similar as possible. The 'pairing' procedure, extensively used in biology, solves the problem of meeting two requirements often conflicting each other: the one achieving as much uniformity as possible in the experimental material, in order to give the experiment maximum sensitivity, and the one bearing as many observations as possible in order to obtain results of maximum reliability. Darwin's experiment consists in choosing 15 pairs of plants, each pair made up of a self-fertilised plant and a cross-fertilised one, and measuring their height on a certain data, taking every precaution in order to avoid any alteration of the experimental conditions. The null hypothesis asserts that any difference in height observed in each pair is accidental and that the limits to which the average height of each type of plant tends are therefore the same. That is to say that the differences in height are distributed normally around the mean value equal to zero, and therefore that the two groups of measurements (the height of the cross-fertilised plants and the height of the self-fertilised ones) are samples of the same normal population. Then we have to decide whether the difference in height observed in the two groups of plants is compatible with the null hypothesis or it is so great to be incompatible with it and thus refuting it.

The mathematical distribution related to this problem was discovered in 1908 by 'Student'[8] and it depends only on the number of independent comparisons (*degrees of freedom*) available for the estimation of errors. The degrees of freedom, which play an important part in all significance tests, give a measurement of the margin of play within which the experimental results are found.[9] Problems very different from this, however, crop up often. A doctor may want to know, for example, if drug A is effective in the treatment of diabetes, or if the contraceptive pill encourages breast cancer, and so on. Obviously, experimental data must undergo a significance test in these cases too; but *Student*'s *t* cannot be used because there are no mean values to be calculated. The χ^2- tests seems suited to the purpose.[10] A very simple example will explain it: a person throws a die on the table sixty times and makes a note of the results: 7 times he gets a one, 9 times a two, 10 times a three, 6 times a four, 15 times a five, and 13 times a six. At this point he stops and asks himself if these results are normal or if there is something wrong with the die. If he happens to be a statistician, he can immediately start the testing of the null hypothesis, asserting that the die is unbiased. He will then proceed to specify exactly the *theoretical* value, that is the mean theoretical frequency of each result; in this case, since 60 throws have been made, with six possible results, the theoretical frequency according to the null hypothesis would require each face of the die to turn up 10 times. The well-known formula for χ^2 will give the value allowing to find the corresponding level of significance in the χ^2-distribution table. In the case in question, the results do not allow to reject the null hypothesis.

The theoretical value, constituted in this example by the ideal mean frequency, can also be offered by a known empirical frequency or can be constituted experimentally by a control group. The effectiveness of a drug, for example, is generally tested on two groups of patients with the same pathology (and, as far as possible, alike in other respects); one group is treated with the drug and the other with a placebo. The latter group will provide the theoretical values. The null hypothesis asserts that the drug and the placebo are equivalent, and that no significant differences between the two groups will be observed.

To sum up, significance tests consist in choosing a function of the stochastic variables under the null hypothesis. Hence, the null hypothesis is rejected if the probability connected with the experimental results is considered sufficiently small.

Significance tests are part of the normal routine of chemical, biological, clinical laboratories, being a substantial part of today's scientific practice. It is important to realize, however, that too often experimental practice applies statistical methods indiscriminately and that there is much confusion between Fisher's significance tests and Neyman-Pearson hypotheses tests, which, though claiming to be 'improvements' on Fisher's methods, differ from them in many essential respects. In the Neyman-Pearson theory, for instance, experiments are always conducted on a pair of opposite hypotheses, for one of which, the so called null hypothesis too, an area of acceptance and an area of refusal are determined. This 'null hypothesis' has nothing in common with Fisher's null hypothesis except the name.

I am only considering Neyman-Pearson method of testing hypotheses in order to point out that it is connected with a pragmatist conception, quite different therefore from Fisher's, which is positivist and objectivist. Neyman himself contrasts Fisher's 'inductive reasoning' – in his view erroneous – with his own 'inductive behaviour', a strategy for choosing the most reasonable action in terms of minimum risk; therefore, in the context of scientific research, Neyman method finds application in a practical-cognitive situation, that is very different from the one contemplated by Fisher. Nevertheless, these methods are often applied indiscriminately, and it is seldom considered that Fisher's ones are significance tests (and have to be considered in the context of the correlation theory), while Neyman's are intended to accept well-determined hypotheses. This confused situation encourages the erroneous use and interpretation of significance tests. As Fisher says, "the ideas that ... and that the purpose of the significance tests is to discriminate or 'decide' between two or more hypotheses has greatly obscured their understanding, when taken not as contingent possibilities but as elements essential to their logic".[11]

3. I mentioned the unexpected analogies between significance tests and a Laplace's use of Bayes theorem. I am examining now these analogies.

In the Laplacean applications of Bayes theorem, aiming at demonstrating the improbability of Chance, there is a constant typical characteristic: while the chance hypothesis had a quantitative determination, the only opposite hypothesis contemplated by Laplace was formulated as a generic universal assertion of causality. In a certain sense, there was no need for any further specification of the causal hypothesis, because Bayes formula could support only a generic determinism, but certainly could not justify a preference for one rather than another causal law, expressed in a strictly universal form. We can suggest that Laplace's real interest is concentrated on the Chance/Cause alternative only in order to affirm the latter by the improbability of the former.

The use of the probability theory made in order to demonstrate the improbability of the Chance was already present before Bayes solution, since De Moivre's and Jacob Bernoulli's works. In 1710 J.Arbuthnot applied the probability theory to birth, to demonstrate that the balance between male and female births was due not to Chance but to Divine Providence: "this probability [of the chance hypothesis] is so small that we can draw the conclusion that it is not a matter of chance that a boy or girl is born".[12] In 1734 Daniel Bernoulli and his father Johan calculated the probability that the inclinations of planes in the orbits of the planets were determined by chance, and found that the probability was so small to refute the chance hypothesis.[13]

Therefore, Laplace has applied Bayes theorem to this reasoning, framing it in the form of inverse inference, more suitable for him. It is difficult to say what actual role Laplace attributed to this proceeding in the domain of science. Perhaps he combined metaphysical instances with a truly scientific spirit. At any rate, it is evident that explanatory hypotheses of phenomena can be neither 'found' nor 'proved', but at most 'justified', by Bayes approach. The formulation and the 'proof', in the strict sense, of hypotheses can only be carried out by the competent science; nevertheless, even the refutation of the chance hypothesis cannot be considered a matter of small account for science, since it provides a basic indication for the specific sciences.

If we now reconsider what Fisher says when he affirms that the importance of significance tests can be understood better if one thinks of them as connected with the formulation of a *single* hypothesis, the null hypothesis, and as being directed exclusively at refuting it, one can see that Fisher and Laplace had in common the need for a method with the only function to 'assay' Chance.

But in order to understand fully the import of the significance tests it is necessary still to retrace the steps of their development.

The first significance tests imitated more or less implicitly Laplace's method. F.Y.Edgeworth, who in 1885 suggested a test for the comparison of two means, imitated it explicitly. But it was the prevalent positivistic view to supply the ideal ground for a precise study of the 'correlation' in Statistics. Karl Pearson, the founder of the correlation theory, expressed his very positivistic conception in the book *The Grammar of Science*: the task of science consists for him in the description of observed perception sequences and in the prediction, founded on the belief in a self-perpetuating order, but he denies resolutely that science has also the task of explaining phenomena. In this conception the causality is reduced to repeated association between perception data and therefore the task of scientific research is to study correlations. So, on the methodological plane, the requirement of tools for discriminating between perceptive sequences, in order to distinguish purely fortuitous coincidences from the ones occurring with a certain regularity, becomes urgent. I see a flash of lightning and immediately afterwards I hear a clap of thunder; repetition of this experience leads me to expect thunder every time I see lightning. But not all experiences that are associated with one another, although repeated many times, can be considered cognitively significant. The doorbell rings when I am in the shower; the telephone rings while I am rushing out of the house, being late for an important meeting: however often these things happen to most people, they cannot be considered as anything but inconvenient coincidences. The theory of correlation, based on the simplest and most instinctive kind of inductive thinking – that is, by repeated association – helps us to bring some degree of order into the chaos of experiences correlated in time and space, setting aside those which are too 'rare' to be learnt from. In this context, the accent does not fall on the rejection of casual correlations, but on the survey of significant correlations: they represent the true objective of scientific research, showing us which regular patterns in nature we can draw reliable conclusions from. Obviously, a statistically significant correlation does not establish anything about the existence of a causal connection between the correlated phenomena. Without wishing to be anything more than the 'description' of what has been observed, the statistically significant correlation claims to be enough in itself to form the basis of future expectations: if *a* and *b* have been closely correlated in the past, we can expect that they will be in the future, and it will be legitimate to base expectation of *b* on the observation of *a*, and *vice versa.*

Too many the limits of so an ingenuous empiricism. Nevertheless, this is the philosophical background of the Pearson's correlation theory and, later, of the significance tests .

If we consider the history of experimental science we cannot, on other hand, fail to be struck by the particular importance that has always been attached to the finding of correlations. Oersted's experiment[14] for example – one of the many experiments which led to new theoretical achievements – revealed the existence of a correlation between electrical and magnetic phenomena; a correlation which did not fit into Newtonian physics and could not be explained except within a new theoretical framework. This particular experiment was guided by precise working hypotheses; but significant correlations can sometimes come to light unintentionally, for example in the course of experiments directed at other objectives, or accidentally, as in the case of puerperal fever.

Everybody knows that Dr. Semmelweis realized with dismay that there was an indisputable significant correlation between the number of deaths from puerperal fever and the first maternity department in the hospital where he worked. Over a period of three years the mortality rate fluctuated between 6.2% and 11.4%, while in the second maternity department in the same hospital it never rose above 2.7%. It was the beginning of a long series of empirical attempts, some of them grotesque, to identify the cause of this abnormal correlation, and each attempt took the form of a search for further, more specific significant correlations. This tormented history makes us reflect about the limits of searching only statistically significant correlations and suggests the necessary connection leading from the correlation theory to the significance tests.[15] The statistical significance of correlations, indeed, cannot warrant their cognitive value: a high correlation does not always indicate that its terms are no-independent! This is the case, for example, of the correlation between the puerperal deaths and the transit, through the first maternity department, of the priest bringing the Extreme Viaticum. In this connection, we can remember also the correlation found in the fifties and which was much talked about between the Stock Exchange quotations and the length of skirts!

Summing up, if the correlation theory works to narrow the field of significant experience, nevertheless statistical significance is too weak a connection between the correlated aspects to warrant their scientific relevance. Therefore, the measure of depth of correlations has to give way to methods able to sound the independence relationship within a correlation. The significance tests elaborated by K. Pearson, and later by Fisher, are therefore a refining of methods to analyse correlations; but they indicate as well the giving up of a merely descriptive point of view towards an active attitude of experiments programming. Consequently, the potential scientific relevance of a correlation which survives the significance tests must not bring to misapprehensions or hurried conclusions. In modern oncological research, for example, epidemiological studies on the connection between diet and cancer seem to suggest the existence of a significant correlation between certain foods rich in vitamin C and a reduced incidence of cancer. This has led some people to the conclusion that vitamin C acts as an anti-cancer agent and has encouraged them to consider the possibility of using it in the prevention and treatment of cancer. However, this conclusion cannot be considered well-founded, since the same foods also contain, among other things, high levels of beta-carotin and fibre, which are also considered to be anti-cancer agents. For this reason the correlation observed has not up to now led to any useful conclusion, but to further research in the prevention and treatment of cancer.

From this discussion emerges an important consideration, connected, if I may look ahead for a moment, with the concluding evaluation that I am about to make: if it seems correct to consider significant correlations as probable signs of cause-and-effect relations still to be determined – and the history of experimental science seems to suggest that it is – it appears, *vice versa*, generally erroneous to make an unqualified identification between the terms of such a causal relation and the terms of the correlation in question. This seems to be the most frequent source of mistaken conclusions drawn from significance tests. It is the error against which Fisher was so anxious to warn people. In doing so, he kept the inheritance of Laplace, as well as that sense of 'correlativeness' which had matured during the development of experimental science and had been expressed in the positivism of K. Pearson.

4. The question at this point seems to present itself in the following terms: if significance tests must be conceived and interpreted only in relation to the null hypothesis, as Fisher claims, and if the reject of null hypothesis cannot be the same as affirming the causal hypothesis drawn from the tested correlation, what function can significance tests still perform in scientific research? The time has come to give a clear exposition of the conclusions to which the foregoing analysis seems to lead us. It is my opinion that this analysis supports the thesis that the Bayesian method and the significance tests method are only seldom, and in exceptional circumstances, both applicable to the same inductive problem.

Now I am considering the two kinds of questions tackled by Laplace using Bayes theorem:

A) what is the probability that the n white balls observed in n extraction come from an urn containing only white balls?

B) what is the probability that the inclinations of the planes of the orbits of the planets are not determined by chance?

Both are inverse questions, yet B is striking in its irreducible indeterminateness. Perhaps it is because of this indeterminateness that Laplace reduces questions of this kind to problems of Chance/Cause opposition, and to the attempt to support the deterministic hypothesis by demonstrating the improbability of Chance. In relation to this surprising Laplacean use of Bayes theorem, the emphasis with which Fisher maintains that significance tests can be considered to be properly interpreted only if they are used exclusively to reject the null hypothesis seems to require attention.

We must now ask ourselves whether behind the striking indeterminateness of cases B does not lie hidden some additional characteristic explaining its oddity. Pursuing this line of thought, I think to have found in this kind of questions a considerable ambiguity leading to the best understanding of significance tests. This ambiguity can be clearly expressed by borrowing a distinction introduced by Reichenbach, the distinction between the context of the discovery and the context of justification of a scientific hypothesis: the B questions are expressed in the form of the context of justification, while they actually imply a real situation in the context of discovery. In other words, we are talking about disguised situations, in which the scientist is not required to produce evidence in support of a given hypothesis, but rather to take his bearings about a preliminary degree of significance in the data and their congruity with the existing theoretical system. The first step in this direction is to try and get the chance hypothesis out of the way. But this is just what significance tests are supposed to do .

More clearly, I think that there are situations to which significance tests and not Bayes theorem are applicable and that it is possible to define an area of research where significance tests play a peculiar role. The situations to which the significance tests seem to be most applicable are those belonging to the orientation phase of research, in which the data are required to supply information necessary to the possible construction of explanatory hypotheses. In this phase there is not yet any specific hypothesis about the data under consideration. It is now that a selection must be made among the data we have gathered, and we must organize and interpret according to their significance degree. Only after we can, in case, working out new explanatory hypotheses or setting the selected data in an existing theoretical framework. This is what is commonly called the 'context of discovery', characterized and directed by a 'Why?' to be pursued, an enigma to be solved.

If the interpretation of significance tests as tools in a context of discovery is correct, the reason of their misapprehensions is more clearly visible. They cannot support either the acceptance or the rejection of non-null hypotheses because their task is simply to signal, by showing correlations, all that is neither normal, or accidental, about the phenomena. By excluding the hypothesis of pure chance, they indicate to the scientist the existence of a problem to be faced and solved.

It goes without saying that, on the contrary, the context peculiar to Bayes theorem can be only the context of 'justification', that is to say a context in which the scientist is required to find evidence in support of clearly defined hypotheses and therefore to know how far the reported data confirm a specific hypothesis. The problem is no longer to extract from the data information useful to orient the following phases of research, so much as to make decision rationally justified, which will depend on an adequate degree of probability.

As a consequence, the Bayesian approach and significance tests seem to assume a more sharply defined identity; and because of this, the relation between them seems to be complementary rather than antagonistic. If the supporters of each of the methodological approaches have kept a quite sterile dispute alive for more than a century, this is due to an inveterate confusion between the plane of discovery and the plane of justification.

Department of Philosophy
University of Florence

NOTES

[1] R.A. Fisher, *Statistical Methods and Scientific Inference*, London 1956, p.39.

[2] Usually 0.05 is taken as the standard level of significance, but in medicine 0.01 is often assumed, or even 0.001. Fisher yet points out that "no scientific worker has a fixed level of significance at which, from year to year and in all circumstances, he rejects hypotheses: he rather gives his mind to each particular case in the light of his evidence and his ideas". *Ibid.*, p.42.

[3] R.A. Fisher, *The Design of Experiments*, London 1949, p.16.

[4] *Ibid.* In line with English empirical tradition, Fisher seems to share the opinion that no observations, however great in number, could be considered enough to 'prove' the validity of a general hypothesis. Inductive reasoning, understood as merely 'passing from the singular to the universal', reveals itself as fallacious. The traditional way of verifying and accepting general hypotheses is founded on false reasoning: induction can only be attempted in the form of *modus tollens*, if one wants to adhere to a logically correct method of reasoning. This seems to be an assumption implied in Fisher's view.

[5] *Ibid.*

[6] R.A. Fisher, *Statistical Methods and Scientific Inference*, cit., p. 42.

[7] *Ibid.*, pp. 43–44.

[8] 'Student' is the pseudonym under which William Sealy Gosset, a chemist who worked for Guinness, the English brewery, published his scientific work. His distribution of *t*, particularly in Fisher's slightly revised form, is one of the normal tools of statistical analysis.

[9] The number of degrees of freedom is equal to *n-1* (*n* being the number of observations reported on the sample). The mean value yielded by *n* observations can be obtained in many different ways (that is, there are many different addends producing the same total), but, once *n-1* have been determined, the last is determined univocally by them.

[10] The chi-squared distribution was discovered in 1876 by the German mathematician Helmert, but passed unnoticed. In 1900 Karl Pearson rediscovered it independently.

[11] R.A. Fisher, *Statistical Methods and Scientific Inference*, cit., p. 42 f.

[12] J. Arbuthnot, *An argument for Divine Providence, taken from the constant regularity observed in the births of both sexes*, 1710–12.

[13] At more or less the same time the French astronomer Louis de Maupertuis (1698–1759), after having observed the appearance of polydactylism in three generations of one family, argued with the help of the probability theory that the anomaly must be hereditary and not a matter of chance (L. de Maupertuis, *Lettres*, in *Ouvres*, Lyon 1768, vol. II, pp. 307–310).

[14] If a wire connected to the poles of a galvanic apparatus is set up parallel with a magnetic needle, when an electric current passes along the wire it will be seen that the needle tends to move, at an angle with its original position. The measurement of the angle depends on the distance between the magnetic needle and the wire. Cf. Ch. Oersted, *Experimenta circa effectum conflictus electrici in acum magneticum,* Annales de Chimie et de Physique, 1820.

[15] This history shows also that the science requirements exceed the mere 'description' of found correlations!

LUIGI ACCARDI

THE QUANTUM PROBABILISTIC APPROACH TO THE FOUNDATIONS OF QUANTUM THEORY: URNS AND CHAMALEONS

INTRODUCTION

The interpretative problems of quantum mechanics motivated a new analysis of classical probability theory leading to the conclusion that some basic probabilistic notions, considered for centuries to be intrinsic to the very notion of probability, are in fact model dependent statements, like Euclid's parallel postulate. This is the basic tenet of quantum probability.

As a consequence of this new analysis a new, model independent, set of axioms for both classical and quantum probability was proposed [3] in which all the axioms have a direct physical meaning and underline what the two theories have in common and where they radically depart from each other. A classification theorem exhibits all the models for these axioms; these include the usual classical and quantum models.

Probably the main success of this axiomatization is that *the superposition principle is deduced from physical axioms* which do not even mention mysterious properties of particles, such as not having a position if not observed. **An implication of this fact is that a consistent physical interpretation of quantum theory can be developed entirely within the conceptual framework of classical physics.**

The EPR correlations are used to argue against the common place according to which quantum theory is intrinsically indeterministic: in some cases *strict quantum determinism* can be experimentally checked. However *experimentally checkable individuation of the deterministic dynamics (whose platonic existence is proved) is ruled out by the Heisenberg principle* (cf. [3]).

The notion of *Einstein (or urn) realism* has been compared with that of *chameleon realism* and it has been argued that the second one is more suitable to give an intuition of the quantum world [1]. The EPR correlations have been reproduced by twin chameleons, thus falsifying the statement that the Bell inequalities are specific of the quantum, or anyway microscopic, world.

In the following we shall describe, following Chapter (IX.7) of [3], a general scheme of measurement which translates into an abstract mathematical language the chameleon metafora and illustrates in a concrete example how

M.L. Dalla Chiara et al. (eds.), Language, Quantum, Music, 95–104.

quantum theory, locality and reality are fully compatible with violation of Bell inequalities.

The basic idea is the following: the standard *counting arguments* from which the Bell inequalities are derived are implicitly based on the *urn model* which is equivalent to the postulate that *the values of all the observables are established a priori.* Given this postulate, the joint probabilities for the triples are nothing but the (platonic) relative frequencies of all triples of values for each particle, which one postulates to be adequately reflected or approximated, by the empirical frequencies. It is well known that these joint probabilities are all what one needs to deduce the Bell inequalities.

In the *chameleon model, the values of all the observables are not established a priori.* What is agreed a priori is only that each particle, if put in interaction with an apparatus x, shall evolve according to a dynamics T_x^t depending, in a completely local and causal manner, on the apparatus x.

This purely local dependence of the dynamical evolution on the measurement apparatus prevents the application of the counterfactual argument on which the standard deductions of Bell's inequality are based: even if particle 1 only feels the instrument x, near to it, and particle 2 only the instrument y, due to the initial singlet constraint, which is a standard conservation law not qualitatively different from those well known in classical physics, the platonic relative frequencies of all the observables of the two particles (i.e. including those which are not actually measured by the apparata x and y), shall depend on both x and y. Therefore there is no physical reason to postulate a priori that, varying x and y one obtains compatible sets of relative frequencies as in the urn model.

The experiments prove that, what was not reasonable to postulate a priori (the existence of joint probabilities) does not in fact take place in nature.

Our conclusion is that it is not possible to concile scientific rationalism with the statement that the Bell (or equivalently two–slit) inequalities provide an *experimental* verification of pretended misterious aspects of quantum theory such as *non–reality, non–locality*, . . .

As strange as this might seem at first sight, a consequence of the above is that the arguments of the Copenhagen interpretration can now be conciled with the everyday practice of physicists and with those of the supporters of hidden variables, but both positions are now cleaned from unnecessary metaphysics, paradoxes, contradictions, . . .

MEASUREMENTS AND LOCALITY FROM A DYNAMICAL POINT OF VIEW

Let us denote:

- S a system
- M a measurement apparatus
- $\mathcal{S}_o(t)$ the set of physical properties that the system S would have at time t if not interacting with the measurement apparatus M
- $M_o(t)$ the set of physical properties that the measurement apparatus M would have at time t if not interacting with the system S
- $\mathcal{S}(t)$ the set of physical properties of the system S *after the interaction* with the measurement apparatus M
- $M(t)$ the set of physical properties of the measurement apparatus M *after the interaction* with the system S.

Following [3] or [4] the set of all physical properties of a system at a given time is called *the Bergsonian state* of that system at that time and has to be distinguished from the notion of *state with respect to a given theory*: the former is a metaphysical and absolute notion, the latter a theory dependent construction. The discussion that follows is constructed so to allow both interpretations. For this reason we shall speak generically of *the state* of the system (cf. [4] and [5] for a detailed analysis of the notion of *physical state*.

Intuitively a dynamical law is something that describes how the physical properties (state) of a system evolve in time. The basic tenet of determinism is that the properties of a system at time t are uniquely determined by the properties of the system at any previous time. Symbolically we can introduce three dynamical laws:

$$T^t_{(S,M)} \quad , \quad T^t_S \quad , \quad T^t_M$$

characterized by the properties:

$$T^t_{(S,M)}(\mathcal{S}_o(0), M_o(0)) = (\mathcal{S}(t), M(t))$$

$$T^t_S\mathcal{S}(0) = \mathcal{S}_o(t)$$

$$T^t_M M(0) = M_o(t)$$

and called respectively *the interacting evolution the free system evolution*, *the free apparatus evolution*. The above formulation implicitly assumes that:

$$\mathcal{S}_o(0) = \mathcal{S}(0) \quad ; \quad M_o(0) = M(0)$$

which is equivalent to the assumption that at the initial time the system S and the apparatus M do not interact.

The fundamental postulate of classical measurement theory can be formulated as follows:

the presence of the measurement apparatus M has a negligible effect on the evolution of the system S, but not conversely.

In symbols:

$$T^t_{(S,M)}(\mathcal{S}(0)) = \mathcal{S}(t) = \mathcal{S}_o(t) = T^t_S \mathcal{S}(0) \tag{1}$$

$$T^t_{(S,M)}(M(0)) = (t) \neq M_o(t) = T^t_M M(0) \tag{2}$$

The two slit experiment, realized with glass beads or billiard balls, illustrate well the content of the relations (1) and (2): a photoelectric cell between the two screens (measurement apparatus) is not activated if there are no balls while it is by their presence. Thus the interacting evolution of the measurement apparatus is different from its free evolution ($T^t_{(S,M)}(M(0)) \neq T^t_M M(0)$). On the other hand the presence of a photoelectric cell does not appreciably affect the trajectory of a billiard ball (system) if random collisions can be neglected: in this case the interacting evolution of the system is equal to the free evolution ($T^t_{(S,M)}(\mathcal{S}(0)) = T^t_S \mathcal{S}(0)$).

An important achievement of quantum theory is the realization of the fact that the fundamental postulate of classical measurement theory is unfounded at a microscopic level. It should be stressed however that, even for macroscopic objects, the negligibility of the effects of the interaction with the measurement apparatus is something that has to be proved case by case and not be given for granted a priori.

For example, in the two–slit experiment, the three possible conditions of the first screen (both slits open, only slit 1 or only slit 2 open) define three different *measurement apparata*:

$$M_{1,2} \quad ; \quad M_1 \quad ; \quad M_2$$

whose corresponding time evolutions shall be denoted

$$T^t_{(S,M_{1,2})} \quad ; \quad T^t_{(S,M_1)} \quad ; \quad T^t_{(S,M_2)}$$

Similarly, in the situation analyzed by Bell and experimentally realized by Aspect and others, S denotes the singlet pair and M measurement apparatus (polarizers along the directions a, b), thus

$$M = M_{a,b}$$

consequently also the dynamical law shall depend on the polarization directions:

$$T^t_{(S,M)} = T^t_{(S,M_{a,b})}$$

In order to introduce **locality** into this scheme, it is convenient to consider the *dynamical generator,* i.e. the *Hamiltonian* H of the system because in it is possible to separate in an easy way the contributions of the free dynamics of S and M, from those of the interaction. Such a separation is usually less intuitively immediate in the dynamical evolution $T^t_{(S,M)}$.

In case of the system (S, M), the dynamical generator is the sum of three pieces:

$$H = H_S + H_M + H_I \tag{3}$$

where H_S generates the free evolution of the system S (i.e. T^t_S); H_M generates the free evolution of the apparatus M (i.e. T^t_M); and H_I describes the interaction between the system S and the apparatus M in the sense that the interacting evolution $T^t_{(S,M)}$ is generated by H given by (3).

In terms of the generator H, it is also very simple to express the locality condition. To fix the ideas let us consider the EPR experiment and recall that both S and M, are made up of two sub–systems: S is composed of the two particles, denoted 1 and 2

$$S = (1, 2)$$

and M is composed of the two, spatially separated, measurement apparata

$$M = (M_1, M_2)$$

The corresponding decomposition of the generators of the free evolutions is:

$$H_S = H_1 + H_2 + H_{1,2}$$

$$H_M = H_{M_1} + H_{M_2} + H_{(M_1,M_2)}$$

where H_1 (resp. H_{M_1}) is the free Hamiltonian of the particle 1 (resp. of the apparatus M_1) and analogously for the particle 2 and the apparatus M_2. The interactions between the two particles – $H_{1,2}$ – and between the two apparata – $H_{(M_1,M_2)}$ – can be neglected because one can assume that these object are far apart. In terms of the Hamiltonians this locality assumption is expressed as follows:

$$H_S = H_1 + H_2 \quad ; \quad H_M = H_{M_1} + H_{M_2} \tag{4}$$

In order to complete the formulation of the locality condition, one should formulate the decomposition, analogue to (4), for the interaction

$$H_{(S,M)} = H_{(1,2),\,(M_1,M_2)}$$

between the particles, i.e. the system $S = (1,2)$, and the apparatus $M = (M_1, M_2)$. In this interaction we recognize various pieces:

- interaction between 1 and M_1 : H_{1,M_1}
- interaction between 2 and M_2 : H_{2,M_2}
- interaction between 1 and M_2 : H_{1,M_2}
- interaction between 2 and M_1 : H_{2,M_1}

Consequently a priori one should have

$$H_I = H_{(S,M)} = H_{1,M_1} + H_{2,M2} + H_{1,M_2} + H_{2,M_1}$$

The locality condition is equivalent to assert that the contributions of the interactions between 1 and M_2 and between 2 and M_1 are negligible because these object are far apart. This is equivalent to claim that, by putting

$$H_{1,M_2} = H_{2,M_1} = 0$$

the error made is negligible. Modulo this error the system 1 interacts only with the apparatus M_1 and the system 2 only with the apparatus M_2. In formulae:

$$H_I = H_{(S,M)} = H_{1,M_1} + H_{2,M_2}$$

In conclusion, up to negligible errors, the dynamical generator is the sum of two pieces:

$$H = \left(H_1 + H_{M_1} + H_{1,M_1}\right) + \left(H_2 + H_{M_2} + H_{2,M_2}\right)$$

one, concerning only the pair *system 1 plus apparatus* M_1; the other one, concerning only the pair *sistema 2 plus apparatus* M_2.

Under these circumstances, *the dynamical law* $T^t_{(S,M)}$, of the interacting system (S, M), *factorizes* into two separated pieces

$$T^t_{(S,M)} = \left(T^t_{(1,M_1)}, T^t_{(2,M_2)}\right) = T^t_{(1,M_1)} \otimes T^t_{(2,M_2)} \tag{5}$$

This expresses tha fact that the particle 1 evolves feeling only the influence of the measurement apparatus M_1, near to it, and similarly for particle 2.

This is the correct formulation of the locality condition.

Now recall that Bell, in one of his formulations of the locality conditions, requires factorization *not of the dynamics, but of the state*). We shall prove the illegitimacy of this condition as well as of the other Bell's assumption, implicit but more fundamental for the proof of his result, that the three probability distributions corresponding to the three (mutually incompatible) pairs of measurements (M_a, M_b), (M_b, M_c), (M_c, M_a) can be deduced from a single probability distribution.

Let σ denote the initial state of the composite system (particles plus measurement apparata). At time zero there can be no relation between the two: the particles cannot suspect which measurement shall be performed on them. This means that

$$\sigma = \sigma_o \otimes \sigma_M \tag{6}$$

The two measurement apparata are also far apart and each experimenter can decide on an entirely local basis what to measure. This means that

$$\sigma_M = \sigma_{M_1} \otimes \sigma_{M_2} \tag{7}$$

On the other hand the two particles are not independent at time zero, since they are in the singlet state σ_o which is not factorizable. Under the dynamics (5), the state of the composite system at time t shall be

$$T^t_{(S,M)}\sigma = \left(T^t_{(1,M_1)} \otimes T^t_{(2,M_2)}\right)\left(\sigma_o \otimes \sigma_M\right) \tag{8}$$

This is by no means a product state even if both the dynamics and the initial state enjoy the factorization properties (5), (6), (7): the two types of factorizations (of the dynamics and of the state) group together different spaces so that the resulting state does not factorize. In the notations of the polarizers, the state of the composite system at time t can be written

$$\sigma_{a,b}(t) = T^t_{(S,M_{a,b})}\sigma = \left(T^t_{(1,M_a)} \otimes T^t_{(2,M_b)}\right)\left(\sigma_o \otimes \sigma_M\right)$$

and similarly for $\sigma_{b,c}(t), \sigma_{c,a}(t)$. None of these states is factorized.

Interpreting the states $\sigma_{a,b}, \ldots$, as probability measures on the same sample space $\Omega_1 \times \Omega_2$ (this is always possible) and denoting S_1, S_2 the projection functions ($S_1(\epsilon_1, \epsilon_2) = \epsilon_1 \in \Omega_1$, $S_2(\epsilon_1, \epsilon_2) = \epsilon_2 \in \Omega_2$), we see that, for any pair of functions f_1, f_2, one has

$$\sigma_{a,b}(t)\big(f_1(S_1)f_2(S_2)\big) = f_1(T^t_{(1,M_a)}S_1)f_2(T^t_{(2,M_b)}S_2)$$

In particular, by taking $f_2 = 1$ (constant function) we obtain

$$\begin{aligned}\sigma_{a,b}(t)\big(f_1(S_1)\big) &= \sigma_{a,b}\big(f_1(T^t_{(1,M_a)}S_1)\otimes 1\big) = \sigma_o\otimes\sigma_{M_a}\otimes\sigma_{M_b}\big(f_1(T^t_{(1,M_a)}S_1)\otimes 1\big) = \\ &= \sigma_o\otimes\sigma_{M_a}\otimes\sigma_{M_b}\big(f_1(T^t_{(1,M_a)}S_1)\otimes(1_2\otimes 1_{M_b})\big) \\ &= \sigma_o\otimes\sigma_{M_a}\big(f_1(T^t_{(1,M_a)}S_1)\otimes 1_2\big)\sigma_{M_b}(1_{M_b}) \\ &= \sigma_o\otimes\sigma_{M_a}\big(f_1(T^t_{(1,M_a)}S_1)\otimes 1_2\big)\end{aligned}$$

In other terms: the probability distribution of the observables of particle 1 does not depend on the measurements one does on the measurements one does on particle 2. This condition (more precisely, a corollary of it) was introduced by Nelson [Nel] who called it *active locality*. The above argument proves that the dynamical locality condition (5) is stronger than active locality.

Notice that the measurement scheme described above is very general: it can refer to a quantum as well as to a classical theory; in the latter case, by *singlet state* one should mean any probability measure on the state space of the corresponding exact theory which is not factorized (e.g. which obeys some conservation rule). Even with this deterministic interpretation there is no reason why the three probability measures

$$\sigma_{a,b}(t) \quad , \quad \sigma_{b,c}(t) \quad , \quad \sigma_{c,a}(t) \tag{9}$$

should be the marginals of a unique probability measure

$$\sigma_{a,b,c}(t)$$

Since the existence of such a measure is the only hypothesis really indispensible for the validity of the Bell inequalities, it follows that:

even for a classical deterministic theory, with a perfectly classical randomness, one can construct thousands of physically meaningful EPR-like examples in which the Bell inequalities are not satisfied but the locality condition is strictly respected.

The *machine* to produce these examples is based on the scheme described above: one fixes one time t (the final time of measurement) and one constructs six transformations

$$T^t_{(1,M_a)} \; , \; T^t_{(1,M_b)} \; , \; T^t_{(1,M_c)} \; , \; T^t_{(2,M_a)} \; , \; T^t_{(2,M_b)} \; , \; T^t_{(2,M_c)}$$

so that the three states (9) do not satisfy the Bell inequalities. One could also add the symmetry condition

$$T^t_{(1,M_a)} = T^t_{(2,M_a)}$$

which reduces the choices to three dynamics, but the free parameters are so many that it is not necessary to make explicit calculations to understand that the solutions are infinitely many.

For example, a simple way to produce within this scheme the EPR correlations in a classical context is to choose three arbitrary probability measures

$$\bar{\sigma}_{a,b} \quad , \quad \bar{\sigma}_{b,c} \quad , \quad \bar{\sigma}_{c,a}$$

on $\{+1,-1\} \times \{+1,-1\}$ such that, denoting S_1, S_2 the projection functions ($S_1(\epsilon_1,\epsilon_2) = \epsilon_1$, $S_2(\epsilon_1,\epsilon_2) = \epsilon_2$), the correlations

$$\bar{\sigma}_{a,b}(S_1 \cdot S_2) \ , \ \bar{\sigma}_{b,c}(S_1 \cdot S_2) \ , \ \bar{\sigma}_{c,a}(S_1 \cdot S_2)$$

are those predicted by the singlet state in the directions a, b, c We also assume that these three directions are those which maximize violation of the Bell inequalities. Now fix arbitrarily three dynamical evolutions

$$T^t_{(1,M_a)} \ , \ T^t_{(1,M_b)} \ , \ T^t_{(1,M_c)}$$

and define three other evolutions by the product rule (5). Finally fix a final time $\bar{t}$ (the time of performance of the experiment) and define the *initial states*

$$\sigma_{a,b} \quad , \quad \sigma_{b,c} \ ; \sigma_{c,a}$$

by the prescription

$$\sigma_{a,b} := \left(T^{-t}_{(1,M_a)} \otimes T^{-t}_{(2,M_b)}\right)\bar{\sigma}_{a,b}$$

(and similarly for the other two). Thus by construction the Bell inequalities are violated. This construction is surely trivial, but notice that it does not guarantee a factorization of the initial state of the form

$$\sigma_{a,b} = \sigma_o \otimes \sigma_{M_a} \otimes \sigma_{M_b}$$

We conjecture that, with an additional effort, also this factorization can be guaranteed.

In the Aspect experiment, and in several subsequent ones, some rapid changes are induced in the apparata M_1, M_2. In the general scheme described above, this amounts to claim that the free evolution of the apparatus, i.e. T^t_M is non trivial. The locality condition implies that it has the form $T^t_{M_a} \otimes T^t_{M_b}$. Thus also the Aspect experiments enter into this scheme.

Practically all the most recent attempts to construct a mathematical model of the measurement process, include a description not only of the system on which the measurement is performed, but also of another system, interacting with it and called, according to the interpretations, *measurement apparatus, environment, thermal bath, reservoir, field, noise, master field, gas,...* (for a broad outline of these approaches the reader may consult the the papers by Belavkin, Ghirardi, Mensky, Namicki and Pascazio, Pearle, Schulman... in the volume [EncIt]).

Just because of its generality, the above scheme does not try to include into the description a differentiation between *macroscopic* and *microscopic* systems. Our goal in fact is to evidentiate why it is reasonable to expect that *the statistics of adaptive systems* should be different from the usual statistics of *passive systems*. To this goal the difference between *macroscopic* and *microscopic* systems is not relevant (in the chameleon example both system and measurement apparatus are macroscopic objects).

The fact that nothing of the above discussion uses specific properties of quantum systems, but everything applies equally well to the classical and to the quantum world shows that it is possible to construct a class of quantum or of classical dynamical models which satisfy the restrictions imposed by locality, formulated in a way that should be acceptable to all physicists, but do not satisfy Bell's inequality.

Faculty of Economics
University of Rome

REFERENCES

[EncIt]: 1994, 'The Interpretation of Quantum Theory: Where Do We Stand?', Acta Enciclopedica, Istituto dell'Enciclopedia Italiana.

[1] Accardi, L.: 1994, 'Einstein-Bohr: One All, in The Interpretation of Quantum Theory: Where Do We Stand?' *Acta Enciclopedica*, Istituto dell'Enciclopedia Italiana, Volterra Preprint N. 174 (1993).

[2] Accardi, L.: 1995, 'Can Mathematics Help Solving The Interpretational Problems of Quantum Theory?', *Il Nuovo Cimento*, 1**10B**, 685–721; cf. also Mesoscopic Physics and Fundamental Problems in Quantum Mechanics, C. Di Castro, F. Guerra, G. Jona–Lasinio (1995).

[3] Accardi, L.: 1997, Urne e Camaleonti: Dialogo Sulla Realtà e l'Interpretazione della Teoria Quantistica, Il Saggiatore.

[4] Accardi L.: 1981, Stato Fisico. *Enciclopedia Einaudi*, **13**, 514–548.

[5] Accardi, L., Reviglio E.: 1992, The Concept of Physical State and the Foundations of Physics, Proceedings Conference 'Scientific Realism', Napoli.

[6] Nelson, L.: 'The Locality Problem in Quantum Mechanics', in New Techniques and Ideas in Quantum Measurement Theory. Preprint January 1986.

DIEDERIK AERTS AND BOB COECKE

THE CREATION-DISCOVERY-VIEW: TOWARDS A POSSIBLE EXPLANATION OF QUANTUM REALITY

1. INTRODUCTION

The creation discovery view and together with it its technically underlying hidden measurement formalism has been elaborated from the early eighties on, and many aspects of it have been exposed in different places [6, 7, 12, 13, 15, 16, 19, 20, 22, 23, 30–37]. In this paper we give an overview of the most important of these aspects.

Quantum mechanics was originally introduced as a non-commutative matrix calculus of observables by Werner Heisenberg [41] and parallel as a wave mechanics by Erwin Schrödinger [43]. These two structurally very different theories could explain fruitfully the early observed quantum phenomena. Already in the same year the two theories were shown to be realizations of the same, more abstract, ket-bra formalism by Dirac [38]. Only some years later, in 1934, John Von Neumann put forward a rigorous mathematical framework for quantum theory in an infinite dimensional separable complex Hilbert space [46]. Matrix mechanics and wave mechanics appear as concrete realizations: the first one if the Hilbert space is l^2, the collection of all square summable complex numbers, and the second one if the Hilbert space is L^2, the collection of all square integrable complex functions. The formulation of quantum mechanics in the abstract framework of a complex Hilbert space is now usually referred to as the 'standard quantum mechanics'.

The basic concepts – the vectors of the Hilbert space representing the states of the system and the self-adjoint operators representing the observables – in this standard quantum mechanics are abstract mathematical concepts defined mathematically in an abstract mathematical space. Several approaches have generalized the standard theory starting from more physically defined basic concepts. John Von Neumann and Garett Birkhoff have initiated one of these approaches [29] were they analyze the difference between quantum and classical theories by studying the 'experimental propositions'. They could show that for a given physical system classical theories have a Boolean lattice of experimental propositions while for quantum theory the lattice of experimental propositions is not Boolean. Similar fundamental structural differences be-

M.L. Dalla Chiara et al. (eds.), Language, Quantum, Music, 105–116.

tween the two theories have been investigated by concentrating on different basic concepts. The collection of observables of a classical theory was shown to be a commutative algebra while this is not the case for the collection of quantum observables [40, 44]. Luigi Accardi and Itamar Pitowski obtained an analogous result by concentrating on the probability models connected to the two theories: classical theories have a Kolmogorovian probability model while the probability model of a quantum theory is non Kolmogorovian [1, 42].

The fundamental structural differences between the two types of theories, quantum and classical, in different categories, was interpreted as indicating also a fundamental difference on the level of the nature of the reality that both theories describe: the micro world should be 'very different' from the macro world. This state of affairs was very convincing also because concrete attempts to understand quantum mechanics in a classical way had failed as well: e.g. the many 'physical' hidden variable theories that had been tried out [45]. The structural difference between quantum theories and classical theories (Boolean lattice versus non-Boolean lattice of propositions, commutative algebra versus non commutative algebra of observables and Kolmogorovian versus non Kolmogorovian probability structure) had been investigated mostly mathematically and not much understanding of the physical meaning of the structural differences had been gained during all these years.

The first step that led to the creation discovery view and its underlying hidden measurement formalism was a breakthrough in the understanding of the physical origin of these mathematical structural differences between quantum and classical theories. Indeed, one of the authors found in the early eighties a way to identify the physical aspects that are at the origin of the structural differences [3, 6, 7]. Let us summarize these findings: it are mainly two aspects that determine the mathematical structural differences between classical and quantum theories in the different categories:

We have a quantum-like theory describing a system under investigation if the measurements needed to test the properties of the system are such that:

(1) The measurements are not just observations but provoke a real change of the state of the system.

(2) There exists a lack of knowledge about the reality of what happens during the measurement process.

The presence of these two aspects is sufficient to render the description of the system under consideration quantum-like. It is the lack of knowledge (2) that is theoretically structured in a non Kolmogorovian probability model. In a certain sense it is possible to interpret the second aspect, the presence of the lack of knowledge on the reality of the measurement process, as the presence of 'hidden measurements' instead of 'hidden variables'. Indeed, if a measurement is performed with the presence of such a lack of knowledge, then this is actually the classical mixture of a set of classical hidden measurements, were for such a classical hidden measurement there would be no lack of knowledge. In an analogous way as in a hidden variable theory, the quantum state is a classical mixture of classical states. This is the reason why we have called the underlying theory of the creation discovery view the hidden measurement formalism. It is possible to illustrate the creation discovery view and the hidden measurement aspect in a very simple way by using a mechanical model that was introduced in [5, 6, 7] and that we have called the quantum machine. This is the subject of next section.

2. THE QUANTUM MACHINE.

Several aspects of the quantum machine have been presented in different occasions [6–10, 15, 19, 20] and we shall therefore introduce here only the basic aspects. The machine that we consider consists of a physical entity S that is a point particle P that can move on the surface of a sphere, denoted *surf*, with center O and radius 1. The unit-vector v where the particle is located on *surf* represents the state p_v of the particle (see Figure 1,a). For each point $u \in surf$, we introduce the following measurement e_u. We consider the diametrically opposite point $-u$, and install a piece of elastic of length 2, such that it is fixed with one of its end-points in u and the other end-point in $-u$. Once the elastic is installed, the particle P falls from its original place v orthogonally onto the elastic, and sticks on it (Figure 1,b). Then the elastic breaks and the particle P, attached to one of the two pieces of the elastic (Figure 1,c), moves to one of the two end-points u or $-u$ (Figure 1,d). Depending on whether the particle P arrives in u (as in Figure 1) or in $-u$, we give the outcome o_1^u or o_2^u to e_u. We can easily calculate the probabilities corresponding to the two possible outcomes.

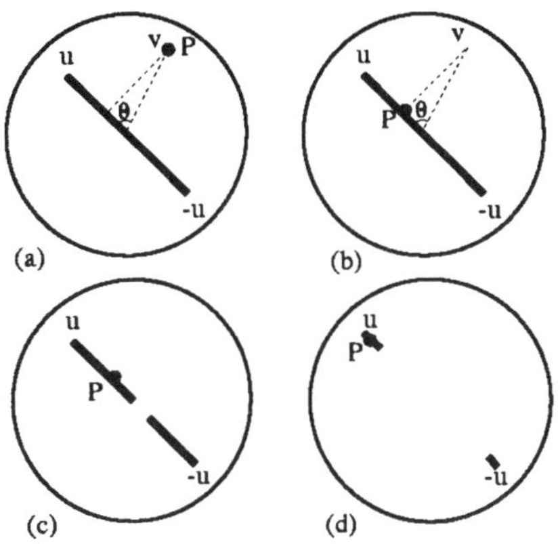

Figure 1. A representation of the quantum machine. In (a) the physical entity P is in state p_v in the point v, and the elastic corresponding to the measurement e_u is installed between the two diametrically opposed points u and $-u$. In (b) the particle P falls orthogonally onto the elastic and stick to it. In (c) the elastic breaks and the particle P is pulled towards the point u, such that (d) it arrives at the point u, and the measurement e_u gets the outcome o_1^u.

The particle P arrives in u when the elastic breaks in a point of the interval L_1 (which is the length of the piece of the elastic between $-u$ and the point where the particle has arrived, or $1 + cos\theta$), and arrives in $-u$ when it breaks in a point of the interval L_2 ($L_2 = L - L_1 = 2 - L_1$). We make the hypothesis that the elastic breaks uniformly, which means that the probability that the particle, being in state p_v, arrives in u, is given by the length of L_1 divided by the length of the total elastic (which is 2). The probability that the particle in state p_v arrives in $-u$ is the length of L_2 (which is $1 - cos\theta$) divided by the length of the total elastic. If we denote these probabilities respectively by $P(o_1^u, p_v)$ and $P(o_2^u, p_v)$ we have:

$$P(o_1^u, p_v) = \frac{1 + cos\theta}{2} = cos^2\frac{\theta}{2} \qquad P(o_2^u, p_v) = \frac{1 - cos\theta}{2} = sin^2\frac{\theta}{2} \tag{1}$$

These transition probabilities are the same as the ones related to the outcomes of a Stern-Gerlach spin measurement on a spin $\frac{1}{2}$ quantum particle, of which the quantum-spin-state in direction $v = (cos\phi sin\theta,\ sin\phi sin\theta, cos\theta)$, denoted by $\bar{\psi}_v$, and the measurement e_u corresponding to the spin measurement in direction $u = (cos\beta sin\alpha,\ sin\beta sin\alpha,\ cos\alpha)$, is described respectively by the vector and the self adjoint operator of a two-dimensional complex Hilbert space.

$$\psi_v = (e^{-i\phi/2}cos\theta/2, e^{i\phi/2}sin\theta/2) \qquad H_u = \frac{1}{2}\begin{pmatrix} cos\alpha & e^{-i\beta}sin\alpha \\ e^{i\beta}sin\alpha & cos\alpha \end{pmatrix} \tag{2}$$

We can easily see now the two aspects in this quantum machine that we have identified in the creation discovery view to give rise to the quantum structure. The state of the particle P is effectively changed by the measuring apparatus (p_v changes to p_u or to p_{-u} under the influence of the measuring process), which identifies the first aspect, and there is a lack of knowledge on the interaction between the measuring apparatus and the particle, namely the lack of knowledge of were exactly the elastic will break, which identifies the second aspect. We can also easily understand now what is meant by the term 'hidden measurements'. Each time the elastic breaks in one specific point λ, we could identify the measurement process that is carried out afterwards as a hidden measurement e_u^λ. The measurement e_u is then a classical mixture of the collection of all measurement e_u^λ: namely e_u consists of choosing at random one of the e_u^λ and performing this chosen e_u^λ. We can see in the example that effectively the final states, after the measurement e_u, hence the states p_u and p_{-u}, are partly created by the measurement. We remark that we have shown in our group in Brussels that such a model can be built for any arbitrary quantum entity [5–7, 14, 30–32, 34, 35]. However, the hidden measurement formalism is more general than standard quantum theory. Indeed, it is very easy to produce quantum-like structures that cannot be represented in a complex Hilbert space. For a presentation of an as general as possible mathematical framework to deal with the hidden measurement ideas we refer to [32, 34–37]. An example of such quantum-like structures is presented in the following section where we define a continuous transition from quantum to classical for the model introduced at the beginning of this section. Hence we claim to solve the old problem of the classical limit.

3. THE QUANTUM CLASSICAL TRANSITION.

If the quantum structure can be explained by the presence of a lack of knowledge on the measurement process, we can go a step further, and wonder what types of structure arise when we consider the original models, with a lack of knowledge on the measurement process, and introduce a variation of the magnitude of this lack of knowledge. We have studied the quantum machine under varying 'lack of knowledge', parameterizing this variation by a number $\epsilon \in [0, 1]$, such that $\epsilon = 1$ corresponds to the situation of maximal lack of knowledge, giving rise to a quantum structure, and $\epsilon = 0$ corresponds to the situation of zero lack of knowledge, generating a classical structure. Other values of ϵ correspond to intermediate situations and give rise to a structure that is neither quantum nor classical [22–25]. We have called this model the ϵ-model and introduce it shortly.

a) The ϵ-Model. We start from the quantum machine, but introduce now different types of elastic. An ϵ, d-elastic consists of three different parts: one lower part where it is unbreakable, a middle part where it breaks uniformly, and an upper part where it is again unbreakable. By means of the two parameters $\epsilon \in [0, 1]$ and $d \in [-1 + \epsilon, 1 - \epsilon]$, we fix the sizes of the three parts in the following way. Suppose that we have installed the ϵ, d-elastic between the points $-u$ and u of the sphere. Then the elastic is unbreakable in the lower part from $-u$ to $(d - \epsilon) \cdot u$, it breaks uniformly in the part from $(d - \epsilon) \cdot u$ to $(d + \epsilon) \cdot u$, and it is again unbreakable in the upper part from $(d + \epsilon) \cdot u$ to u (see Figure 2).

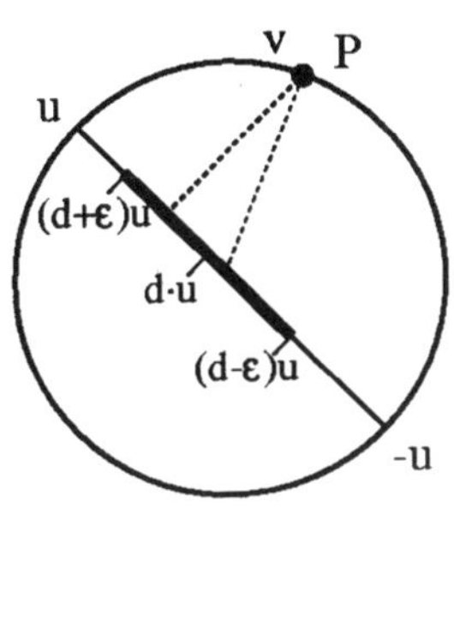

Figure 2. A representation of the measurement $e^{\epsilon}_{u,d}$. The elastic breaks uniformly between the points $(d - \epsilon)u$ and $(d + \epsilon)u$, and is unbreakable in other points.

An e_u measurement performed by means of an ϵ, d- elastic shall be denoted by $e^{\epsilon}_{u,d}$. We have the following cases:

(1) $v \cdot u \leq d - \epsilon$. The particle sticks to the lower part of the ϵ, d-elastic, and any breaking of the elastic pulls it down to the point $-u$. We have $P^{\epsilon}(o_1^u, p_v) = 0$ and $P^{\epsilon}(o_2^u, p_v) = 1$.

(2) $d - \epsilon < v \cdot u < d + \epsilon$. The particle falls onto the breakable part of the ϵ, d-elastic. We can easily calculate the transition probabilities and find:

$$P^{\epsilon}(o_1^u, p_v) = \frac{1}{2\epsilon}(v \cdot u - d + \epsilon) \qquad P^{\epsilon}(o_2^u, p_v) = \frac{1}{2\epsilon}(d + \epsilon - v \cdot u) \tag{3}$$

(3) $d + \epsilon \leq v \cdot u$. The particle falls onto the upper part of the ϵ, d-elastic, and any breaking of the elastic pulls it upwards, such that it arrives in u. We have $P^{\epsilon}(o_1^u, p_v) = 1$ and $P^{\epsilon}(o_2^u, p_v) = 0$.

We are now in a very interesting situation from the point of view of the structural studies of quantum mechanics. Since the ϵ-model describes a continuous transition from quantum to classical, its mathematical structure should be able to learn us *what are the structural shortcomings of the standard Hilbert space quantum mechanics*. Therefore we have studied the ϵ-model in the existing mathematical approaches that are more general than the standard quantum mechanics: the lattice approach, the probabilistic approach and the *-algebra approach.

b) The Lattice Approach. In this lattice approach exists a well known axiomatic scheme that reduces the approach to standard quantum mechanics if certain axioms are fulfilled. For intermediate values of ϵ, that is $0 < \epsilon < 1$, we find that 2 of the 5 axioms needed in the lattice approach to reconstruct standard quantum mechanics are violated. The axioms that are violated are the weak-modularity and the covering law, and it are precisely those axioms that are needed to recover the vector space structure of the state space in quantum mechanics [22, 23, 25].

c) The Probabilistic Approach. If we take the case of vanishing fluctuations ($\epsilon = 0$), do we obtain the Kolmogorovian theory of probability? This would be most interesting, since then we would have constructed a macroscopic model with an understandable structure (i.e. we can see how the probabilities arise) and a quantum and a classical behavior. We [15] proposed to test the polytopes for a family of conditional probabilities. The calculations can be found in [26] and the result was what we hoped for: a macroscopic model with a quantum and a Kolmogorovian limit. For intermediate values of the fluctuations ($0 < \epsilon < 1$) the resulting probability model is neither quantum nor Kolmogorovian: we have identified here a new type of probability model, that is quantum-like, but not really isomorphic to the probability model found in a complex Hilbert space.

*d) The *-Algebra Approach.* The *-algebra provides a natural mathematical language for quantum mechanical operators. We applied the concepts of this approach to the epsilon model to find that an operator corresponding to an ϵ measurement is linear if and only if $\epsilon = 1$ [21]. This means that for the classical and intermediate situations the observables cannot be described by linear operators.

CONCLUSION: Quantum theory and classical theories appear as special cases $(\epsilon = 1, \epsilon = 0)$ *and the general intermediate case, although quantum- like, cannot be described in standard Hilbert space quantum mechanics.*

e) The Measurement Problem and the Schrödingers Cat Paradox. The result stated in the conclusion means that all the paradoxes of standard quantum theory that are due to the fact that quantum theory is used as a universal theory, also being applied to macroscopic system, for example the measuring apparatus, are not present in our hidden measurement formalism. We explicitly have in mind the 'measurement problem' and the 'Schrödinger cat paradox'. Indeed, the measurement apparatus should be described by a classical model in our approach, and the physical system eventually by a quantum model. The problem of the presence of quantum correlations between physical system and measuring apparatus, as it presents itself in the standard theory, takes a completely different aspect. We are working now at the elaboration of a concrete description of the measurement process within the hidden measurement formalism [23]. Our result also shows that it is possible in the hidden measurement formalism to formulate a 'classical limit', namely as a continuous transition from quantum to classical [25].

4. NON-LOCALITY: A GENUINE PROPERTY OF NATURE.

The measurement problem and paradoxes equivalent to Schrödingers cat paradox disappear in the hidden measurement formalism, because standard quantum mechanics appears only as a special case, the situation of maximum fluctuations. All quantum mysteries connected to the effect of 'non-locality' remain (Einstein Podolsky Rosen paradox and the violation of the Bell inequalities). It is even so that non-locality unfolds itself as a fundamental aspect of the creation discovery view. This is due to the fact that if we explain the quantum structures as it is done in this view, a quantum measurement has two concrete physical effects: (1) it changes the state of the system and (2) it produces probability due to a lack of knowledge about the nature of this change. With the quantum machine we have given a macroscopic model for a spin measurement of a spin $\frac{1}{2}$ particle. If we apply the hidden measurement formalism to the situation of a quantum system described by a wave function $\psi(x)$, and to a position or a momentum measurement performed on this system, we also have the two mentioned effects. For example in the case of a position measurement: the detection apparatus changes the state of the quantum system, in the sense that it 'localizes' the quantum system in a specific place of space, and the probabilities that are connected with this measurement are due to the fact that we have a lack knowledge about the specific way in which this localization takes place. This means that 'before the detection has taken place' the quantum system was in general not localized: it was not present in a specific region of space. With other words, quantum systems are fundamentally non-local systems: the wave function $\psi(x)$ describing such a quantum state is not interpreted as a wave that is present in space, but as indicating these regions of space were the particle can be localized, where $\int_R |\psi(x)|^2 dx$ is the probability that this localization will happen in region R. A similar interpretation must be given to the momentum measurement of a quantum entity: the quantum entity has no momentum before the measurement, but the measurement creates partly this momentum. In [25] we have calculated the ϵ- situation for a quantum system described by a wave function $\psi(x)$, element of the Hilbert space of all square integrable complex functions, and we have found the following very simple procedure. Suppose that ϵ is given, and the state of the quantum system is described by the wave function $\psi(x)$ and $\phi(x)$ is the corresponding probability distribution (hence $\phi(x) = |\psi(x)|^2$). We cut, by means of a constant function ϕ_Ω, a piece of the function $\phi(x)$, such that the surface contained in the cutoff piece equals ϵ (see step 1 of Figure 3). We move this piece of function to the x-axis (see step 2 of Figure 3). And then we renormalize by dividing by ϵ (see step 3 of Figure 3). If we proceed in this way for smaller values of ϵ, we shall finally arrive at a delta-function for the classical limit $\epsilon \to 0$, and the delta-function is located in the original maximum of the quantum probability distribution. We want to point out that the state $\psi(x)$ of the physical system is not changed by this ϵ-procedure, it remains always the same state, representing the same physical reality. It is the regime of lack of knowledge going together with the detection measurement that changes with varying ϵ. For $\epsilon = 1$ this regime is one of maximum lack of knowledge on the process of localization, and this lack of knowledge is characterized by the spread of the probability distribution $\phi(x)$. For an intermediate value of ϵ, between 1 and 0, the spread of the probability distribution has decreased (see Figure 3) and for zero fluctuations the spread is 0. Let us also try to see what becomes of the non-local behavior of quantum entities taking into account the classical limit procedure that we propose. Suppose that we consider a double slit experiment, then the state p of a quantum entity having passed the slits can be represented by a probability function $p(x)$ of the form represented in Figure 4. We can see that the non-locality presented by this probability function gradually disappears when ϵ becomes smaller, and in the case where $p(x)$ has

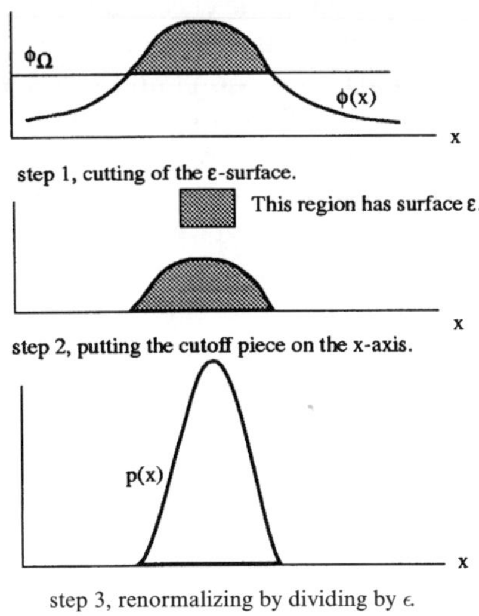

Figure 3.

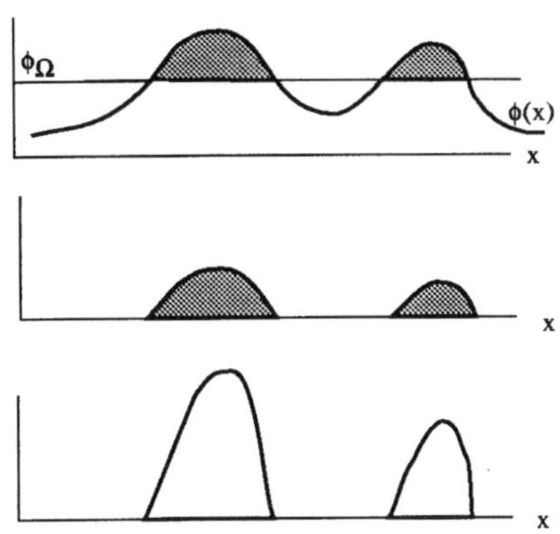

Figure 4. The classical limit procedure in the situation of a non-local quantum state.

only one maximum finally disappears completely. When there are no fluctuations on the measuring apparatus used to detect the particle, it shall be detected with certainty in one of the slits, and always in the same one. If $p(x)$ has two maxima (one behind slit 1, and the other behind slit 2) that are equal, the non-locality does not disappear. Indeed, in this case the limit- function is the sum of two delta-functions (one behind slit 1 and one behind slit 2). So in this case the non-locality remains present even in the classical limit. If our procedure for the classical limit is a correct one, also macroscopic classical entities can be in non-local states. How does it come that we don't find any sign of this non-locality in the classical macroscopic world? This is due to the fact that the set of states, representing a situation where the probability function has more than one maximum, has measure zero, compared to the set of all possible states, and moreover these states are 'unstable'. The slightest perturbation will destroy the symmetry of the different maxima, and hence shall give rise to one point of localization in the classical limit. Also classical macroscopic reality is non-local, but the local model that we use to describe it gives the same statistical results, and hence cannot be distinguished from the non-local model.

5. BELL INEQUALITIES AND HIDDEN CORRELATIONS.

It is interesting to consider the violation of Bell's inequalities within the creation discovery view. The quantum machine, as presented in section 3, delivers us a macroscopic model for the spin of a spin $\frac{1}{2}$ quantum entity and starting with this model it is possible to construct a macroscopic situation, using two of these models coupled by a rigid rod, that represents faithfully the situations of two entangled quantum particles of spin $\frac{1}{2}$ [11]. The 'non-local' element is introduced explicitly by means of a rod that connects the two sphere-models. We also have studied this EPR situation of entangled quantum systems by introducing the ϵ-variation of the amount of lack of knowledge on the measurement processes and could show that one violates the Bell-inequalities even more for classical but non-locally connected systems, that is, $\epsilon = 0$. This illustrates that the violation of the Bell inequalities is due to the non-locality rather then to the indeterministic character of quantum theory. And that the quantum indeterminism (for values of ϵ greater than 0) tempers the violation of the Bell inequalities [18]. This idea has been used by one of the authors to construct a general representation of entangled states (hidden correlations) within the hidden measurement formalism [33, 36]. More precisely, it is possible to prove that for a collection of quantum entities described in a tensor product of Hilbert spaces there always exists one unique hidden correlation representation, i.e., a representation as a collection of individual entities which are correlated in such a way that a change of state of one of the entities (due to a measurement) induces in a well-defined change of state of the other entities (for more details we refer to [36]). Structurally, these correlations are definitely mechanistic. Nonetheless, they are non-local.

CLEA
Free University of Brussels

REFERENCES

[1] Accardi, L.: 1982, *Nuovo Cimento*, **34**, 161.

[2] Accardi, L.: 1984, 'The Probabilistic Roots of the Quantum Mechanical Paradoxes', in S. Diner et al., eds., *The Wave-Particle Dualism*. Reidel Publishing Company, Dordrecht.

[3] Aerts, D.: 1982, *Lett Nuovo Cimento*, **34**, 107.

[4] Aerts, D.: 1984, 'How Do We Have To Change Quantum Mechanics In Order To Describe Separated Systems', in S. Diner et al., eds., *The Wave-Particle Dualism*. Reidel Publishing Company, Dordrecht.

[5] Aerts, D.: 1985, 'A Possible Explanation for The Probabilities of Quantum Mechanics and a Macroscopic Situation that Violates Bell Inequalities', in P. Mittelstaedt et al., eds., *Recent Developments in Quantum Logic*; in *Grundlagen der Exacten Naturwissenschaften*, **6**, Wissenschaftverlag, Bibliographisches Institut, Mannheim, 235.

[6] Aerts, D.: 1986, *J Math Phys*, **27**, 202.

[7] Aerts, D.: 1987, 'The Origin of the Non-Classical Character of the Quantum Probability Model', in A. Blanquiere et al., eds., *Information, Complexity, and Control in Quantum Physics*. Springer-Verlag.

[8] Aerts, D.: 1988a, 'The Physical Origin of the EPR Paradox and How to Violate Bell Inequalities by Macroscopic Systems', in P. Lahti et al., eds., *Symposium on the Foundations of Modern Physics*. World Scientific, Singapore.

[9] Aerts, D.: 1988b, 'The Description of Separated Systems and a Possible Explanation of the Probabilities of Quantum Mechanics', in van der Merwe et al., eds., *Microphysical Reality and Quantum Formalism*. Kluwer Academic Publishers.

[10] Aerts, D.: 1991a, 'A Macroscopic Classical Laboratory Situation with Only Macroscopic Classical Entities Giving Rise to a Quantum Mechanical Probability Model', in L. Accardi, ed., *Quantum Probability and Related Topics*, Vol. VI. World Scientific, Singapore.

[11] Aerts, D.: 1991b, *Helv Phys Acta*, **64**, 1.

[12] Aerts, D.: 1992, *Int J Theor Phys.*, **31**, 1815.

[13] Aerts, D.: 1993, *De Muze van het Leven: Quantummechanica en de Aard van de Werkelijkheid*, Pelckmans, Kapellen, Belgium (in this book the creation discovery view is exposed in its most complete form, an English translation is being prepared).

[14] Aerts, D.: 1994, *Found Phys*, **24**, 1227.

[15] Aerts, D.: 1995, *Int J Theor Phys*, **34**, 1165.

[16] Aerts, D.: 1997, *Int J Theor Phys*, **37**, 291.

[17] Aerts, D., Aerts, S.: 1996, 'The Hidden Measurement Approach to Quantum Mechanics', in M. Ferrero and A. Van der Merwe, eds., *New Developments on Fundamental Problems in Quantum Physics*. Kluwer Academic, Dordrecht.

[18] Aerts, D., Aerts, S., Coecke, B., Valckenborgh, F.: 1996, *The Meaning of the Violation of Bell's Inequalities: Non-Local Correlation or Quantum Behavior?*, Preprint CLEA. Free University of Brussels.

[19] Aerts, D., Coecke, B., Durt, T., Valckenborgh, F.: 1997a, *Tatra Mt Math Publ*, **10**, 225.

[20] Aerts, D., Coecke, B., Durt, T., Valckenborgh, F.: 1997b, *Tatra Mt Math Publ*, **10**, 241.

[21] Aerts, D., D'Hooghe, B.: 1996, *Int J Theor Phys*, **35**, 2241.

[22] Aerts, D., Durt, T.: 1994a, *Found Phys*, **24**, 1353.

[23] Aerts, D., Durt, T.: 1994b, 'Quantum, Classical and Intermediate; A Measurement Model', in K.V. Laurikainen, C. Montonen, K. Sunnar Borg, eds., The Proceedings of the *Symposium on the Foundations of Modern Physics*, Helsinki, 1994. Editions Frontieres, Gives sur Yvettes, France.

[24] Aerts, D., Durt, T., Van Bogaert, B.: 1992, *Tatra Montains Math Publ*, **1**, 5.

[25] Aerts, D., Durt, T., Van Bogaert, B.: 1993, 'Quantum Probability, The Classical Limit and Non-Locality', in T. Hyvonen, ed., *Symposium on the Foundations of Modern Physics*. World Scientific, Singapore.

[26] Aerts, S.: 1996, *Int J Theor Phys*, **35**, 2245.

[27] Bell, J.: 1964, *Physics*, **1**, 195.
[28] Bohm, D.: 1951, *Quantum Theory*, Prentice-Hall.
[29] Birkhoff, G., Von Neumann, J.: 1936, *J Ann Math*, **37**, 823.
[30] Coecke, B.: 1995a, *Int J Theor Phys*, **34**, 1313.
[31] Coecke, B.: 1995b, *Found Phys Lett*, **8**, 447.
[32] Coecke, B.: 1995c, *Hidden Measurement Systems*, Doctoral Dissertation. Free University of Brussels.
[33] Coecke, B.: 1996a, *Int J Theor Phys*, **35**, 2371.
[34] Coecke, B.: 1997a and b, *Helv Phys Acta*, **70**, 442 and 462.
[35] Coecke, B.: 1997c, *Tatra Mt Math Publ*, **10**, 63.
[36] Coecke, B.: 1998a and b, *Found Phys*, **28**, 1109 and 1347.
[37] Coecke, B., Valckenborgh, F.: 1998c, *Int J Theor Phys*, **37**, 311.
[38] Dirac, P.A.M.: 1958, *The Principles of Quantum Mechanics*. Clarendon Press, Oxford.
[39] Einstein, A., Podolsky, B., Rosen, N.: 1935, *Phys Rev*, **47**, 777.
[40] Emch, G.G.: 1984, *Mathematical and Conceptual Foundations of 20th Century Physics*. North-Holland, Amsterdam.
[41] Heisenberg, W.: 1925, *Zeitschr Phys*, **33**, 879.
[42] Pitovski, I.: 1989, *Quantum Probability – Quantum Logic*. Springer Verlag.
[43] Schrödinger, E.: 1926, *Ann der Phys*, **79**, 1926.
[44] Segal, I.E.: 1947, *Ann Math*, **48**, 930.
[45] Selleri, F.: 1990, *Quantum Paradoxes and Physical Reality*. Kluwer Academic Publishers, Dordrecht.
[46] Von Neumann, J.: 1955, *Mathematical Foundations of Quantum Mechanics*. Princeton University Press, Princeton.

GUIDO BACCIAGALUPPI AND PIETER E. VERMAAS

VIRTUAL REALITY: CONSEQUENCES OF NO-GO THEOREMS FOR THE MODAL INTERPRETATION OF QUANTUM MECHANICS

1. INTRODUCTION

The modal interpretation of quantum mechanics is a no-collapse interpretation that provides a rule for assigning properties to quantum mechanical systems, depending on their (reduced) quantum state, whether this state be pure or mixed. The modal interpretation was originally developed by Kochen (1985), Dieks (1989) and Healey (1989) (with slight differences) modifying ideas by Van Fraassen (1973, 1991), and was further developed by Vermaas and Dieks (1995) and Clifton (1995). In the original version, the rule for assigning properties is applied to all quantum systems. However, a recent series of no-go theorems by Bacciagaluppi (1995), Clifton (1996) and Vermaas (1997) has shown that this unrestricted application of the rule leads to contradictions. This has prompted some authors (Bacciagaluppi and Dickson, 1999; Dieks, 1998) to restrict application of the rule only to 'elementary' or 'atomic' systems. We call this version the atomic modal interpretation. In this paper, we shall describe the no-go results, and discuss the picture yielded by the atomic modal interpretation. We shall proceed as follows.

In Section 2, we introduce the rule for assigning properties in the original modal interpretation, together with natural requirements and further results on joint property ascriptions to different systems. In Section 3, we describe the no-go theorems mentioned above. Then, in Section 4, we present the atomic modal interpretation. In Section 5 we discuss the ontological picture provided by the atomic modal interpretation. Because the rules are applied only to atomic systems and because systems which are composites of atoms are assigned only properties which are products of properties assigned to atoms, composite systems possess strictly speaking no holistic properties in the atomic modal interpretation. On the other hand, as remarked by Dieks (1998), there is the possibility of collective or holistic effects of a collection of atoms on other systems. A similar point is made by Clifton (1996), who talks of dispositions of a composite system. We stress this distinction between 'real' properties of a system, arising from application of the rules of the modal interpretation to the atoms of the system, and 'virtual' properties of the system, which can be

M.L. Dalla Chiara et al. (eds.), Language, Quantum, Music, 117–128.

manifested in appropriate contexts via collective effects. In a sense, this is a revisitation of Kochen's concept of 'witnessing'. Finally, in Section 6, we mention how previous discussions of the measurement problem in the modal interpretation need to be modified if one adopts the atomic modal interpretation. We further point out that the relevance of virtual properties depends on the possibility of their manifestation in realistic contexts, in particular in realistic contexts of measurement.

2. THE KOCHEN–DIEKS INTERPRETATION

The rule of the original modal interpretation for assigning properties to quantum mechanical systems is as follows. Take a system α with associated Hilbert space $\mathcal{H}^\alpha$ and suppose that the state of α is given by a density operator ρ^α defined on $\mathcal{H}^\alpha$. Let the spectral resolution of the state be[1]

$$\rho^\alpha = \sum_j \lambda_j^\alpha P_j^\alpha. \tag{1}$$

The projections $\{P_j^\alpha\}_j$ are the *eigenprojections* of ρ^α and project onto the eigenspaces corresponding to the non-zero eigenvalues of ρ^α. The first step in assigning properties to α is to assign definite value 1 to one of the eigenprojections P_j^α (written $[P_j^\alpha] = 1$) with probability $\mathrm{Tr}^\alpha(\rho^\alpha P_j^\alpha)$. We call this the *core property ascription*. The full property ascription is then as follows (see Vermaas, 1998a).

$$\begin{array}{l} \text{With probability } \mathrm{Tr}^\alpha(\rho^\alpha P_j^\alpha) \text{ it is the case that:} \\ [Q^\alpha] = 1 \text{ for all projections } \{Q^\alpha | Q^\alpha P_j^\alpha = P_j^\alpha\}, \\ [Q^\alpha] = 0 \text{ for all projections } \{Q^\alpha | Q^\alpha P_j^\alpha = 0\}. \end{array} \tag{2}$$

The full set of properties $\mathcal{F}(P_j^\alpha)$ with definite values given a specific core property ascription $[P_j^\alpha] = 1$ has the structure of a faux-Boolean algebra (Dickson, 1995). The intersection of all the $\mathcal{F}(P_j^\alpha)$ is obviously the set of properties that are assigned definite values irrespective of the specific core property ascription. This intersection $\cap_j \mathcal{F}(P_j^\alpha)$ is also a faux-Boolean algebra, which we denote by $\mathcal{F}(\{P_j^\alpha\}_j)$, and we shall refer to it as the set of *definite properties* of the system α.

This rule of property ascription refers just to one system α. We can now consider a set of such systems $\{\alpha_k\}_k$, and apply the rule to all the systems simultaneously. For the time being we do not restrict the set $\{\alpha_k\}_k$. It could be denumerable or non-denumerable, consisting of systems in one specific factorisation of the universe (that is, a set of disjoint systems plus their composites) or consisting of systems from all possible factorisations. One may or may not want to consider the union of the sets $\mathcal{F}(\{P_j^{\alpha_k}\}_j)$ as a set of projections in the total Hilbert space, thus identifying properties P of a subsystem with properties of the form $P \otimes \mathbf{I}$ of the total system (the universe).

(Clifton (1995, 1996) discusses this point under the heading of 'property composition'.) This set is no longer a faux-Boolean algebra, but one can take its closure under partial Boolean algebra (PBA) operations. We can now require the following.

(i) If one performs the closure under PBA operations, require that the assigned values of the definite properties respect this algebraic structure.

(ii) Whether or not one performs the closure under PBA operations, require that there exist a *joint probability distribution* over the total set of properties.

As shown by Vermaas and Dieks (1995), a partial specification of the joint probabilities required in (ii) can be given as follows. Take a collection of *disjoint* systems $\alpha, \beta \ldots \kappa$ and call their composite $\omega := \alpha \otimes \beta \otimes \ldots \otimes \kappa$. If one requires that joint probabilities are consistent with the predictions of quantum mechanics (in the standard interpretation) with respect to joint measurements, the joint probability that the core property ascriptions assign simultaneously value 1 to $P_j^\alpha, P_k^\beta, \ldots$ and P_m^κ must be equal for all states ρ^ω to

$$\mathrm{p}(P_j^\alpha, P_k^\beta, \ldots, P_m^\kappa) = \mathrm{Tr}^\omega(\rho^\omega P_j^\alpha \otimes P_k^\beta \otimes \ldots \otimes P_m^\kappa). \tag{3}$$

We shall now see how requirement (i) leads to the no-go theorems of Bacciagaluppi and of Clifton and how requirement (ii) together with (3) leads to the no-go theorem of Vermaas.

3. THE NO-GO THEOREMS

3.1. Bacciagaluppi's theorem

Bacciagaluppi's (1995) no-go theorem is based directly on Kochen and Specker's (1967) construction of a finite partial Boolean subalgebra of the algebra of projections of a 3-dimensional Hilbert space that does not allow a homomorphism onto the Boolean algebra $\{0, 1\}$ (that is, it does not allow an assignment of values that respects the algebraic structure of the projections). He reproduces a finite subalgebra of the PBA of definite properties isomorphic to the finite subalgebra of Kochen and Specker by taking the union of the definite properties of systems from different (3×3)-factorisations of a 9-dimensional Hilbert space.

Such factorisations can be defined as follows.

Definition 3.1 *Let $n = pq$, and let $\mathcal{H}$ be an n-dimensional Hilbert space. A* $(p \times q)$-factorisation Φ *of $\mathcal{H}$ is an equivalence class of isomorphisms*

$$f : \mathcal{H} \rightarrow \mathcal{H}_1 \otimes \mathcal{H}_2$$

(where $\mathcal{H}_1$ and $\mathcal{H}_2$ are Hilbert spaces of dimension p and q, respectively), defined by the following equivalence relation:

$$\tilde{f} \sim f \quad :\Leftrightarrow$$
$$\exists V_1 : \mathcal{H}_1 \rightarrow \mathcal{H}_1,\ V_2 : \mathcal{H}_2 \rightarrow \mathcal{H}_2, \text{ both unitary, such that } \tilde{f} = V_1 \otimes V_2 f.$$

And with respect to a factorisation we can define:

Definition 3.2 *A unitary transformation $U : \mathcal{H} \rightarrow \mathcal{H}$ is* factorisable with respect to a factorisation Φ *iff there is an $f \in \Phi$ and $U_1 : \mathcal{H}_1 \rightarrow \mathcal{H}_1$ and $U_2 : \mathcal{H}_2 \rightarrow \mathcal{H}_2$ both unitary, such that*

$$fU = U_1 \otimes U_2 f.$$

One can now show that one passes from one factorisation Φ to a different one, precisely by means of unitary transformations $U : \mathcal{H} \rightarrow \mathcal{H}$ that are *not* factorisable with respect to Φ. That is, given an isomorphism $f \in \Phi$, iff U is not factorisable with respect to Φ the isomorphism $\tilde{f} := fU$ induced by U will not be in Φ.

One can further show that under such a non-factorisable transformation, the definite properties (considered as projections in the total Hilbert space $\mathcal{H}$) of the systems defined by the factorisation Φ, will be mapped to the definite properties of the systems defined by the new factorisation induced by U (the details are in Bacciagaluppi (1995)). By choosing an appropriate sequence of transformations U_j, one can, indeed, reproduce the finite subalgebra constructed by Kochen and Specker.

Bacciagaluppi's theorem shows that given the assumption that definite properties are assigned by the modal rule to systems defined by all possible factorisations of the universe, and the requirement (i) that the values of these properties respect the algebraic structure of the projections in the total Hilbert space, the modal interpretation faces a contradiction. This could be lifted by either giving up (i) or by restricting to a smaller set of systems the applicability of the rule for assigning properties.

3.2. Clifton's theorem

Clifton's theorem applies precisely to a situation in which the applicability of the rule for property assignment is restricted, namely to a set of disjoint systems together with all their composites. This corresponds to the choice of one particular factorisation of the universe into 'atomic' systems and their 'molecular' composites (this can be independently justified by consideration of space-time symmetries, see the discussion in Dieks (1998)). As a matter of fact,

Clifton's proof applies already to the simple case of two 8-dimensional systems α and β and their tensor product $\alpha \otimes \beta$. Essential to his proof are the conditions of property composition, closure under PBA operations, and the requirement of a homomorphic assignment of values.

Clifton's theorem is based on the following crucial lemma.

Lemma 3.3 *If $\alpha \otimes \beta$ is in a pure state $|\Psi^{\alpha\beta}\rangle$ and if P is a projection in the set of definite properties of $\alpha \otimes \beta$, then so is the projection onto $P|\Psi^{\alpha\beta}\rangle$.*

Consider now a composite $\alpha \otimes \beta$ with state $|\Psi^{\alpha\beta}\rangle$, and let α have a reduced state with spectral resolution $\rho^\alpha = \sum_j \lambda_j^\alpha P_j^\alpha$ so that the $\{P_j^\alpha\}_j$ are in the set of definite properties of α. From requirement (i) it follows that the projections $\{P_j^\alpha \otimes \mathbf{I}^\beta\}_j$, as well as their sums, are in the set of definite properties of $\alpha \otimes \beta$. This, of course, is unproblematic. However, the lemma implies that also the projections onto the vectors $P_j^\alpha \otimes \mathbf{I}^\beta|\Psi^{\alpha\beta}\rangle$, $P_k^\alpha \otimes \mathbf{I}^\beta|\Psi^{\alpha\beta}\rangle$ and $(P_j^\alpha + P_k^\alpha) \otimes \mathbf{I}^\beta|\Psi^{\alpha\beta}\rangle$, $j \neq k$, will be in the set of definite properties of $\alpha \otimes \beta$. But now, these projections will generally not commute with one another. Thus, the lemma leads very easily to a non-trivial PBA structure.

Clifton now chooses a state $|\Psi^{\alpha\beta}\rangle$ such that the definite properties obtained via the lemma lead to a Kochen–Specker contradiction (Clifton does not reproduce Kochen and Specker's original contradiction, but adapts a simpler one by Kernaghan (1994)).

3.3. Vermaas's theorem

Vermaas's (1997) no-go theorem does not focus on the algebraic structure of the set of definite properties but is based on requirement (ii) that there exist joint probability distributions on the set of definite properties, in particular a joint probability distribution on the core property ascriptions to all possible systems. Like Clifton, Vermaas considers atomic and molecular systems in a preferred factorisation. But unlike Clifton's, Vermaas's theorem does not depend on assuming the existence of a homomorphic value assignment, nor indeed on the closure of the set of definite properties under PBA operations.

Vermaas proves that there does not exist in general a classical joint probability distribution for the simultaneous ascription of the core properties to collections of atomic and molecular systems that is consistent with (3). He considers systems α, $\alpha \otimes \beta$ and $\alpha \otimes \beta \otimes \gamma$. A candidate joint probability $\mathrm{p}(P_j^\alpha, P_k^{\alpha\beta}, P_l^{\alpha\beta\gamma})$ for the simultaneous core property ascription to these systems wil be a classical probability distribution if and only if it obeys a set of so-called Bell–Wigner inequalities (see Pitowsky, 1989). One of these inequalities is

$$\mathrm{p}(P_k^{\alpha\beta}) - \mathrm{p}(P_j^\alpha, P_k^{\alpha\beta}) - \mathrm{p}(P_k^{\alpha\beta}, P_l^{\alpha\beta\gamma}) + \mathrm{p}(P_j^\alpha, P_l^{\alpha\beta\gamma}) \geq 0 \quad (4)$$

(which is an elementary theorem of probability theory). All terms in this inequality can be calculated by means of the joint probability (3). In fact, by introducing a fourth system δ such that $\alpha \otimes \beta \otimes \gamma \otimes \delta$ is in a pure state $|\Psi^{\alpha\beta\gamma\delta}\rangle$, one has for instance

$$\mathrm{p}(P_j^\alpha, P_k^{\alpha\beta}) = \mathrm{p}(P_j^\alpha, P_k^{\gamma\delta}), \tag{5}$$

because by (3) there exist one-to-one correlations between the properties $P_k^{\alpha\beta}$ of $\alpha \otimes \beta$ and $P_k^{\gamma\delta}$ of $\gamma \otimes \delta$. The right-hand side of equation (5) can now be calculated using (3) since α and $\gamma \otimes \delta$ are disjoint systems.

It is now possible to choose $|\Psi^{\alpha\beta\gamma\delta}\rangle$ such that the above Bell–Wigner inequality is violated, proving that in general there do not exist classical probabilities for simultaneous core property ascriptions to three or more systems.

4. THE ATOMIC MODAL INTERPRETATION

In the light of the theorems described in the previous section, it is manifest that the original version of the modal interpretation is difficult to defend. (Notice that giving up requirement (i) is no longer a possible option because Vermaas's theorem is independent of this assumption.) The most natural response seems to be that one should reduce further the scope of the rule for property ascription; for instance, and quite drastically, to include only *atomic* systems in a preferred factorisation, and none of the molecular ones. This suggestion has explicitly been made by Bacciagaluppi and Dickson (1999) (in the context of the definition of a dynamics for the modal interpretation) and by Dieks (1998). We shall call this version of the modal interpretation the *atomic modal interpretation.*

The rules of the atomic modal interpretation are as follows. One assumes a preferred (finite or denumerably infinite) factorisation of the universe into atomic systems. The ascription of properties to these atomic systems is as above in Section 2. The ascription of properties to any molecular system ω, composed, say, of the atoms $\alpha, \ldots, \kappa$ is as follows. With probability $\mathrm{Tr}^\omega(\rho^\omega P_j^\alpha \otimes \ldots \otimes P_m^\kappa)$ the system ω possesses the property $P_j^\alpha \otimes \ldots \otimes P_m^\kappa$. This is the core property ascription. This means that the core properties of the molecular systems are just the products of the core properties of the atomic systems. Thus, one does not ascribe holistic properties to ω. The core properties of ω are determined by property composition from the ones of the atomic systems, and the corresponding probabilities are the joint probabilities (3).

The full property ascription is now as follows.

$$\begin{array}{l}\text{With probability } \mathrm{Tr}^\omega(\rho^\omega P_j^\alpha \otimes \ldots \otimes P_m^\kappa) \text{ it is the case that:}\\ [Q^\omega] = 1 \text{ for all projections } \{Q^\omega | Q^\omega P_j^\alpha \otimes \ldots \otimes P_m^\kappa = P_j^\alpha \otimes \ldots \otimes P_m^\kappa\},\\ [Q^\omega] = 0 \text{ for all projections } \{Q^\omega | Q^\omega P_j^\alpha \otimes \ldots \otimes P_m^\kappa = 0\}.\end{array} \tag{6}$$

As before, this set is a faux-Boolean algebra $\mathcal{F}(P_j^\alpha \otimes \ldots \otimes P_m^\kappa)$, and the intersection of all these sets is again taken as the faux-Boolean algebra of the definite properties of ω.

One can now show that the atomic modal interpretation has several welcome features. First of all, property composition (as well as the converse relation of property decomposition) always holds. Since a faux-Boolean algebra is a PBA, the set of definite properties is, indeed, closed under PBA operations, and it admits many homomorphisms onto the Boolean algebra $\{0, 1\}$. Any such homomorphism in fact corresponds uniquely to a core property ascription. Further, there is a joint probability distribution for all properties in the set of definite properties, in particular for all core property ascriptions (see e.g. Vermaas, 1998b).

Finally, Bacciagaluppi and Dickson (1999) have developed a framework for stochastic dynamics in the modal interpretation, that is, a framework for the evolution of the *actually* possessed properties. This framework yields possible dynamics once a general expression for joint probabilities is given, as is the case in the atomic modal interpretation. Thus, the dynamics for the atomic modal interpretation can be explicitly constructed. It takes a particularly interesting form, and in many ways it resembles the dynamics of the Bohm theory.

The fact that the core property ascription in the atomic modal interpretation is in terms of *product* properties is responsible for these welcome features, in particular for allowing the interpretation to escape the conclusions of the no-go theorems. On the other hand, this very fact can be viewed as strongly counter-intuitive, because it rids quantum mechanics altogether of *holistic* properties. Thus, two electrons will never, in the atomic modal interpretation, possess a property corresponding to the quantum mechanical singlet state, or any other entangled state, not even if the quantum state of the electron pair is, indeed, such an entangled state, and thus any measurement of the projection on that state will yield a positive outcome with probability 1. In the next section, we shall discuss precisely this kind of objection.

5. VIRTUAL PROPERTIES

Clifton (1996), in the discussion of his no-go theorem, points out that such an absence of holistic properties is present from the start also in the Bohm theory (Pagonis and Clifton, 1995). In the Bohm theory, especially in the first-order formulation of the theory, one can say that the only properties possessed by the particles are their positions, i. e. configurational properties. The concept of a quantum mechanical observable is applicable only in describing measurement situations, due to the special dynamics of the theory. Similarly, the concept of a quantum state of the particle is applicable only in certain situations, as an effective state, call it $|\psi\rangle$. Namely, whenever the particle will, in the context of a measurement of the projection onto $|\psi\rangle$, elicit with probability 1 a positive response from the apparatus. As Pagonis and Clifton put it, the particle has a *disposition* to elicit a positive response in this measurement context. This disposition is, in fact, grounded in the configurational properties of the particle and in the Bohmian dynamics, but it can be conveniently described by assigning an effective state vector $|\psi\rangle$ to the particle.

Clifton transposes this idea from the Bohm theory to the atomic modal interpretation. He distinguishes between *categorical* properties, namely the ones that are assigned by the rules of the modal interpretation (the analogues of the positions in the Bohm theory), and *dispositional* properties, described by those projections whose measurements will yield a positive result with probability 1.

Similarly, Dieks (1998) discusses the question of holistic properties in terms of *collective effects* of the atomic systems constituting a molecular system. Dieks emphasises the dynamical aspects of these collective effects. He considers the case in which the molecular system interacts as a whole with its environment rather than as a collection of individual atomic systems. That is, the interaction Hamiltonian is a function of some non-factorisable operators on the Hilbert space of the molecular system (and of some operators on the Hilbert space of the environment). In this case, Dieks argues that the molecular system behaves 'as if' it were an atomic system. That is, a coarse-grained version of the atomic modal interpretation, in which the molecular system is treated as atomic, is adequate to describe the effects of the molecular system on its environment. Thus Dieks also distinguishes between the properties that are assigned to molecular systems by assigning properties to the individual atomic systems, and the holistic properties, corresponding to collective effects, assigned to the molecular systems when a different, coarse-grained, level of description is appropriate, which in turn depends on the nature of the interactions.

Both Clifton and Dieks seem to acknowledge that the basic ontology of the atomic modal interpretation is somewhat too spartan, and they try to enrich it by adding a second layer of properties. This would lead to the following picture. In the atomic modal interpretation, there is a layer of properties, which we would like to call *real*, that are determined by the rules of the modal interpretation applied to the atomic systems, as in Section 4. On top of these one can reintroduce a layer of holistic properties, which we would like to call *virtual*, because they will be manifested only in specific contexts. This does not drive the atomic modal interpretation back to the problems caused by the no-go theorems. In fact, because these properties are manifested only in specific contexts, one cannot argue that they should satisfy any of the requirements (i) and (ii).

One should notice that the introduction of holistic properties is not essential to the explanatory aspects of the atomic modal interpretation. Bacciagaluppi and Hemmo (1998) have argued that the Bohmian analogy, that is, the properties of the atomic systems and their dynamics, is perfectly adequate from the explanatory point of view. On the other hand, the advantage of introducing these virtual properties lies in the resulting visualisability of explanations in the atomic modal interpretation, for the virtual properties of one system can be taken to explain the occurrence of real properties in another system. For example, an ideal measurement of the projection onto $|\psi\rangle$ performed on a molecular system in the state $|\psi\rangle$, yields a positive result in terms of a real possessed measurement outcome because the molecular system has $|\psi\rangle$ as a virtual property. This explanation of the measurement result recovers the idea stressed by Healey (1989, Ch. 5) that quantum correlations in the case of entangled states can be explained in terms of holistic properties. (The above presupposes that the modal interpretation guarantees that measuring apparatuses always have real pointer readings, as we shall mention again in the last section.)

This example also shows the limitations of the relevance of virtual properties: they have meaning only in so far as they have a manifestation in terms of real properties. For any virtual property of a system, one can of course construct a context in which this property is manifested. Namely, one can always think of the system as embedded in a context in which the environment is itself an atomic system and the total system is in a pure state. In this case, the virtual property of the system is in one-to-one correlation with a real property of its environment. In general, however, the environment of a molecular system will be another molecular system, so that the virtual properties of a system will be in one-to-one correlation only with virtual properties of its environment. Thus, although the virtual properties of a system and its environment are always 'witnessed' by each other (in Kochen's (1985) sense), they need not be manifested in terms of real properties. Whether a virtual property will be manifested is thus a dynamical question of how a system further interacts with its environment. However, if virtual properties are not manifested, the introduction of a layer of holistic properties produces but a sterile realm of virtual reality.

6. FINAL REMARKS

We have already mentioned that the relevance of virtual properties depends on an adequate solution to the measurement problem. We shall close by pointing out how the measurement problem itself is modified in the atomic modal interpretation. In general, the measurement problem in the modal interpretation is formulated as the problem of whether, after a measurement, the measuring apparatus exhibits definite pointer readings, that is, whether the properties corresponding to readings are in the set of definite properties of the apparatus after the measurement is completed.

The status of the measurement problem in the original modal interpretation has been considered in detail in a number of recent papers.[2] The results seem to depend crucially on the models of measurement taken into consideration. In the case of discrete, low-dimensional models, the definite properties of the apparatus will tend to be extremely close to the idealised pointer eigenstates of the model, while in the case of continuous, infinite-dimensional models, the definite properties of the pointer will tend to correspond to delocalised states. These results, however, cannot simply be applied to the case of the atomic modal interpretation, because the above discussions apply the modal rule to the apparatus as a whole, and thus the properties of the pointer that are singled out are holistic properties. In general, the discussion applies to what in the atomic modal interpretation are merely virtual properties and not real ones. And this can lead at best to an infinite regress.

In fact, recall that the relevance of virtual properties lies in the possibility of their manifestation in terms of real properties in specific contexts. In principle, such contexts always exists, as seen in the previous section. But in practice, these virtual properties will need to be *made* manifest in specific measurement contexts. If now the atomic modal interpretation only yields that after measurements the readings are virtual properties, then, as a result, virtual properties of a system are manifested only as virtual properties of a pointer, which, in their turn, are manifested only as virtual properties of a second pointer, and so on, *ad infinitum*, and a sterile realm of virtual reality is established.

In order to avoid this regress, one needs to discuss anew the measurement problem in the atomic modal interpretation, namely in terms only of *real* properties, that is, in terms of the properties of the atomic constituents of the apparatus, and in terms of collective properties of the apparatus that will be emergent in some unobjectionable sense (e.g. centre-of-mass observables). The question of whether the atomic modal interpretation is able to do so, is a subject for further investigation (see Vermaas, 1998b).

In this paper, we have reviewed the status of the modal interpretation with respect to the no-go theorems. Virtual reality might be an imaginative response to these theorems. On the other hand, even if it modifies its formulation, the atomic modal interpretation does not directly address the measurement problem (nor the related problem of instability, discussed by Bacciagaluppi, Donald and Vermaas (1995)). Only a solution to the measurement problem will qualify it as a successful interpretation of quantum mechanics.

G. Bacciagaluppi
Balliol College and Sub-Faculty of Philosophy
University of Oxford
P.E. Vermaas
Department of Philosophy
Delft University of Technology

NOTES

[1] In this spectral resolution the sum runs over all different non-zero eigenvalues λ_j^α of ρ^α. The P_j^α are the eigenprojections of ρ^α and are multi-dimensional iff the corresponding eigenvalues λ_j^α are degenerate. The spectral resolution is unique in the sense that the sets $\{\lambda_j^\alpha\}_j$ and $\{P_j^\alpha\}_j$ are uniquely determined by ρ^α.

[2] See in particular Bacciagaluppi and Hemmo (1996), and Bacciagaluppi, Donald and Vermaas (1995), Donald (1998), Ruetsche (1998), Vermaas (1998) and Bacciagaluppi (1999, Ch. 6).

REFERENCES

Bacciagaluppi, G.: 1995, 'Kochen–Specker Theorem in the Modal Interpretation of Quantum Mechanics', *International Journal of Theoretical Physics*, **34**, 1206–1215.

Bacciagaluppi, G.: 1999, *Modal Interpretations of Quantum Mechanics.* Cambridge: Cambridge University Press, forthcoming.

Bacciagaluppi, G., Dickson, M.: 1999, 'Dynamics for Modal Interpretations', forthcoming in *Foundations of Physics.*

Bacciagaluppi, G., Donald, M.J., Vermaas, P.E.: 1995, 'Continuity and Discontinuity of Definite Properties in the Modal Interpretation', *Helvetica Physica Acta,* **68**, 679–704.

Bacciagaluppi, G., Hemmo, M.: 1996, 'Modal Interpretations, Decoherence and Measurements', *Studies in History and Philosophy of Modern Physics*, **27**, 239–277.

Bacciagaluppi, G., Hemmo, M.: 1998, 'State Preparation in the Modal Interpretation', *Minnesota Studies in the Philosophy of Science*, **17**, 95–114.

Clifton, R.: 1995, 'Independently Motivating the Kochen–Dieks Modal Interpretation of Quantum Mechanics', *British Journal for the Philosophy of Science*, **46**, 33–57.

Clifton, R.: 1996, 'The Properties of Modal Interpretations of Quantum Mechanics', *British Journal for the Philosophy of Science*, **47**, 371–398.

Dickson, M.: 1995, 'Faux-Boolean Algebras, Classical Probability, and Determinism', *Foundations of Physics*, **8**, 231–242.

Dieks, D.: 1989, 'Resolution of the Measurement Problem Through Decoherence of the Quantum State', *Physics Letters*, **A 142**, 439–446.

Dieks, D.: 1998, 'Preferred Factorizations and Consistent Property Attribution', *Minnesota Studies in the Philosophy of Science*, **17**, 144–159.

Donald, M.J.: 1998, 'Discontinuity and Continuity of Definite Properties in the Modal Interpretation', in D. Dieks and P.E. Vermaas, eds, *The Modal Interpretation of Quantum Mechanics.* Dordrecht: Kluwer, 213–222.

van Fraassen, B.: 1973, 'Semantic Analysis of Quantum Logic' in C.A. Hooker, ed., *Contemporary Research in the Foundations and Philosophy of Quantum Theory.* Dordrecht: Reidel, 180–213.

van Fraassen, B.: 1991, *Quantum Mechanics: An Empiricist View.* Oxford: Clarendon Press.

Healey, R.: 1989, *The Philosophy of Quantum Mechanics: An Interactive Interpretation.* Cambridge: Cambridge University Press.

Kernaghan, R.: 1994, 'Bell–Kochen–Specker Theorem with 20 Vectors', *Journal of Physic,* **A 27**, L829–L830.

Kochen, S.: 1985, 'A New Interpretation of Quantum Mechanics', in P. Lahti and P. Mittelstaedt, eds, *Symposium on the Foundations of Modern Physics 1985: 50 Years of the Einstein–Podolski–Rosen Gedankenexperiment.* Singapore: World Scientific, 151–169.

Kochen, S., Specker, E.P.: 1967, 'On the Problem of Hidden Variables in Quantum Mechanics', *Journal of Mathematics and Mechanics*, **17**, 59–87.

Pagonis, C., Clifton, R.: 1995, 'Unremarkable Contextualism: Dispositions in the Bohm Theory', *Foundations of Physics*, **25**, 281–296.

Pitowsky, I.: 1989, *Quantum Probability – Quantum Logic.* Berlin: Springer.

Ruetsche, L.: 1995, 'Measurement Error and the Albert–Loewer Problem', *Foundations of Physics Letters*, **8**, 327–344.

Vermaas, P.E.: 1997, 'A No-Go Theorem for Joint Property Ascriptions in the Modal Interpretations of Quantum Mechanics', *Physical Review Letters*, **78**, 2033–2037.

Vermaas, P.E.: 1998a, 'Expanding the Property Ascriptions in Modal Interpretations of Quantum Theory', *Minnesota Studies in the Philosophy of Science*, **17**, 115–143.

Vermaas, P.E.: 1998b, 'The Pros and Cons of the Kochen–Dieks and the Atomic Modal Interpretation', in D. Dieks and P.E. Vermaas, eds, *The Modal Interpretation of Quantum Mechanics*. Dordrecht: Kluwer, 103–148.

Vermaas, P.E., Dieks, D.: 1995, 'The Modal Interpretation of Quantum Mechanics and Its Generalization to Density Operators', *Foundations of Physics*, **25**, 145–158.

ROBERT C. BISHOP AND FREDERICK M. KRONZ

IS CHAOS INDETERMINISTIC?

1. INTRODUCTION

Although classical mechanics (CM) is a successful, mature and stable theory, our understanding of CM has undergone substantial growth and refinement over the last twenty-five years. The recent explosion in the study of chaotic dynamics has uncovered a rich variety of behavior in simple macrophysical systems previously unsuspected by the vast majority of the physics community (Hilborn, 1994). One of the essential features of chaos is sensitivity to initial conditions. In CM the behavior of simple physical systems is described using models such as the harmonic oscillator which capture the main features of the systems in question (Giere, 1988). Some models of CM evolve in much the same way for all nearly identical initial states. Chaotic models, however, evolve in radically different ways in a relatively short time period and may be extremely sensitive to quantum fluctuations. Jesse Hobbs (1991) and Stephen Kellert (1993) have argued that quantum events can influence chaotic macrophysical systems through sensitivity to initial conditions. Such arguments raise questions about whether chaos is indeterministic and what types of models are appropriate for capturing the dynamics of such systems.

The semantic view of theories (Giere, 1988) is a working presupposition of this essay, as opposed to the syntactic view which was favored by the logical positivists. The latter has fallen out of favor in recent philosophy of science. As Ian Hacking (1983) points out, the practice of physics and other sciences requires that we move beyond the realm of first-order languages and incorporate higher-order quantifiers and indexical expressions. Such extensions raise difficult questions regarding completeness, decidability and other deduction properties (Hacking, 1983, 104–8). Simple examples show why it is reasonable to make these extensions. In order to represent properties of matter such as mass or charge as intrinsic, the use of second-order logic is required. Additionally many properties of CM take on a continuum of possible physical values requiring a formal language powerful enough to describe continuous functions of real numbers (Earman, 1986, 20).

A more complex example involves the mathematical language of physics itself. Although what mathematicians call real analysis appears to be expressible in terms of second-order predicates with identity, physics also makes use of mathematical concepts such as manifolds and the calculus of manifolds,

M.L. Dalla Chiara et al. (eds.), Language, Quantum, Music, 129–141.

both of which require going beyond second-order languages. The situation is similar in quantum mechanics (QM) when we shift from using separable Hilbert spaces to nonseparable ones or to even more abstract spaces (Hellman, 1989, 108–15; Thirring, 1983, 29–31). For example the *canonical commutation relations* are one of the defining features of QM and are closely associated with the Heisenberg relations (Merzbacher, 1970, 146–8). These fundamental relations, however, cannot be represented by operators in a separable Hilbert space in an unproblematic way, but such problems disappear in a rigged Hilbert space (Reed and Simon, 1972, 270–6; Dubin and Hennings, 1990, 10–13).

2. UNIQUE EVOLUTION

Textbook models of CM will repeat their behaviors exactly if the same initial conditions (ICs) and boundary conditions (BCs) are specified. For example the equations of motion for a frictionless pendulum will produce the same solution for the motion as long as the same initial velocity and initial position are chosen. Both Hobbs and Kellert explicate this feature of determinism in CM.[2] They use John Earman's formulation of what they call *unique evolution*:

> Letting W stand for the collection of all physically possible worlds, that is, possible worlds which satisfy the natural laws obtaining in the actual world, we can define the Laplacian variety of determinism as follows. The world $w \in W$ is *Laplacian deterministic* just in case for any $w' \in W$, if w and w' agree at any time, then they agree for all times. By assumption, the world-at-a-given-time is an invariantly meaningful notion and agreement of worlds at a time means agreement at that time on all relevant physical properties (Earman, 1986, 13).

Earman's formulation is meant to be a metaphysical statement about our world which says in brief that all identical copies of our world will have the same history.

Although avoiding some of the deep problems of possible-worlds analysis by quantifying over only those worlds that have the same physics as ours, Earman's formulation is still problematic. It is not clear whether physical laws are metaphysical expressions about our world or methodological expressions about our models of the world. To read unique evolution in the former way requires spelling out the relationship between the laws and models of mathematical physics and the world of our experience through structural similarities, isomorphisms or something else. In addition agreement on 'all relevant physical properties' is not enough to establish the metaphysical type of determinism intended. Earman needs agreement on all properties. The more plausible and less controversial interpretation is to follow the latter alternative and regard physical laws as statements about models and their properties. After all, in physics our models are not considered to be identical with the world, but only capture those elements we think are physically relevant to understanding the way things work. We will follow the model interpretation.

There is an additional problem in whether the notion of 'agreement at a time' really delivers the right condition for the type of determinism Earman intends to capture. First if time is not a property of a given model, then to what time is Earman referring when comparing the various models? Suppose that we had two models with identical physical laws except that the relevant properties of model *A* agree with those of model *B* two minutes ago. In other words model *A* is always two minutes behind *B* in terms of its evolution. There is never a time where the two models agree on all their properties, but clearly they are *isomorphic* to one another. That is to say, we can easily translate between the two models by shifting the time forward or backward two minutes.

If, moreover, talk of agreement on 'all relevant physical properties' really amounts to agreement on all properties, then Earman's version of unique evolution appears subject to criticism based on Leibniz's law of the identity of indiscernibles. There is no way to distinguish different models from one another or to state that there is more than one model involved. This criticism is too quick, however. For the various models could differ as to the potential functions used to describe the gravitational, electrostatic or other potential fields. Since some of the relevant physical properties of the models are derived mathematically *via* gradient or divergence operations on the potential functions, the models could differ with regard to their potential functions up a differentiable scalar function (Jackson, 1975, 220). For example we can add any differentiable function of time to the electrostatic potential and the equation for the charge distribution remains the same. The models could all have electrostatic potentials differing from each other by various differentiable functions of time on sets of measure zero, and would still have the same charge distribution. In this way the models would be discernible to an omniscient Laplacian demon having access to the exact forms of the potential functions in each model.

These considerations suggest that a modification to Earman's formulation may serve to provide a better definition of 'unique evolution'. Let *M* stand for the collection of all models sharing the same set *L* of physical laws, and suppose that *P* is the set of relevant physical properties for specifying the time evolution of a model recognized by *L*. The above term is then defined as follows:

A model $m \in M$ exhibits *unique evolution* iff every $m' \in M$ isomorphic to m with respect to *P* undergoes the same time evolution as m.

Roughly the idea is that if we take a model to be the equations of motion, every time we return the model to the same initial state, it will undergo the same history of transitions from state to state. In other words the evolution of the model will be unique with respect to a particular specification of the ICs and BCs. Furthermore we can choose *any* state in the history of state transitions as the initial starting point and the model's history will remain unchanged. The 'collection' of models is just all the isomorphisms allowed by the translation properties of the equations of motion. Exploring whether a given model has the property of unique evolution is the study of the equations of motion and their solution properties under the allowed translations.

Although this is a strong requirement to make, it is important if physical determinism is to be a meaningful concept applicable to either models or physical systems. Imagine a typical physical model m as a film. Unique evolution means that if we were to start the film over and over at the same frame (returning the model to the same initial state), then m would repeat every detail of its total history over and over again and identical copies of the film would produce the same sequence of pictures. So no matter whether we always start *Jurassic Park* at the beginning frame, the middle frame or any other frame, it plays the same. No new frames are added to the movie nor is the sequence of the frames changed simply by starting it at an arbitrary frame. In other words, any m' isomorphic to m (each having the same evolution equations and the same ICs and BCs) would follow the same sequence of state transitions as m.

3. THE PRECARIOUS NATURE OF UNIQUE EVOLUTION

Earman (1986) argues that unique evolution, and hence determinism, turns out to be problematic even in the case of CM. Consider the case of CM in a Newtonian spacetime. The spacetime structure is Euclidean, that is to say, it has zero curvature (two infinitely long parallel lines would never get any nearer to nor farther from one another). In addition space and time are both absolute quantities. No matter what frame of reference is chosen, two different events are always a fixed spatial distance from one another and are always measured to occur in the same temporal order. Newtonian spacetime also admits Galilean transformations – linear time and space translations and rigid rotations – as its basic symmetries.

Earman examines the case of noncolliding point particles under the influence of Newton's law of gravity. Under the assumptions of an infinite Euclidean spacetime domain and unbounded but finite particle velocities, it is possible that as $t \to t^*$ (i.e. as t increases toward some time t^*), $x(t) \to \infty$. Under such conditions Earman claims the particle 'disappears from the universe' (1986, 34). On the other hand, consider the mirror image of our disappearing particle. This second particle begins its journey immediately after t^* and as t increases it flies into our Euclidean universe from the spatial infinity to which our first particle flew off. Unique evolution fails because there is no way to specify that two copies of a given model may start in the same initial state or follow the same history of state transitions because of the possibility of invasion from outside influences in the form of particles from spatial infinity.

Furthermore value definiteness is problematic in that the value of the position function $x(t)$ is mathematically undefined at t^* in our model. There are two interesting possibilities for interpreting the status of the particle. On the one hand, we could say the particle has left the spacetime manifold and now has some indefinite properties. On the other hand, we could say that the particle no longer exists. The latter interpretation seems more problematic due to violations of conservation principles of mass, energy, charge and momentum. The former may allow for the preservation of at least some of those principles; so it is the one we adopt. In any case, the appearance of space invaders violates unique evolution, hence the types of CM models Earman has in mind clearly cannot be deterministic.

There are questions that can be raised about the interpretation Earman assigns to the behavior of particles as $x(t) \rightarrow \infty$. For example are events such as leaving the infinite Euclidean universe and space invaders from beyond to be taken as predictions of CM or as indications that CM is breaking down because the function $x(t)$ becomes undefined? Furthermore such behavior may be artifacts of the type of spacetime Earman is using. Suppose we switch from CM models with infinite Euclidean spatial domains to CM models with spatially finite Euclidean domains. Now $x(t)$ always remains finite because the spatial domain is finite. Such models would be appropriate for spatially finite universes and in such universes particles would never leave the space nor would invaders from beyond ever appear.

Earman does consider a finite cylindrical universe and briefly discusses two cases (Earman, 1986, 45–7). The first case he considers are field theories, where fields are taken to the basic physical quantities of the models. In that case there would clearly be no space invaders. On the other hand, Earman thinks that particle theories, where particles are taken to be the basic physical quantities of the models, would still be subject to space invaders. His reasoning is somewhat abbreviated. It appears that Earman takes the case of an infinite Euclidean plane where space invaders occur, rolls it up into a finite cylinder (connecting the east and west 'edges') and finds that the trajectories of the space invaders do not go away. But this approach to answering the question ignores several crucial features about the differences between the case of the infinite Euclidean plane and the finite cylinder. The most obvious feature is the fact that the equations of motion for particles in a finite cylinder *differ* from those in a Euclidean plane. Furthermore the process of rolling up a Euclidean plane means that the east and west 'edges' of the plane will have to be identified with each other when they are connected. This introduces a BC very different from that found in the case of the Euclidean plane. Earman is silent on what type of BCs he is using in his analysis, though it is likely he intends for there to be periodic BCs. Finally one might ask why Earman does not consider the case of a finite Euclidean plane instead of attempting to transform an infinite Euclidean plane into a finite cylinder. The crucial case regarding the effect different spacetimes have on space invaders is the comparison between infinite Euclidean planes and finite ones. Since there is no spatial infinity in the latter and nothing in the equations of motion to suggest that particles could ever reach spatial infinity, it is difficult to see how space invaders could be a problem for finite Euclidean spaces.

4. THE SENSITIVE DEPENDENCE ARGUMENT

Even if unique evolution can be preserved in CM, Hobbs and Kellert both present arguments that if quantum mechanics (QM) is taken into account, microscopic events will influence macroscopic chaotic systems *via* sensitive dependence (Hobbs, 1991, 157; Kellert, 1993, 69–75). Hobbs' formulation can be stated succinctly:

(1) Macroscopic chaotic models exhibit sensitivity to initial conditions. That is to say, trajectories starting out in a highly localized region of phase space will begin to diverge exponentially from one another during the model's relaxation time.

(2) QM limits the precision with which macroscopic physical models can be specified to a patch in phase space no less than the size of Planck's constant (10^{-27} in cgs units).

(3) Given enough time and the quantum mechanical bound on the neighborhood ϵ for the ICs, two trajectories of the same chaotic model will have final states localizable to a much larger region δ in phase space (from (1) and (2)).

(4) Therefore, QM will affect the outcomes of macroscopic chaotic models leading to a violation of unique evolution.

The relaxation time referred to in (1) is the time it takes for the exponential divergence to render predictions based upon the specified ICs inaccurate. We will call (1)–(4) a 'sensitive dependence argument'.

Briefly, the reasoning runs as follows. Because of sensitive dependence, chaotic models whose initial states can be located only within a patch of phase space of finite size will have final states that can be located only within a patch of much larger size. Therefore, two isomorphic models whose initial states are localized to a very small patch in phase space will have final states that can be localized only within a huge patch. Since QM sets a lower bound on the size of the initial patch of ICs, unique evolution must be failing for chaotic models.

An obvious criticism of this argument is that to have any force against unique evolution, it must presuppose that the two isomorphic models are in identical initial states. In order to falsify unique evolution, two isomorphic models must have the same ICs, but differ in their history of state transitions. The point of the criticism is that the terms 'initial conditions' and 'unique evolution' are problematic. From a classical point of view, it only makes sense to speak of identical initial states for some proposed solutions to the measurement problem such as particle-based interpretations involving non-local quantum potentials. From a quantum point of view, the concept of initial state is unambiguous, but it only makes sense to speak of unique time evolution for some solutions to the measurement problem such as the particle-based interpretation mentioned above, but not for others such as spontaneous collapse of the wave packet.

In addition complications arise in the analysis of how QM affects the phase space representation of a macroscopic chaotic model in the Hobbs and Kellert sensitive dependence arguments. From Kellert's more detailed discussion (Kellert, 1993, 69–72), it is clear that he confuses the notions of classical phase space and Hilbert space. If we attempt to carry out the analysis completely in classical phase space, then we need to switch to the Wigner distribution formulation of QM (Wigner, 1932; Kim and Noz, 1991). The essential idea is to represent quantum mechanical properties using the tools of classical phase space and to carry out calculations on a distribution function in this space instead of on the state vector in Hilbert space. Eugene Wigner originally constructed the distribution function that bears his name to make it easy to calculate the correction terms contributed by quantum processes to thermodynamic equilibrium. The Wigner distribution is a complicated function of the state vector, so in principle this distribution function contains the same content, but is defined on classical phase space.

This approach raises a number of problems, however. First the patch in phase space representing quantum indeterminacy now becomes a probability density represented by the Wigner distribution function. As constructed this function looks like a probability distribution for simultaneous values of position and momentum, but it cannot have this interpretation. Not only do the Heisenberg relations preclude simultaneously precise values for the position and momentum variables in some versions of QM, the Wigner distribution may take on negative values (Wigner, 1932, 751) which violates one of the postulates of standard probability theory that probability distributions be strictly non-negative. At best the Wigner distribution can be viewed as an auxiliary function for carrying out calculations. When integrated with respect to position, it gives the correct quantum probabilities for the momentum and when integrated with respect to momentum, it gives the correct quantum probabilities for the position (Wigner, 1932, 750).

In addition there are questions about how to interpret the patch in phase space represented by the Wigner distribution. For example the patch could be regarded as a matter of ignorance due to experimental limitations (which is primarily an epistemological interpretation) or it could represent an actual indefiniteness in the classical state of the system (which is primarily an ontological interpretation). There is a further question about whether QM as formulated in this phase space picture now becomes a statistical theory on a par with statistical mechanics (i.e. having an underlying classical structure) or is still an irreducibly indeterministic theory (i.e. involving an ineliminable chance element). The supposed failure of unique evolution in this phase space picture stands or falls based how these questions are answered.

On the other hand, if we attempt to carry out the analysis completely in Hilbert space, then we need to use a formulation of CM like that introduced by B.O. Koopman (Koopman, 1931) and are forced to use unitary operators for CM and will loose the ability to model chaotic behavior because such operators do not admit exponentially diverging solutions. This is because the Koopman view focuses on the dynamics of distribution functions like the Liouville equation description in classical statistical mechanics, rather than on the dynamics of individual trajectories, to describe the time evolution of a distribution function in phase space. The Liouville equation is linear in the distribution function, so the evolution of the distribution function does not exhibit sensitive dependence on ICs even though the individual trajectories do (Hilborn, 1994, 324).[3]

Independent of the technical difficulties regarding the state space to be used in analyzing the sensitive dependence argument, there is a more fundamental problem on which we will focus: the success of that argument depends on the view of QM and the measurement problem one takes.[4] Suppose state vectors describing quantum systems are specified as vectors in Hilbert space, where the tip of the vector is a point traveling continuously on the surface of a unit hyper-sphere. Given the same initial preparation of the quantum state, the state vectors will undergo the same Schrödinger evolution (the tip of the arrow will trace out the same history of state transitions) until interacting with a macroscopic system.

Now, consider the time-evolution of a compound system consisting of a photon and a damped, driven pendulum of macroscopic dimensions. Suppose that the pendulum parameters are set so that the pendulum's time-evolution is chaotic. We want to consider how the photon's interaction with the pendulum might affect the character of the pendulum's time-evolution. It is appropriate to construe the interaction as a measurement-like interaction due to the sensitive dependence characteristic of chaotic pendula, which results in an irreversible act of amplification of the photon's transfer of momentum to the pendulum.

On interpretations of the quantum formalism in which measurement processes do not collapse the state vector (i.e. in which there is only Schrödinger evolution), there will be no indeterminism at the quantum level for sensitive dependence to amplify to the macro level. Whereas on other interpretations of the quantum formalism, measurement processes do lead to indeterministic collapse of the state vector to one sharp value from the superposition of possible values represented by the state vector. The collapse results in potential violations of unique evolution (at least for this simple case).

What is regarded as identical sets of ICs under the collapse interpretations for the pendulum plus photon system depends on the time at which the compound system is initially examined; two key times are the time of preparation of the photon and the time of measurement of the photon.

Specifying the pendulum plus photon system before the interaction takes place as the IC, leads to the same IC every time the experiment is run (provided the initial quantum states are prepared identically). This is because the photons start out in the same pure state in a superposition of possible energy and frequency states and it is this state vector as a whole that is taken as the IC. Under this specification the system is indeterministic due to the later collapse of the photon state vector as a result of the measurement-like interaction between the photon and the pendulum. Unique evolution fails for that case because, although the system always starts in the same initial state, the photon state vector collapses at a different location on the pendulum transferring a different amount of momentum each time the experiment is run producing different trajectories for the pendulum. Hence we have the same initial states, but different final states which explicitly violates the conditions for unique evolution.

On the other hand specifying the moment of collapse as the IC results in different ICs every time the experiment is run. This amounts to treating the photon as an indeterministic perturbation to the motion of the pendulum. The photons may be prepared in the same initial state each time the experiment is run, but an indeterministic change in the pendulum's position and momentum state results because of the indeterministic collapse of the photon's state vector upon interacting with the pendulum arm. There would then be a different IC (a different indeterministic perturbation) for the pendulum's motion.[5] Hence we have different initial states and different final states which is consistent with the conditions for unique evolution. Therefore the purported failure of unique evolution depends on how one 'cuts up' the system even in cases where the quantum state vectors collapses.

This line of argument leads to a related problem of specifying initial states in the sensitive dependence argument that is related to a fundamental problem of QM noted by Bell:

> The problem is this: quantum mechanics is fundamentally about 'observations'. It necessarily divides the world into two parts, a part which is observed, and a part which does the observing. The results depend in detail on just how this division is made, but no definite prescription for it is given. All we have is a recipe which because of practical human limitations is sufficiently unambiguous for practical purposes (1987, 124).

Essentially QM allows several different ways of 'cutting up the system' between observed and unobserved. For the case of our pendulum and photon, the first specification of the initial state corresponds to choosing to observe the photon plus pendulum system as a whole. The second specification corresponds to the pendulum as measuring apparatus which in turn responds to the perturbation imparted by the detected photon. So the question of choice of ICs is related to the problem of how to divide the system into observer and observed.[6]

To remove the ambiguity noted above with regards to the collapse interpretation in the photon plus pendulum experiment, it suffices to consider a slightly more complicated case. Suppose that in addition to the initially prepared photon, we add another photon prepared well after the first. Once the initial photon has been registered by the pendulum, any further interactions between the pendulum and the second photon lead to violations of unique evolution. One might attempt to account for the effect of the second photon by means of a BC, but the net effect of this move is to add an irreducibly indeterministic element to the model of the pendulum plus photon system. Such an addition obviously takes this model outside the domain of CM.

Kellert seems to be aware of the general problem with identifying identical ICs in QM, but misses the point when he tries to appeal to the possibility of quantum chaos as the source of the confusion on the question of identical initial states (1993, 72–3). The lack of determinate values in the ICs still poses the same problem for understanding what it means to have two models in identical initial states even if quantum chaos exists. It is obvious that the sensitivity argument depends crucially on the interpretation of how quantum systems interact with macro systems. The measurement theory – and in concrete cases such as the photon and pendulum, what counts as the ICs – makes a difference as to whether unique evolution exists for chaotic systems.[7]

5. CONCLUSIONS.

The sensitive dependence arguments of Hobbs and Kellert are much more naive than the authors realize. In cases where quantum effects are amplified to the classical level *via* chaos for situations like the damped, driven pendulum, the question of whether such effects lead to chaotic phenomena that violate unique evolution depends upon how one interprets the quantum formalism, the measurement problem and, in concrete cases, what counts as the ICs for the quantum plus chaotic system under investigation. Under some of these interpretations, the quantum fluctuations amplified by sensitive dependence *will not* lead to violations of unique evolution. For those interpretations where amplified quantum effects *do* influence macrophysical systems in an indeterministic way leading to violations of unique evolution, classical chaotic models would be inadequate descriptions of the chaotic systems in question. In the case of the damped, driven pendulum, textbook classical chaotic models would fail to describe the indeterministic behavior of the pendulum due to quantum influences. Such a result would not force us to revise our conception of physical determinism as Kellert suggests (1993, 69–76), but, rather, would serve as evidence that not all physical systems traditionally thought to be in the domain of CM can be described using deterministic CM models.

This limitation has important consequences for our approach to macroscopic chaotic systems. The dynamical approach to chaos focuses on the properties of individual trajectories as described by equations of motion. As noted above, however, when we shift to the Koopman formulation and focus on descriptions of systems using distribution functions, chaos at the level of individual trajectories gets masked. The equations and the operators involved in these latter models are linear and the behavior of trajectories does not enter into the description, so the global behavior of the models does not exhibit sensitive dependence. Distribution models are very useful for statistical mechanics (SM) where we are dealing with systems that have a large number of components. It seems that, on the one hand, at the level of description of individual trajectories in nonlinear SM models, there is sensitive dependence and, hence, the possibility for quantum events to influence such macrophysical systems. On the other hand, at the level of description of distributions in those same SM systems, there is no sensitive dependence in the evolution of the global properties and, hence, no way to capture possible quantum influences on macrophysical systems.

In order to capture the possible influences of quantum events on the global properties of such systems, we need to shift to different types of descriptions. One possibility is to extend the Koopman formalism to more abstract, extended spaces than a separable Hilbert space, such as the Prigogine school's most recent approach which involves rigged Hilbert spaces (Antoniou and Prigogine, 1993; Antoniou and Tasaki, 1993). Such an approach might provide models that could describe a form of chaos or indeterminism at the macrophysical level that SM models cannot describe. Another possibility is Bohm's theory of QM, where there are exact particle positions and momenta as well as nonlinear quantum forces (1952a,b; Bohm and Hiley, 1993). Although usually negligible at the macrophysical level – Holland (1993, 224–9) argues that quantum forces are most likely constant or vary slowly in space for many macrophysical situations – quantum forces could be amplified up *via* sensitive dependence in nonlinear SM models in a fashion similar to the photon and pendulum experiment described above with the additional feature that the quantum force is nonlocal and would act on all components of the system being modeled. Such possibilities for modeling indeterminism or chaos in the global properties of macrophysical systems call for careful examination.

Department of Philosophy
University of Texas at Austin

NOTES

[1] We would like to thank Harald Atmanspacher for many useful comments and criticisms on the penultimate draft of this essay.

[2] Kellert (1993, 50–62) extends Mark Stone's analysis of determinism (1989) by identifying four different levels or aspects of determinism: (1) the equations of motion including ICs and BCs where no irreducible probability is involved; (2) unique evolution; (3) the existence of precise values for physical quantities; and (4) the possibility of predicting any observational quantities. We take unique evolution to be the crucial feature for determinism, unlike Kellert.

[3] Distribution functions exhibit other chaotic features (Hilborn, 1994, 356–8).

[4] Neither Hobbs nor Kellert explicitly state their preferred interpretation on these issues.

[5] There is a continuum of possible position and momentum states for the photon that could be actualized in interaction with the pendulum. Therefore, the probability of achieving the same ICs on multiple runs under this second construal is of measure zero.

[6] Primas (1990) has suggested making a distinction between epistemic and ontic states, thus providing a more formal characterization of different features of the situation considered above. In particular, measurement can be treated as a transition from an ontic to an epistemic level of description. See also Atmanspacher (1994).

[7] Barone *et al.* (1993) have recently performed computer simulations of classical electrons orbiting in the field of a charged sphere with a uniform magnetic field superimposed. They try to argue that for parameter settings leading to chaos, macroscopic indeterminism results from the fact that QM limits the precision available in phase space for specifying ICs. Their approach, however, is subject to the same problems as those identified for the sensitivity arguments put forward by Hobbs and Kellert.

REFERENCES

Antoniou, I.E., Prigogine, I.: 1993, 'Intrinsic Irreversibility and Integrability of Dynamics', *Physica A*, **192**, 443–64.

Anonious, I.E., Tasaki, S.: 1993, 'Generalized Spectral Decomposition of Mixing Dynamical Systems', *International Journal of Quantum Chemistry,* **46**, 425–73.

Atmanspacher, H.: 1994, 'Is the Ontic/Epistemic-Distinction Sufficient to Represent Quantum Systems Exhaustively?', in K.V. Laurikainen, C. Montonen, and K. Sunnarborg, eds, *Symposium on the Foundations of Modern Physics 1994*. Gif-sur-Yvette: Editions Frontieres.

Barone, S.R., Kunhardt, E.E., Bentson, J., Syljuasen, ?: 1993, 'Newtonian Chaos + Heisenberg Uncertainty = Macroscopic Indeterminacy', *American Journal Of Physics*, **61**, 423–7.

Bell, J.S.: 1987, *Speakable and Unspeakable in Quantum Mechanics.* Cambridge: Cambridge University Press.

Bohm, D.: 1952a, 'A Suggested Interpretation of the Quantum Theory in Terms of 'Hidden Variables' I', *Physical Review*, **85**, 166–79.

Bohm, D.: 1952b, 'A Suggested Interpretation of the Quantum Theory in Terms of 'Hidden Variables' II', *Physical Review*, **85**, 180–93.

Bohm, D., Hiley, B.J.: 1993, *The Undivided Universe: An Ontological Interpretation of Quantum Mechanics.* London: Routledge.

Dubin, D.A., Hennings, M.A.: 1990, *Quantum Mechanics, Algebras and Distributions.* New York: Longman Scientific & Technical.

Earman, J.: 1986, *A Primer on Determinism.* Dordrecht, The Netherlands: D. Reidel Publishing Company.

Giere, R.: 1988, *Explaining Science.* Chicago: University of Chicago Press.

Hacking, I.: 1983, *Representing and Intervening: Introductory Topics in the Philosophy of Science.* Cambridge: Cambridge University Press.

Hellman, G.: 1989, *Mathematics without Numbers: Towards a Modal-Structural Interpretation.* Oxford: Clarendon Press.

Hilborn, R.C.: 1994, *Chaos and Nonlinear Dynamics: An Introduction for Scientists and Engineers.* Oxford: Oxford University Press.

Hobbs, J.: 1991, 'Chaos and Indeterminism', *Canadian Journal of Philosophy*, **21**, 141–64.

Holland, P.R.: 1993, *The Quantum Theory of Motion: An Account of the de Broglie-Bohm Causal Interpretation of Quantum Mechanics.* Cambridge: Cambridge University Press.

Jackson, J.D.: 1975, *Classical Electrodynamics.* New York: John Wiley & Sons.

Kim, Y.S., Noz, M.E.: 1991. *Phase Space Picture of Quantum Mechanics: Group Theoretical Approach.* Singapore: World Scientific.

Kellert, S.H.: 1993, *In the Wake of Chaos.* Chicago: University of Chicago Press.

Koopman, B.O.: 1931, 'Hamiltonian Systems and Transformations in Hilbert Space', *Proceedings of the National Academy of Sciences*, **17**, 315–18.

Merzbacher, E.: 1970, *Quantum Mechanics.* New York: John Wiley & Sons.

Primas, H.: 1990, 'Mathematical and Philosophical Questions in the Theory of Open and Macroscopic Quantum Systems', in A.I. Miller, ed., *Sixty-Two Years of Uncertainty*. New York: Plenum.

Reed, M., Simon, B.: 1972, *Methods of Modern Mathematical Physics I: Functional Analysis*. New York: Academic Press.

Stone, M.A.: 1989, 'Chaos, Prediction and Laplacian Determinism', *American Philosophical Quarterly*, **26**, 123–31.

Thirring, W.: 1983, *A Course in Mathematical Physics 4: Quantum Mechanics of Large Systems*, tr. Evans M. Harrell. New York: Springer-Verlag.

Wigner, E.P.: 1932, 'On the Quantum Correction for Thermodynamic Equilibrium', *Physical Review*, **40**, 749–59.

GIOVANNI BONIOLO

WORMHOLES AND TIMELIKE CURVES: IS THERE ROOM FOR THE GRANDFATHER PARADOX?

1. INTRODUCTION

Since W.J. van Stockum, K. Gödel and F.J. Tipler proposed their solution to the Einsteinian field equations and de Felice *et al.* discussed Kerr's ones, the time-machines have turned from fiction to physical objects. Immediately, philosophers of science, such as Earman (1972), Malament (1985) and Horwich (1987), began to analyse them on account of the several paradoxes implied by backward causation.

Notwithstanding their rigorous and eye-opener approach, these philosophers have limited themselves almost exclusively to discussing the philosophical implications of Gödelian time-machines, and whether spacetime was Gödelian or not. However, on the one hand, Gödel's solution to the Einsteinian field equations is not the only one which implies closed timelike curves (CTCs). On the other hand, besides this kind of Einsteinian time-machines, due to the quantisation of spacetime on Plank's scale, there is the possibility of another class of CTCs, that is, the one implied by the so-called wormholes. Therefore, also these results, obtained by K.S. Thorne and I. Novikov's pioneering studies, have to be considered by the philosopher of science, as Earman (1995) shows in a brilliant way.

In what follows, I will try to discuss the philosophical implications of the quantum time machines. To reach this goal, I will act as a new Carneades. Firstly, I will assume that the proposal given by Friedman *et al.* (1990) has truly to do with the grandfather paradox. So I will analyse it and arrive at *the principle of self-consistency* and at a more restrictive principle: *the principle of the unique time version*. Then (Section 5), I will discuss whether the mentioned assumption is really correct.

However, before starting, I would like to stress a point that must not be forgotten. Many authors have maintained, and have been sustaining, that the physical possibility of CTCs inevitably implies the grandfather paradox, in one of its several more or less elaborated versions. It concerns the possibility that someone went back in his past and killed his grandfather before his father's birth. In this case, the killer would not be born and he could not have gone back.

M.L. Dalla Chiara et al. (eds.), Language, Quantum, Music, 143–157.

However, I suggest drastically abandoning such a kind of anthropomorphisms which may dazzle, owing to the ambiguity of the everyday language by means of which they are formulated. Consequently, I propose to consider only the physical systems, that is, the very source of the question. The implications connected with CTCs may be discussed without appealing to grandchildren who kill their grandfathers, or other strange anthropomorphic situations loved by some scholars in backward causation. This does not imply a reductive form of positivism. Anyway we have two possibilities:

(1) The philosophical discussions on time travels are completely independent of any real physical problem. In this case, they appear to me academic exercises on science fiction.

(2) The philosophical discussions on time travels start from real physical problems. In this case, we have to take into consideration the limitation of physical analysis. To discuss beings with free will which go back in their past is not a province of physics. Physics concerns only more or less idealised objects, of course without free will, which have to conform to certain laws. Physics studies the fall of a body along an inclined plane and does not concern the question of a skier endowed with free will who decides on stopping during his descent. *This is not a physical problem because free will is not a physical problem.*

2. QUANTUM TIME-MACHINES AND TIME PARADOXES

It should be noted that even if it is certain that Classical General Relativity formally implies CTCs, this is not the case at the level of Quantum Gravity, especially after the results obtained by Kim-Thorne (1991). On the other hand, many authors (e.g. Thorne and Novikov) have warned that if Quantum Gravity does not imply QTMs, their papers have to be considered only as an exercise. Actually, even if Quantum Gravity does not imply QTMs, what has been said by these authors is relevant because it spurs us both to better comprehend what the CTCs imply and to try to disentangle some aspects of time paradoxes.

In particular Friedman *et al.* (1990) have simply taken a wormhole and they have put it inside an otherwise flat and empty 4-dimensional spacetime, then they have discussed the evolution of a scalar field in three cases: (1) the two mouths of the wormhole are both at rest with respect to a Lorentzian frame; (2) one mouth is accelerated as in the twin paradox with respect to the other; (3) one mouth is moving inertially with respect to the other. The first case can be exemplified by means of a very intuitive and fascinating example which enables to intuitively and immediately grasp the core of the problem.

It concerns a semiclassical scenario: there are a billiard ball and a static wormhole in an otherwise flat and empty 4-dimensional spacetime. The ball, moving with non-relativistic velocity, enters the mouth B of the wormhole at the time t^0 indicated by the clock of the Lorentzian frame in which the wormhole is at rest (from now on, it is called the 'observer's clock'). It comes out from the mouth A of the same wormhole at the time $t^0 - \delta t$, where δt is the time delay, measured by the observer's clock. Therefore, such a wormhole - consisting of two spherical holes (the mouths A and B) in a 3-dimensional space – works like a time-machine. It means that a CTC passes through it. The mouths are connected by a handle whose length l is arbitrary small and independent of their distance r in the external space. We may suppose that $l << r$.

The paradox arises (Figure 1a) from the fact that the ball, after covering the trajectory Y_1 and entering mouth B at the time t^0, comes out from A at the time $t^0 - \delta t$; that is, before entering B. It means that, after coming out from A, it covers the trajectory O_1 and collides in Z with its younger self, which, after the collision, covers the trajectory Y_3. In such a case, the latter cannot enter mouth B. If it cannot enter B, how can it come out from A? Here there should be the time paradox and the causality violation, and in this case *the physical system would have a self-inconsistent evolution.*

Echeverria *et al.* (1991) have proposed a theorem by means of which the 'paradox' is solved and, consequently, it seems that many problems regarding backward causation may be solved without any dangerous anthropomorphism. For the simple case just shown, they have demonstrated that for each set of initial conditions, which gives self-inconsistent solutions, there are also self-consistent solutions. That is, under the same initial conditions, the physical system may evolve in a self-inconsistent way or in a self-consistent way. In the first case, there is a paradox; in the second one, there is not. It should be noted that such a theorem is possible only by imposing constraints to the initial conditions of a system evolving in a region with CTCs. However such constraints are 'mild in the sense that in some neighbourhood of every event one is free to specify initial data arbitrarily, with the initial data elsewhere being adjusted to guarantee self-consistent evolution' (Friedman *et al.*, 1990, p. 1915).

What is a self-consistent solution? Suppose that the initial position and velocity of the ball are the same as before. Also the trajectory of the ball, before Z' – the new collision point – is the same as before. However (Figure 1b), now the ball enters the mouth B following a slightly altered trajectory. This means that it does not enter B at the same point as before, but at a slightly shifted point. Let Y_2' be such a new altered trajectory after the collision point Z'. In this case the ball comes out from A always at the time $t^0 - \delta t$ and it covers the trajectory O_1'. Following this trajectory, the older version of the ball, after coming out from A, collides in Z' with its younger version *exactly in such a way that this latter is deflected from its original trajectory,* Y_1*, to the new altered trajectory,* Y_2'. Therefore, there is no paradox because the ball can enter B, even if at a slightly shifted point. In such a way, the evolution of the physical system is completely self-consistent and there is no violation of causality. *It happens*

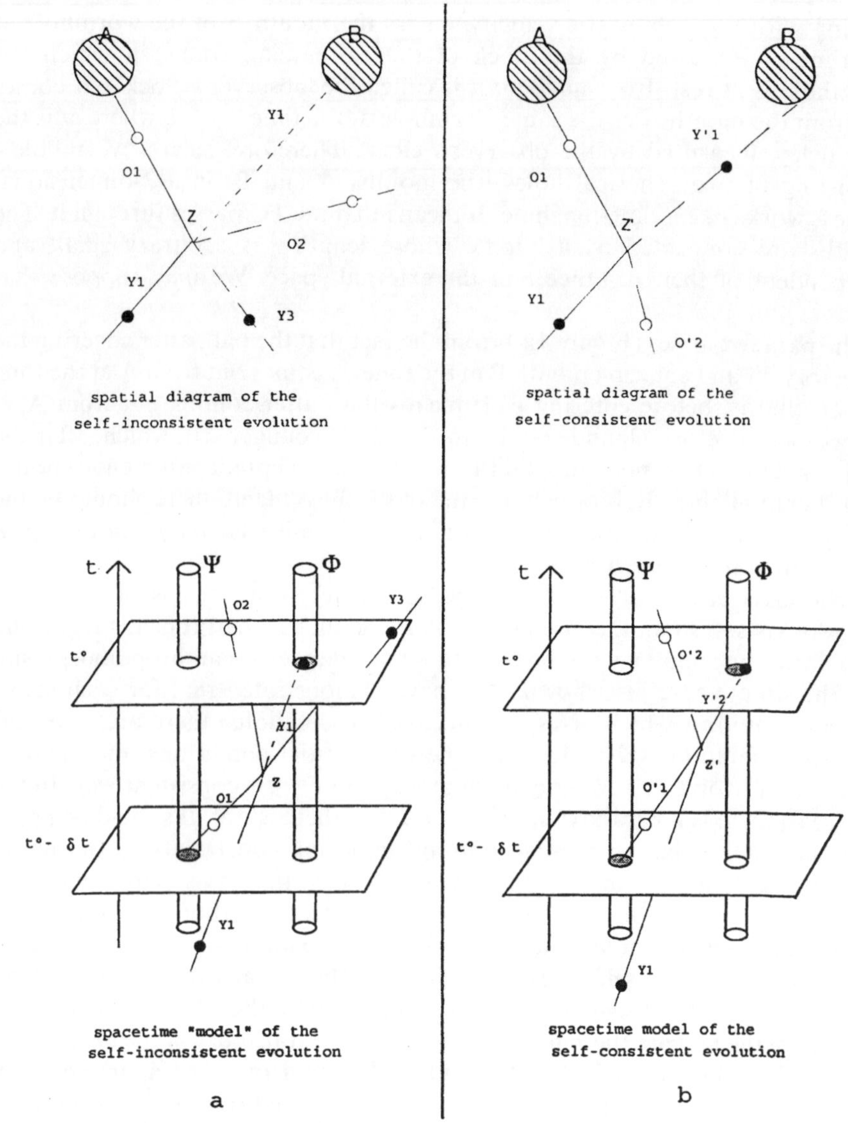

The ball enters B at observer's time t° and comes out from A at observer's time t-δt

○ older ball Ψ world tube of the mouth A
● younger ball Φ world tube of the mouth B

Figure 1.

that, in this case, self-consistency is a consequence of the self-adjustment of the physical system itself!

The authors underline that the paradox (case a) was due to the fact that the collision of the older ball with its younger self has not been taken into consideration and only the evolution of the younger one has been analysed. That is, *the influence of the future on the past has not been taken into consideration.*

The fact that these authors (especially Novikov 1991, pp. 1989–1990) use the term 'influence' amazingly agrees with the difference highlighted by some philosophers (e.g. Horwich 1987, Ch. VII) between '*changing* the past and *influencing* the past'. According to Horwich, 'time travel would allow one to influence the past but not to change it. Changing the past is indeed logically impossible. However, no such contradictions are involved in the idea of influencing the past. There is nothing tricky about this'. It seems that physicists have formalised exactly what philosophers have said. But only now we possess a clear notion of what 'changing the past' and 'influencing the past' mean.

Thus we have a *theorem of self-consistency* which states that *besides self-inconsistent solutions, according to which the past is changed, there is the possibility of having self-consistent solutions, according to which the past is influenced.*

It seems that self-inconsistent evolution is the formal representation of the grandfather paradox. Therefore, self-consistent evolution would be its solution, as Friedman *et al.* (1990) and Novikov (1991) explicitly propose and as Earman (1995) seems to suggest.

However, it should be noted that if the above-mentioned theorem certainly holds for the simple case of the billiard ball and for more general 4-dimensional flat and empty spacetimes with single wormholes and scalar fields, it need not hold in general. In more complex systems, the constraints to the initial conditions, to obtain self-consistent solutions, are not so trivial as in the case shown (cf. Maudlin, 1990 and Novikov, 1991, sects. II–IV).

These limitations notwithstanding, it would seem that, at this point, a first important step, concerning causality, towards the disentanglement of the question about time travels has been made. We have reached the conclusion that certain time-machines influence the past and do not change it.

3. THE PRINCIPLE OF SELF-CONSISTENCY

Friedman *et al.*, beyond the above-mentioned theorem, proposed also a *principle of self-consistency* by means of which they rule out the self-inconsistent solutions and therefore they forbid changing the past. The *principle of self-consistency* states that '*the only solutions to the laws of physics that can occur locally in the real Universe are those which are globally self-consistent.* This principle allows one to build a local solution to the equations of physics only if that local solution can be extended to be part of a (not necessarily unique) global solution, which is defined throughout the non-singular regions of the spacetime' (Friedman *et al.* 1990, pp. 1916–1917). Therefore one has to expunge the self-inconsistent solutions of any equation or set of equations from physics and the principle becomes an *ukase* stating that the self-inconsistent solutions are physically insignificant.

Besides this negative heuristic, there is also a positive heuristic according to which, for each class of self-inconsistent solutions, that is, solutions which imply changes of the past, we should look for the class of corresponding self-consistent solutions, that is, solutions which imply the influence of the past. However, as stressed, this is not a simple question, but a difficult and still open question.

At this point, the question is: What is the philosophical status of this principle? Is it valid? According to Friedman *et al.* (1990), the principle is

(1) imposed *by fiat* in such a way that the changes of the past are ruled out;

(2) 'non-totally tautological' because it implicitly affirms that it is necessary to eliminate a new form of local physics, whenever it is different from global physics; that is, whenever it implies self-inconsistent solutions.

No doubts on the fact that it is imposed *by fiat*; but what about the second point? There are authors, e.g. Deutsch (1991), who object that the principle of self-consistency is totally tautological because it admits only physics we have. If it were really tautological, it would be totally void. Nevertheless, I do not think that this objection is right because the principle of self-consistency does not simply affirm that we have the physics we have, but that we have, and we should have, a form of physics in which there cannot be a contradiction between local and global level.

Actually, the real question is: is there any good reason to support it? Therefore, the very problem is in justifying its introduction.

If it were interpreted as a metaphysical principle it would be the formal translation of an ontological belief on the structure of the world in itself. But the supporters of this way have the hard *onus argomentandi* for the ontological self-consistency of the world in itself.

If the principle were interpreted in a methodological way, it would state that the laws of physics do not admit self-inconsistency owing to the physics we have. We may argue inductively for it by affirming that 'Up to now, the negation of the principle of self-consistency has never occurred in the real world; therefore I feel justified to reject any solution which violates it. Perhaps, such a violation might be the clue for new physics, but I do not believe this because I have no good reason to support this idea'. Yet, this argumentation raises the same doubts that are raised by any kind of inductive confirmation.

Nevertheless, the methodological interpretation of the principle might be also argued by considering the consequences of its negation and showing that they contradict other strongly accepted physical statements. However, we may also find a more restrictive principle that, by forbidding certain formal possibilities, implicitly forbids self-inconsistent solutions without appealing to the principle of self-consistency. It is exactly this path which I will move along. But before arriving at this point, I want to better analyse the billiard ball problem.

4. ANALYSIS OF THE BILLIARD BALL PROBLEM

The main point of what follows lies in the fact that I will turn from causality to time identity of physical objects, by stressing what happens when spacetime regions with CTCs (i.e. chronological-violating regions, CVRs) are embedded into the usual spacetime regions (i.e. chronological-respecting regions, CRRs). After considering the question of the time delay, I will dwell on the differences between the consequences arising from having closed spatial trajectories and those arising from having open spatial trajectories.

4.1 The problem of the choice of δt

The time delay, δt, has been considered such that the younger version of the ball collides with its older self in Z'. From now on, I indicate this particular time delay with δT. However we have to consider all the possibilities and thus we have to consider what happens for a time delay $\delta t \neq \delta T$.

Of course, there are two classes of possibilities:

1. $\delta t < \delta T$; the younger ball passes through Z' when its older self has already passed.

2. $\delta t > \delta T$; the younger ball passes through Z' when its older self has not yet passed.

In both cases, there is no collision. In the first case, the younger ball is in the past of its older self. On the contrary, in the second case, the younger ball is in the future of its older self. It means that *these two classes concern only self-consistent solutions and therefore the principle of self-consistency is no longer necessary.*

Moreover, each class has the extreme possibility:

1'. $\delta t \to 0$; at the same time as the ball enters B, it comes out from A.

2'. $\delta t \to \infty$; the ball enters B, but it never comes out from A. We have a strange time-machine: the ball vanishes after entering B.

4.2 CTC with closed spatial trajectories: a spacetime niche

Suppose that the spatial trajectory external to the two mouths of the wormhole is closed. By means of simple reasoning on the physical consequences of the opposite case, the older version of the ball can never reach its younger self and therefore *there is no collision.* What occurs is that the older ball becomes, step by step and in a continuous way, its younger self and enters the mouth B. The same situation can be described in another way. The younger ball enters B at time t^0 but its older self has come out at time $t^0 - \delta t$, where δt is the time delay. Therefore, it never collides with its younger self and, in the time δt in which it reaches the mouth B, it becomes its younger version. As soon as we relate the CVR to the CRR in which it has embedded, we have interesting consequences. We know that the observer's clock indicates t^0 when the younger version enters B and $t^0 - \delta t$ when it comes out from A. What happens in this case? We have the ball's proper time trapped between t^0 and $t^0 - \delta t$. And this for ever. We are in the presence of a sort of *spacetime niche*, inside which time flows in a circular way while outside time flows regularly.

4.3 CTC with open spatial trajectories

We have seen that we have collision between the younger ball and its older self if $\delta t = \delta T$. However, in case of open spatial trajectories, this is only one of the necessary conditions for the collision. In fact, in order to have it, beyond $\delta t = \delta T$, the open spatial trajectories of the older and younger balls have to intersect. That is, there has to be at least a spacelike hypersurface t^* with $t^* \in [t^0 - \delta t, t^0]$, such that in it the two world lines of the two versions of the ball intersect (therefore if $\delta t = \delta T, t^*$ is the time of the collision).

We suppose now that $\delta t = \delta T$ and that there exists a point in the 3-dimensional space in which the two spatial trajectories of the two versions intersect. Since we are considering CVR embedded in CRR, let us analyse what happens in the time interval $[t^0 - \Delta t, t^0 + \Delta t]$, with $\Delta t \supset \delta T$, measured by the observer's clock. That is, we are dealing with a time interval longer than the time delay.

Since we are considering only open spatial trajectories, at $t^0 - \Delta t$, in the scenario (given by the spacelike hypersurface $t^0 - \Delta t$) there is only the younger ball which is moving towards the mouth B. Instead, at $t^0 + \Delta t$, the situation has changed: in the scenario (given by the spacelike hypersurface $t^0 + \Delta t$) there is only the older ball which is moving away from the mouth A, which it has come out from.

If we group together the 'pictures' of the four spacelike hypersurfaces, $t^0 - \Delta t$, $t^0 - \delta t$, t^0, and $t^0 + \Delta t$ (Figure 2), we have four sections of the spacetime evolution of the system during the time interval $[t^0 - \Delta t, t^0 + \Delta t]$.

Now, we may correctly grasp in what sense, if the spatial trajectories are open, intersecting and with $\delta t = \delta T$, CTCs are 'closed'. In order to reach this

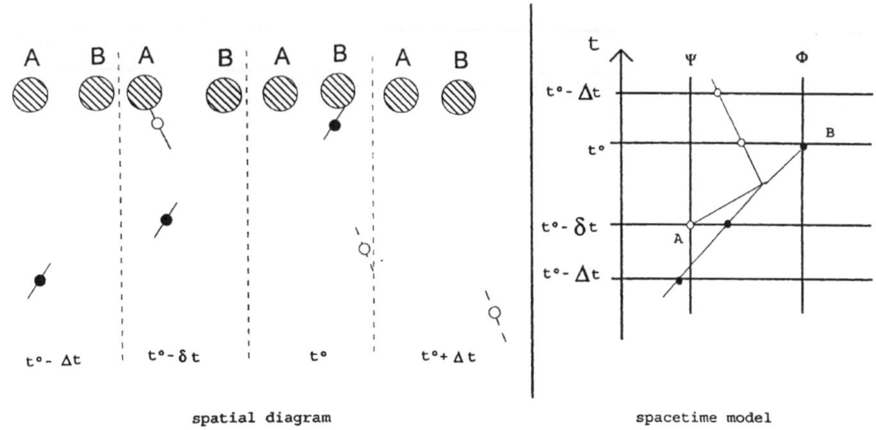

Figure 2.

goal, it is sufficient to take into consideration the whole evolution of the system in the interval $[t^0 - \Delta t, t^0 + \Delta t]$. In this case, it is easily graspable that *the closed timelike curve actually is a sort of 'open timelike loop'* . That is, the ball 'goes back' only once (from t^0 to $t^0 - \delta T$ of the observer's clock), but, then, its proper time flows as usual.

Notice that*there is a real never ending circularity in time only and uniquely in the case of spatial circularity. That is, the timelike curve is indefinitely closed if and only if the system is moving along a closed spatial trajectory.*

At this point, we have to observe more closely what happens in such a case. From what has been said, it should be clear, as noted also by Deutsch (1991), that there is an observer's time interval, between the coming out of the older ball from A and the entering of the younger self in B, in which two different time versions of the same ball are present. This has to happen, otherwise there could be no collision owing to the fact that one version of ball would be in the future (or in the past) of the other version. That is, if we relate the transition from a CRR to a CVR with the transition from a CVR to a CRR, we pass from a situation with only one ball (for $t < t^0 - \delta T$, where t^0 is the entering time of the younger ball into B as to the observer's frame) to a situation with two balls (for each t such that $t^0 - \delta T < t < t^0$), and, finally, to a situation with only one ball again (for $t > t^0$). In other words, there exists a multiplicity of spacelike hypersurfaces, between the observer's spacelike hypersurface $t^0 - \delta T$ and the one t^0, in which there are two time versions of the same ball.

What has been said above may be schematised in the following way:

	time delay	*collisions*	*self-consistency*	*two time versions*
Closed spatial trajectories	δt→0	no	always	no
	0<δt<∞	no	always	yes
	δt→∞	no	always	no
Open intersecting spatial trajectories	δt→0	no	always	no
	δt <δT	no	always	yes
	δt =δT	yes	it depends	yes
	δt >δT	no	always	yes
	δt→∞	no	always	no

Scheme

From the above scheme, it is clear that the usually discussed case, that is, the one in which there are CTCs with open intersecting spatial trajectories and with a time delay $\delta t = \delta T$ such that there is collision, is only one of the many possible time-machines we may have from physical solutions. Moreover, we see that only in this case there are problems with causality! In addition, the theorem of self-consistency showed that for $\delta t = \delta T$ there can be self-inconsistent solutions but also self-consistent solutions, which do not violate causality because they do not imply changing the past, but influencing the past.

However, what is relevant from the above scheme is the fact that the time-machines which imply possible causality violations imply also a time interval in which there are two time versions of the same physical object.

As soon as we forbid this state of affairs, automatically we forbid causality violation without involving any discussion on whether, and when, there are self-consistent evolutions and on the problem of the constraints to the initial conditions. In this way, we turn from the problem of causality to the problem of time identity.

4.4 The principle of the unique time version

In order to forbid the above-mentioned state of affairs, we might introduce the *principle of the unique time version* stating that *the only solutions to the laws of physics that can occur in the real Universe are those which do not admit observer's time intervals in which there is more than one time version of the same physical object.*

I would not consider this principle as the formal translation of an ontological belief on the structure of the world but as a methodological principle on physical propositions. That is, a principle which demarcates those which are physically meaningful from those which are not physically meaningful, even if they are formally meaningful.[2]

At this point, if we take a look at the scheme, we grasp that the only time-machines permitted by the principle are the ones which imply a time delay $\delta \neq \delta T$ with $\delta t \rightarrow 0$ or $\delta t \rightarrow \infty$.

It means that in the case of CTCs with closed spatial trajectories and of CTCs with open intersecting spatial trajectories such that the time delay is $\delta \neq \delta T$ with $\delta t \rightarrow 0$ or $\delta t \rightarrow \infty$, are we dealing with physically meaningful solutions.[3] *The other solutions have to be considered only mathematical explorations of the formal power of the theory, but nothing more.*

Again, it should be noted that the physically meaningful cases do not need the principle of self-consistency owing to the fact that they have nothing to do with collisions. That is, *on the one hand, we are not forced to introduce an ukase to prevent me from killing my infant self, or my grandfather, etc., and, on the other hand, we can leave these possibilities to fiction without any impoverishment of our physical knowledge of nature and of our philosophical knowledge of knowledge.*

5. IS THERE A PARADOX?

Up to now, I have examined the time evolution of the system from the point of view of an 'external' observer who knows that there is a wormhole, that the second ball is the older time version of first ball, etc. From now on, I want to analyse the time evolution from the point of view of an internal observer. That is, I am behaving exactly as Carneades did.[4]

We started by considering the 'paradox' from the observer's time t^0 in which the ball enters B. Then, we went on to see what happens at $t^0 - \delta t$, the time of the coming out of the ball from A. Finally, by discussing the possible ways of collision between the older ball and its younger self, we arrived at differentiating the case of self inconsistency (the past is changed) from the one of self-consistency (the past is influenced). But is this quite correct?

Let us try to analyse once more the evolution with the collision from a different approach. That is, let us analyse the entire evolution from the point of view of the observer at rest in the Lorentzian frame (Figure 3).

At a certain t^1 (equal to the old $t^0 - \Delta t$), there is one billiard ball in our Minkowskian spacelike hypersurface.

At $t^2 > t^1$ (t^2 is equal to the old $t^0 - \delta t$), something strange happens: in a certain point A of the spacelike hypersurface, a new ball suddenly appears.

At $t^3 > t^2$ (t^3 is equal to the old $t^* > t^0 - \delta t$), the first ball and the new one collide.

At $t^4 > t^3$ (t^4 is equal to the old t^0), two different evolutions may be observed:

(α) nothing strange happens and we have two balls (this is the old case a);

(β) again something strange happens; one of the two balls suddenly vanishes in a certain point B of our spacelike hypersurface.

At $t^5 > t^4$ (t^5 is equal to the old $t^0 - \Delta t$), we continue observing either the two balls (case α) or only one ball (case β). And this *ad infinitum*.

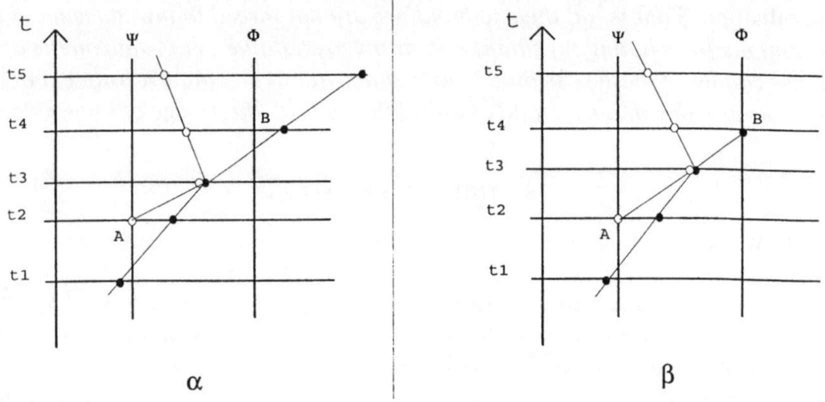

Figure 3.

Neither case α, nor case β, is self-inconsistent, neither there is a grandfather-like paradox. We simply have a collision between two bodies which, after the collision and according to their initial conditions, move along two different spatial trajectories. Certainly, there is something strange in these situations. In both cases, a new ball, of the same physical dimensions as the other one is, suddenly appears at a certain point of spacetime. Moreover (*but only in the case* β), there is another strangeness: one of the two balls suddenly vanishes at a certain point (B) of spacetime.

From this point of view, why should we grant a privilege to case β and rule out case α? After all, it is exactly the latter which is less strange than the former. In it there is only one strange thing (the appearance of a new ball), instead in the other one there are two strange things (the appearance and the vanishing).

Case α might be considered self-inconsistent (that is equal to the old case a) only provided that we know (1) that the younger ball is going to enter mouth B of a wormhole through which a CTC passes; (2) that the new ball is actually the older version of the first ball and that it (the new one) has come out from mouth A of the same wormhole.

However, we would to know too much and all of this cannot be inserted into the physics of the problem. The only observer in the playground is the one at rest in the Lorentzian frame and it knows only what it observes during its time evolution. At t^1, this observer cannot know what is going to happen in its future. And, at t^2, it cannot know that the ball which has suddenly appeared is the older version of the other ball which, at t^4, is going to disappear into the mouth of a wormhole.

Therefore, to apply the interpretation given in sections 3 and 4 means to apply a knowledge which cannot belong to the physical observer at rest in the Lorentzian frame. We should introduce a new non-interacting observer who knows the entire future evolution of the system from the beginning. That is, we should introduce a sort of transcendent observer who knows the future. Only in this case would it know that the ball which is appearing in A is the older version of the ball which, later, is going to disappear in B and that A and B are the two mouths of a same wormhole connecting two different spacetime regions. Ironically, it is worth noting that this is also the knowledge of the author of the papers on CTCs and time-machines. But, usually, in physics what the author knows does not interfere with what is writing in his paper.

To conclude, if what has been said in this section is correct, that is, if we consider only the physical evolution of the system without any kind of transcendent knowledge, neither of the two situations (α and β) is self-inconsistent, even if some strange things happen. And the time paradox vanishes. Moreover, even if a real wormhole existed in the actual world (and if we were the observer at rest in the Lorentzian frame), we could not be aware of it. In fact, what we would observe, would only be the appearance of a new object physically similar to one that already exists, and in certain cases (i.e. in case β) the vanishing of one of the two objects.

Only an observer who is external to the evolution in spacetime could say something more. But this would be God, and – in my opinion – nowadays there is no room for God in physics. Or it would be the author of the paper concerning time-machines. But, so far as I know, there is no physical link between the knowledge of the author and the topics of the paper he is writing.

6. CONCLUSIONS

The conclusion is that if we accept that *inside physics there is no room for a transcendent observer* therefore we have to accept the second approach (the one given in Section 5). It follows that in this case there is neither self-inconsistent evolution, nor – *a fortiori* – a grandfather paradox, which even implies free will. Time travels exist only in science fiction because the authors and the readers know what is happening and what is going to happen. Only in fiction have we time travellers endowed with free will. But not in the examined physics.

Therefore, what is the value of what has been said in Sections 3–4? With respect to the result by Friedman, Novikov and colleagues, one is forced to say that they do not at all tackle the grandfather paradox, neither do they solve it. What they have proposed is a physical-mathematical problem in which they explore the possibility of having more then one evolution of a system in a Minkowskian spacetime in which they suppose that there are strange things such as the appearance and vanishing of bodies.

And what is the value of my analysis in Sections 3–4? It would be correct if the problem existed. But I believe that this is not the case. Therefore, it is only a demonstration that it is possible to raise an apparently correct philosophical discussion of a non-existent problem. And this is only an example!

Department of Philosophy
University of Padova

NOTES

[1] I would like to express my sincere thanks to H. Brown, F. de Felice, F. Gonella and S. Saunders for the discussions on this topic. Of course, they need not completely agree with my rather drastic conclusions.

[2] I would point out that a physically meaningful solution of a physical theory tells us something about certain aspects of the world around us, whereas a non-physically (but formally) meaningful solution tells us only something about the power of the formal and mathematical language of a physical theory.

[3] We should include also the case of open and non-intersecting spatial trajectories with $\delta t \to 0$ or $\delta t \to \infty$. Instead we should rule out all the other cases of open non-intersecting spatial trajectories: we would have two time versions of the same physical ball in $[t^0 - \delta t, t^0]$.

[4] Once (around I cent. BC), the Greek philosopher Carneades went to Rome. The Romans, knowing he was there, asked him for some lectures. In the first, he explained to the Roman students how well-founded and untouchable Plato's doctrine was, and the students went away strongly convinced about its validity. In the second day, Carneades meticulously criticised Plato's doctrine and showed how badly founded and without any great relevance it was. The students went away completely disconcerted. In the same way, up to now, I have analysed the matter as if there were a paradox (self-inconsistency). Now, I want to discuss if this is right.

REFERENCES

Clarke, C., de Felice, F.: 1982, 'Globally Non Causal Space-Times', *J Phys A*, **15**, 2415.

Clarke, C., de Felice, F.: 1984, 'Globally Non Causal Space-Times II. Naked Singularities and Curvature Conditions', *Gen Rel & Grav*, **16**, 139.

de Felice, F., Calvani, M.: 1979, 'Causality Violation in the Kerr Metric', *Gen Rel & Grav*, **10**, 335.

de Felice, F.: 1981, 'Time-Like Non Geodesic Trajectories Which Violate Causality. A Rigorous Derivation', *Il Nuovo Cimento*, **65B**, 224.

Deutsch, D.: 1991, 'Quantum Mechanics Near Closed Timelike Lines', *Phys Rev D*, **44**, 3197.

Earman, J.: 1972, 'Implication of Propagation Outside the Null-Cone', *Austral J Phil*, **50**, 223.

Earman, J.: 1995, *Bangs, Crunches, Whimpers, and Shrieks*. Oxford: Oxford University Press.

Echeverria, F., Klinkhammer, G., Thorne, K.S.: 1991, 'Billiard Balls in Wormhole Spacetimes with Closed Timelike Curves. Classical Theory', *Phys Rev D*, **44**, 44.

Friedman, J., Morris, S.M., Novikov, I.D., Echeverria F., Klinkhammer, G., Thorne, K.S., Yurtsever, U.: 1990, 'Cauchy Problems in Spacetimes with Closed Timelike Curves', *Phys Rev D*, **42**, 1915.

Horwich, P.: 1987, *Asymmetries in Time*. Cambridge: The MIT Press.

Kim, S.W., Thorne, K.S.: 1991, 'Do Vacuum Fluctuations Prevent the Creation of Closed Timelike Curves?', *Phys Rev D*, **43**, 3939.

Malament, D.: 1985, 'Time Travel in the Gödel Universe', *PSA*, **II**.

Maudlin, T.: 1990, 'Time-Travel and Topology', *PSA*, **I**, 303.

Thorne, K.S.: 1994, *Black Holes and Time Warps. Einstein's Outrageous Legacy.* New York: W.W. Norton & Company.

Novikov, I.D.: 1991, 'Time Machine and Self-Consistent Evolution in Problems with Self Interaction', *Phys Rev D*, **45**, 1989.

[illegible] ... [illegible] Fragments from the Gospel and [illegible]

[illegible] 1997

[illegible] [illegible] in the [illegible]

THOMAS BREUER

JOHN VON NEUMANN MET KURT GÖDEL: UNDECIDABLE STATEMENTS IN QUANTUM MECHANICS

INTRODUCTION

John von Neumann met Kurt Gödel. It was at a 1930 conference in Königsberg in a discussion of von Neumann's talk that Gödel for the first time in public mentioned his incompleteness theorem. Von Neumann in his 1932 book on quantum mechanics introduced the quantum measurement problem to the physics literature. The main question of this paper is: Is Gödel's incompleteness theorem related in any way to the quantum measurement problem? Is perhaps self-reference at the heart of both?

1. THE MEASUREMENT PROBLEM

The quantum mechanical measurement problem is the incompatibility of the following four assumptions:

A There is a one-to-one correspondence between the measurable quantities and self-adjoint operators on the Hilbert space.

B Closed systems always evolve according to the Schrödinger equation.

C In an individual system a physical quantity has an unambiguous value if and only if the system is in an eigenstate of the corresponding operator.

D Each single measurement has some – perhaps unknown – result.

Assumptions A, B, and C are consequences of assuming traditional quantum mechanics to be universally valid. They require that all physical systems are described by the traditional quantum mechanical formalism. Assumption D is justified by everyday experience: after every experiment the pointer displays a certain number.

D refers to some measurement result; this should not be taken to imply that the experimenter records free of doubt one result. This is usually not the case since experimenters specify error bars for their results. D requires that the physical quantity 'pointer reading' has *some* value, even if the experimenter

M.L. Dalla Chiara et al. (eds.), Language, Quantum, Music, 159–170.

does not know it exactly. It postulates the factual reality of measurement results but it does not require that the experimenter's knowledge be perfect.

Why are A, B, C, and D incompatible? The standard argument goes as follows. If quantum mechanics is universally valid, it applies also to the measurement apparatus. By assumption A there is a self-adjoint operator Z on the Hilbert space H_A of the apparatus corresponding to the physical quantity 'pointer reading'. Let us assume that the measured observable and the pointer observable are represented by operators with discrete, non-degenerate spectrum. This assumption is an idealisation, but it is not the reason of the measurement problem. Denote by: A the measurement apparatus, O the observed system, o_n the eigenstates of the measured observable, z_n the eigenstates of the pointer observable, i the initial state of the measurement apparatus.

Assume that the time evolution carries an initial state $o_n \otimes i$ into a final state $o_n \otimes z_n$. If we know that after the measurement the pointer observable has value n, then we know that before and after the measurement the measured observable has the value n as well. Therefore this process is called a 'measurement'.

There are several unrealistic assumptions made here. Firstly it is assumed that the pointer value n displys directly the measured value n. In general the value of the measured quantity rather will be calculated from the pointer reading. Secondly, it is assumed that the apparatus is always in the same initial state i. In realistic experiments only a few characteristic properties of the apparatus will indicate that it is ready for the experiment. These few properties will not in general determine a unique ready state. Thirdly it is assumed that the measurement is repeatable. All these idealisations can be removed but it turns out that the measurement problem does not disappear. If the idealisations are not the reason of the measurement problem, we can stick to them for the sake of simplicity.

Now, the incompatibility of A, B, C, and D arises in the following way. If $A\&O$ is closed, it evolves by assumption B according to the Schrödinger equation. (If $A\&O$ is not closed, it is always possible to extend A in such a way that $A\&O$ is closed. If necessary $A\&O$ has to be the whole universe. Since we assumed quantum mechanics to be universally valid, this is not a problem. But I will come back in Section 2 to the question whether it helps to treat $A\&O$ as an open system.) Since the time evolution carries $o_n \otimes i$ into $o_n \otimes z_n$, we infer from the linearity of the Schrödinger equation that the initial state $\sum c_n o_n \otimes i$ is carried into

$$s_1 := \sum c_n o_n \otimes z_n.$$

Therefore, if the initial state of O was not an eigenstate of the measured observable, the state of the apparatus after the experiment will not be an eigenstate of the pointer observable. According to assumption C in an individual system a quantity has an unambiguous, well-defined value only if the system is in an eigenstate. Thus we conclude: *if in an individual system the measured observable does not have an well-defined value before the experiment, the pointer will not have an well-defined value after the experiment.* For individual systems this contradicts assumption D.

And what about measurements of an ensemble? For ensembles C is not directly applicable. But C and D imply restrictions on the final statistical state of the compound system. D requires that after each individual measurement the pointer observable has a well-defined value. According to C this is only the case if the final state after each individual measurement is an eigenstate of $\mathbb{1} \otimes Z$. For the time evolution considered the state of the compound system after an individual experiment is one of the $o_n \otimes z_n$. For the ensemble the mixture of such states yields a density matrix

$$s_2 := \sum_n |c_n|^2 |o_n \otimes z_n\rangle\langle o_n \otimes z_n|.$$

This state is not pure but mixed, and in the basis $o_n \otimes z_n$ it is diagonal. Now if the initial statistical state is $\sum c_n o_n \otimes i$ one infers from A, B, and C, that the statistical final state is s_1, whereas A, C, and D imply that the final state should be s_2. This is a contradiction, because s_1 and s_2 are different states: there are operators which take different expectation values in s_1 and s_2, and by assumption A these operators are observable. s_1 and s_2 can be distinguished by an interference experiment.

What is a solution to the measurement problem? Solutions aim at avoiding the contradiction between A, B, C, and D by dropping one of them, or weakening it, or arguing that it does not apply to real measurements. This can be done in many ways, and which solution seems acceptable, if any, depends on one's esthetical and metaphysical proclivities. Any of the assumptions is more dear to some than it is to others. No wonder that no solution proposed up to now gained unanimous support.

Followers of Bohr would deny the universal applicability of A, B, and C because they would not treat the apparatus as quantum mechanical system. In the superselection appproach one direction of A is denied: some self-adjoint operators do not represent observables. The modal interpretation and the Bohm interpretation deny one direction of C: some observables (position in the case of the Bohm theory) may possess a well-defined value even if the system is not in an eigenstate. The decoherence approach and the stochastic reduction programme treat the apparatus as an open system and thereby deny the applicability (not the truth) of B. GRW deny B even for closed systems by assuming a stochastic evolution of closed systems.

With his projection postulate von Neumann (1932) introduced a second kind of time evolution which occurs during measurement processes and independent of the Schrödinger evolution. In this way he drops assumption B in order to render A and C compatible with D. As recipe for the prediction of measurement results the projection postulate is marvellously effective. But it is difficult to believe that nature 'knows' which interaction is a measurement interaction. This and other objections against the projection postulate are often motivated esthetically or metaphysically. They have to be taken serious, but they are not logically compelling.

From D together with C it follows that the final state of a statistical experiment is s_2. But the converse does not hold. Even if the final statistical state is s_2 we cannot be sure that in each individual case the the pointer has a well-defined value. There are many ensembles consisting not of eigenstates of the pointer which are assigned the statistical state s_2. Only if the pointer has a well-defined value in all individual systems of *all* ensembles which are described by the density matrix s_2, or if one can give a good reason why exactly such an ensemble should be realised, we say that an *ignorance interpretation* of s_2 with respect to the pointer observable is admissible.[1] Then we can be confident that indeed the pointer observable has a well-defined—albeit unknown—value. The system is described by a state s_2 without sharp pointer value just because the pointer value is unknown, not because it is objectively undefined.

2. THE APPARATUS AS OPEN SYSTEM

The incompatibility of A, B, C, and D has been demonstrated under the assumption that *A&O* is a closed system. By B it then develops according to the Schrödinger equation. But if *A&O* is an open system it does not evolve according to the Schrödinger equation. Then the evolution is not deterministic any more: depending on the initial state of the environment, a given state of *A&O* can evolve into different directions. Also, the equations of motion are not linear if *A&O* is open. So some of the essential ingredients for showing that A, B, C, and D are incompatible are not at hand if *A&O* is open. There is hope that the measurement problem can be solved by assuming *A&O* to be open. Is it legitimate to assume that *A&O* is open?

To this there are two reasonable answers, one contradicting the other. On the one hand macroscopic systems such as measurement apparatuses from a practical point of view never can be isolated from their environment; therefore *A&O* has to be treated as an open system. On the other hand, one can always chose for *A* the rest of the universe; then *A&O* is the whole universe and thus closed.

Which of the two answers is the right one? If the question is why we have the *impression* that measurements have results, then it is probably legitimate to neglect the unobserved environment of the apparatus. From this point of view it is justified to assume that *A&O* is open. If the question is why measurements *objectively have* results – if they do – then one may not neglect the environment of the apparatus. The ontological reality of results should not be explained by the epistemic fact that we do not observe the environment. With this reservation in mind, let us discuss briefly whether D can be explained by assuming $A\&O$ to be an open system.

Dynamical Decoherence. Zeh (1970), later Zurek (1982) and others propose the following solution of the measurement problem. The apparatus is in interaction with an environment E, therefore $A\&O$ has to be considered as an open system. Since the environment is not observed, what matters is only the state of $A\&O$ resulting from taking the partial trace over E. The decisive point is: *If the free Hamiltonian of $A\&O$ commutes with the interaction between $A\&O$ and E, then there is a basis in which those coefficients of the density matrix of $A\&O$ which are not on the diagonal go to zero very quickly.* Zurek defines the pointer observable to be one whose eigenvectors are the vectors of this basis.

Dynamical decoherence faces two problems. Firstly, it takes place just if the Hamiltonian of the interaction between $A\&O$ and E commutes with the free Hamiltonian of $A\&O$. If this condition is not satisfied, there is no basis in which the density matrix of $A\&O$ approximates a diagonal one. In contradistinction to the interaction between measured system and apparatus, the interaction between apparatus and environment cannot be chosen by the experimenter. Therefore the experimenter cannot make sure that the condition for dynamical decoherence is satisfied.

Secondly, it remains to be argued that the almost diagonal final state admits an ignorance interpretation. Zurek argues that the non-diagonal terms are so small that practically they cannot be observed. An approximate superselection rule holds for the pointer observable. But for the purpose of an ignorance interpretation only strict superselection rules are good enough. An observable which does not strictly commute with all other observables does not have a well-defined value in some pure states, and it is always possible to decompose any mixed state into such pure states.

Stochastic Reduction. The influence of the environment on $A\&O$ can be described on the level of density matrices by taking the partial trace over E of the Schrödinger evolution of $A\&O\&E$. It is no problem to write down Markovian master equations which carry the initial state $\sum c_n o_n \otimes i$ into s_2. But, again, there is the problem of the ignorance interpretaion.

To avoid this problem there are attempts (e.g. Gisin, 1984; Diósi, 1988; Belavkin, 1990; Holevo 1991 and others) to describe the time evolution on the level of individual systems, i.e. for pure states. This is usually done with the help of stochastic diffusion equations. These technically sophisticated results among other things show the following: there is a stochastic diffusion equation for the pure states the solutions of which fulfil – after averaging over the stochastic process – a Markovian master equation carrying the initial state into s_2. Under some additional conditions the Markovian master equation determines the corresponding stochastic equation even uniquely.

The use of stochastic diffusion equations seems to solve the problem of the ignorance interpretation. The Markovian master equation determines *uniquely* a stochastic process and thus a decomposition of mixed states into pure ones. Therefore one could propose a modification of quantum mechanics saying that mixed states have to be decomposed according to the measure of the stochastic process determined by the time evolution. This would be a real modification of quantum mechanics because traditionally all decompositions are admitted.

But even this modification does not help much. Replacing the Markovian master equation by a stochastic diffusion equation for the pure states is just a simulation which yields the right statistics for $A\&O$. But it is not possible to say that $A\&O$ is in a pure state at every moment. If this were so, $(A\&O)\&E$ would always be in a product state. But the only time evolutions carrying product states into product states are the ones generated by product Hamiltonians $H_{A\&O} \otimes H_E$. The reduced time evolution of $A\&O$ then is the one generated by $H_{A\&O}$, which is not stochastic but deterministic. The dilemma is that the evolution of the total system cannot both carry pure states of $A\&O$ into pure states and be stochastic.

3. SELF-REFERENCE PROBLEMS FOR INTERNAL OBSERVERS

In this section I am going to report some results on restrictions on the experimental accessibility of systems for internal observers: *an internal observer cannot distinguish all states of a system in which he is contained.* These restrictions are due to problems of self-reference. A measurement on a system in which the observer is contained yields information also about the observer. In this sense measurements from inside are self-measurements. The requirement that this self-reference has to be consistent is the main reason for the restrictions.

Since the argument can be found in Breuer (1995, 1996), I will only give a condensed version. For a more detailed discussion of the rôle of self-reference in the argument, and of the question of how we have to conceive of universally valid theories, I refer to (1995), and for a discussion of some consequences in quantum mechanics to (1996).

Description of measurements. A measurement performed by an apparatus A on some observed system O establishes certain correlations between the states of A and of O. After a measurement, we infer information about the state of O from information we have about the state of A. To describe this inference, let us use a map I from the power set $P(S_A)$ of the set S_A of apparatus states into the power set $P(S_O)$ of the set S_O of system states. The inference map I characterises the kind of measurement performed and is defined by:

I assigns to every set X of apparatus states (except the empty set) the set $I(X)$ of object states compatible with the information that the apparatus after the experiment is in one of the states in X.

I is different in different measurement situations. But when the observer chooses the experimental set-up, he also chooses a map I describing how he is going to interpret the pointer reading after the experiment. This map is fixed throughout the measurement.

I will say that an experiment with inference map I is able to *distinguish between the states* s_1, s_2 if there is one set X_1 of final apparatus states referring to s_1 but not to s_2, and another set X_2 referring to s_2 but not to s_1: $I(X_1) \ni s_1 \notin I(X_2)$ and $I(X_1) \not\ni s_2 \in I(X_2)$.

Measurements from inside. Now consider the case where the apparatus A is measuring a system $A\&O$ in which it is contained. The state of $A\&O$ determines the state of the subsystem A. To describe this we use a map $R : S_{A\&O} \to S_A$: if $A\&O$ is in the state s, the subsystem A is in the state $R(s)$. In classical mechanics, for example, a map R is defined by discarding coordinates which refer to O. In quantum mechanics, one can take R to be for example the partial trace over O. For our purposes it is enough to take an arbitrary but fixed map R. If O has at least one degree of freedom

$$(\exists s, s' \in S_{A\&O}) : R(s) = R(s'), s \neq s'.$$

In the case of quantum mechanics it is problematic to restrict a state of $A\&O$ to the subsystem A. Usually this is done by taking R to be the partial trace over O. Now the problem is that if $A\&O$ is in a pure non-product state, then the partial trace yields as reduced state of A a density matrix, not a pure state. When taking only pure states as possible states of a single system, then we have to conclude that A is not in a state in which a single system can be, although it is a subsystem of a single system. This is one of the great riddles of quantum mechanics. Fortunately we do not have to solve it. All we need in the sequel is that the state of $A\&O$ determines the state of A, and we describe this by the restriction map R. Whether the resulting states of A are described by density matrices or wave functions, and how we interpret these, does not concern us here.

If the apparatus is contained in the observed system a circle of reference arises. Starting from a state of A, I refers to some states of $A\&O$, and by restriction R these again determine states of A. This is because measurements from inside are self-referential: when you know something about a system in which you are contained, you also know something about yourself. Inference maps I for which this self-reference is consistent fulfil

$$R(I(\{s_A\})) = \{s_A\}$$

for all states s_A of A. This means that each apparatus state s_A can refer only to states of $A\&O$ whose restriction to A is s_A. An inference map not satisfying this gives rise to contradictions.

Theorem 1 *There exists no consistent inference map* $I : P(S_A) \rightarrow P(S_{A\&O})$ *which can distinguish states* s_1, s_2 *of A&O for which* $R(s_1) = R(s_2)$.

A proof of Theorem 1 can be found in Breuer (1995).

4. MEASUREMENTS FROM INSIDE AND QUANTUM MECHANICAL MEASUREMENTS

I come back to the question whether self-reference is at the heart not only of Gödel's incompleteness theorem but also of the quantum mechanical measurement problem. This expectation is nourished by several parallels between measurements from inside and quantum mechanical measurements.

Firstly, both for measurements from inside and for quantum mechanical measurements the observer cannot be separated from the observed system. For measurements from inside this is a consequence of the observer being contained in the observed system. A cut around the observer would affect the observed system as well. In quantum mechanics the observed system cannot be separated from the observer because generically the two are entangled.

Secondly, both, self-reference and the quantum mechanical measurement problem lead to the conclusion that standard quantum mechanics cannot be a universally valid theory which is at the same time experimentally fully accessible. Self-reference problems prevent any universally valid theory from being experimentally fully accessible: according to Theorem 1 no observer can distinguish all states of the universe, since every observer is contained in it. The measurement problem is the contradiction between the universal validity of quantum mechanics (assumptions A, B, C) and the possibility of measurements (assumption D). Therefore quantum mechanics cannot be universally valid and experimentally fully accessible. Both, problems of self-reference and the measurement problem, can be avoided if we renounce the universal validity of quantum theory and do not apply it to the observer. For an observer outside the quantum world looking at a quantum system neither the measurement problem nor problems of self-reference arise.

Thirdly, there is this remarkable analogy between von Neumann's (1932) chain of quantum mechanical observers and the hierarchy of object theory, meta-theory, meta-metatheory etc. enforced by problems of self-reference. Apart from that there is the provocative remark of London and Bauer (1939) that it is the mind's faculty of *introspection* which causes the wave packet to collapse. Introspection involves self-reference. Thus the claim of London and Bauer is: the reduction of the wave packet is an effect of self-reference.

Before I investigate whether these three parallels really justify the claim that the measurement problem is a problem of self-reference let us look at self-reference and the measurement problem from a different point of view.

5. PHYSICS-FROM-INSIDE AND PHYSICS-FROM-OUTSIDE

Primas (1990) suggests to view the measurement problem as the problem of compatibility of *endo-* and *exophysics*. These terms are used by Rössler, Svozil (1993), and Finkelstein (1988) in a different way. To avoid these terminological discussions I will use the terms 'from inside' and 'from outside'. As physics-from-inside I dub the observations and the formalism used by an observer contained in the observed or described system, as physics-from-outside the physics of an outside observer.

A description from outside presupposes a separation of observer and observed system. This is not deep philosophy or exciting physics, it is a logical necessity. When we speak about measurements from outside we have to specify outside what. So the observed system must be conceivable as something distinct from the observer. This does not mean that there may not be any interaction between the two, nor that they must not be correlated, nor that there is an absolute observer outside all systems described by physics-from-outside. But since measurements without interaction are impossible, systems which are observed from outside have to be *open* systems. Only open systems can be subject of physics-from-outside. The universe cannot be subject of physics-from-outside, so physics-from-outside cannot be universally valid. But in physics-from-outside there are no restrictions on measurability since Theorem 1 does not apply.

Closed systems, as for example the universe, can be observed only from inside. Thus only physics-from-inside is able to deal with closed systems, and only physics-from-inside can be universally valid. The price to pay for potential universality is the failure of full measurability. In physics-from-inside full measurability fails for problems of self-reference (Theorem 1).

This is true in general but there are some unexpected consequences for quantum mechanics. Quantum-mechanics-from-inside refers to closed systems and according to assumption B their time evolution is governed by the Schrödinger equation. They evolve linearly, deterministically, and reversibly. The experimental investigation of such systems is problematic because they can be observed only from inside. Firstly, this gives rise to problems of self-reference (Theorem 1). Secondly, the measurement problem arises since an observer being part of the observed quantum mechanical system must be described quantum mechanically as well.

In contradistinction, quantum-mechanics-from-outside, avoids problems of self-reference and the measurement problem. The measurement problem does not arise because quantum mechanics is only applied to the observed open system but not to the observer. Neither do problems of self-reference arise for observations from outside. But the universe cannot be described by quantum-mechanics-from-outside. The Copenhagen interpretation is the paradigm example of quantum-mechanics-from-outside. It requires that quantum mechanics must not be applied to the observer. According to the Copenhagen interpretation assumption B is: inapplicable to the observer; and violated by quantum systems because during a measurement the wave function collapses.

The measurement problem can be viewed as a problem of compatibility of quantum-mechanics-from-inside with quantum-mechanics-from-outside. Is it possible for quantum mechanics both to be universally valid, as is quantum-mechanics-from-inside, and to allow for measurements, as does quantum-mechanics-from-outside in the Copenhagen interpretation? The two can be compatible only if there is no contradiction between the validity of the Schrödinger equation for closed systems and the occurrence of a collapse during a measurement, while the system is open. Is it consistent to apply the projection postulate to quantum systems observed from outside and at the same time to believe that the wave function of the universe follows the Schrödinger equation? This is the measurement problem.

The alleged contradiction lies not in the fact that the evolution during the collapse is non-linear and stochastic. This does not necessarily contradict the validity of the Schrödinger equation for closed systems since during the collapse the observed system is open. The incompatibility of quantum-mechanics-from-inside and quantum-mechanics-from-outside, and the measurement problem, is that if the Schrödinger equation holds for closed systems the evolution during the collapse cannot be explained by the observed system being open during the measurement.

6. HIERARCHIES OF FORMAL SYSTEM AND VON NEUMANN'S CHAIN OF OBSERVERS

In his theory of the quantum measurement process von Neumann (1932) introduced a chain of observers. If A makes a measurement on O the values of the pointer observable will be related to the values of the measured observable. Knowing the value of the pointer we know the value of the measured observable. But if the measured observable did not have a well-defined value before the measurement, the pointer observable will not have a well-defined value afterwards. Now a second observer A_2 can try to measure the value of the pointer observable. But this measurement performed by A_2 on A faces the same problem: the value of the second pointer observable can be related to the one of the first and thus to the one of the measured observable, but if the first pointer did not have a well-defined value the second will not have one either. The same will be true if a third observer A_3 tries to determine the value of the second pointer, etc. etc.

What precisely is the analogy of this situation with the hierarchy of object theory, meta-theory, meta-theory etc.? In a formal system there are propositions like the Gödel formula which cannot be proved or refuted by the rules of the system. Correspondingly, a quantum measurement of A on O does in general not have a result if $A\&O$ is closed. In the meta-theory there are propositions which cannot be decided by the rules of the meta-theory.

Correspondingly, a measurement of A_2 on A does not have a result if $O\&A\&A_2$ is closed. Is there an analogy to the decidability of the Gödel formula in the meta-theory? Yes, if the measurement problem could be solved by assuming $A\&O$ to be open. Then measurement of A on O do not have a result for A but they do have a result for an observer outside $A\&O$.

This is how far the analogy carries in a traditional analysis of the measurement problem. It was based on an analogy of 'having a proof of a statement' with 'having a result of a measurement'. This is not entirely mistaken. After all, measurement and proof are both semantic concepts in that they establish a relation between a formalism physical or mathematical, and what is referred to by the formalism. Still, in this form the analogy is vague and wanting: there is no analogy to the decidability of the Gödel formula on the meta-level if, as I argued in section 2, the occurrence of results cannot be explained by assuming $A\&O$ to be open.

With the results of Section 3 one can spell out the analogy more explicitly and formulate undecidable propostions. Remember from Theorem 1 that A cannot distinguish between states s, s' of $A\&O$ for which $R(s) = R(s')$. This is the case for the pure state s_1 and the decohered s_2: $R(s_1) = R(s_2) = \sum_n |c_n|^2 |z_n\rangle\langle z_n|$. Thus a statement like

$$\text{The state of } A\&O \text{ after the measurement is } s_1 \text{ and not } s_2. \qquad (1)$$

or 'In the final statistical state the interference terms between A and O vanish' is undecidable for A.[2] There is no consistent inference map and no possible measurement outcome $X \subset S_A$ such that just one but not both s_1, s_2 is in $I(X)$.

A proposition which is undecidable on the meta-level, i.e. for A_2, is: 'The interference terms between A_2 and $A\&O$ vanish.' The system $O\&A$ corresponds to the object theory, $O\&A\&A_2$ to the meta-theory etc. Similar to the language of the meta-theory being richer than the one of the object theory and containing it, the physical system of the meta-level contains the one of the object level.

Is (1) decidable for A_2? Yes, because there are many consistent inference maps $I : P(S_{A_2}) \to P(S_{A\&O})$ which can distinguish s_1 from s_2. An outside observer like A_2 can distinguish s_1 from s_2 simply by measuring the interference terms between A and O. But things are not that easy. As long as the measurement problem is not solved we cannot assume that any measurement is possible. The existence of an inference map does not imply that the measurement problem has been solved.

Department of Philosophy
University of Salzburg

NOTES

[1] This is an ignorance interpretation with respect to some physical quantity, namely the pointer value. In a different terminology, introduced by d'Espagnat (1966) the term ignorance interpretation is used in the following way: a density matrix admits an ignorance interpretation in the sense of d'Espagnat if all single systems of all ensembles described by this density matrix are in a pure state. In this case the system is described by a mixed state only because the pure state it is actually in is unknown, not because the system is actually in a mixed state. D'Espagnat showed: an ignorance interpretation in his sense is not admissible for mixed states of a subsystem resulting from taking the partial trace over a pure entangled state of a bigger system.

[2] Primas (1990) formulates a different quantum mechanical undecidability theorem: *The statement 'The cat is either dead or alive' is endophysically undecidable (i.e. not provable and irrefutable), even it were true.* Unfortunately he does not give an argument for his undecidability theorem. If one translates 'endophysically' with 'for the cat' and does not refer to a single experiment but rather to a series then Primas' undecidability theorem reads: *The statement 'In the final statistical state the interference terms between cat and observed system vanish' is undecidable for the cat.* This is my (1). But it seems that Primas' statement is not undecidable if read as referring to single experiments. States of *A&O* with different restrictions to *A* (like definite life-state and superpositions of definite life states) can be distinguished by some inference map.

REFERENCES

Albert D.Z.: 1987, *Philosphy of Science*, ***57***, 577.

Breuer T.: 1995, *Philosophy of Science*, ***62***, 197.

Breuer T.: 1996, *Synthese*, ***107***, 1.

Belavkin V.P.: 1990, *Journal of Mathematical Physics*, ***31***, 2930.

de Espagnat B.: 1966, *Nuovo Cimento Suppl*, ***4***, 828.

Diósi L.: 1988, *Journal of Physics A*, ***21***, 2885.

Finkelstein D.: 1988, 'Finite Physics', in R.Herken, ed., *The Universal Turing Machine. A Half-Century Survey*. Oxford, Oxford University Press.

Gisin N.: *Physical Review Letters* ***52***, 1657.

Holevo A.S.: 1991, in C. Bendjaballah *et al.* 1991, *Quantum Aspects of Optical Communication*, Lecture Notes in Physics, Vol. 378. Heidelberg, Springer; 127–137.

London F., Bauer E.: 1939, *La Théorie de L'Observation en Mécanique Quantique*. Paris, Hermann.

von Neumann J.: 1932, *Mathematische Grundlagen der Quantenmechanik*. Berlin, Springer.

Primas H.: 1990, in J. Audretsch, K. Mainzer, eds, *Wieviele Leben hat Schrödingers Katze?*. Mannheim, BI-Wissenschaftsverlag; 209–243.

Rössler O.E.: 1987, 'Endophysics', in J.L.Casti, A. Karlqvist, eds., *Real Brains – Artificial Minds*. New York, North-Holland; 25–46.

Svozil K.: 1993, *Randomness and Undecidability in Physics*. Singapore, World Scientific.

Zeh H.D.: 1970, *Foundations of Physics*, ***1***, 69.

Zurek W.H.: 1982, *Physical Review D*, ***26***, 1862.

NEWTON C.A. DA COSTA AND DÉCIO KRAUSE

SET-THEORETICAL MODELS FOR QUANTUM SYSTEMS

1. INTRODUCTION

The general mathematical framework underlying the usual physical theories is naïve set theory, but it is obvious that every construction can be performed in an axiomatized set theory like Zermelo-Fraenkel or Kelley-Morse. In fact, we could consider the axiomatization of physical theories by means of Suppes' predicates (or Bourbaki's species of structures), which are formulas of set theory, and note that the models of such predicates are also set-theoretical structures [6]. In particular, the usual formulations of quantum mechanics (henceforth, QM) use a fragment of the language of functional analysis; so, they are based on set theory.[1]

Several authors like Heisenberg, Bohr and Schrödinger have observed that in QM some basic concepts like velocity, position, observation, physical object should be sharply distinguished from the corresponding concepts of classical physics [3], [22, Cap. 1] [37]. They have also noted that several new, non-classical concepts should be introduced in such a context, for instance the concept of complementarity [24, pp. 337ff]. Furthemore, the classical concepts cannot be applied, so to say, 'freely' as in classical physics: some restrictions in their range of applicability should be imposed, such as the uncertainty relations. As recalled by M. Jammer,[2] Heisenberg pointed out that the disparity between formalism and intuitive conceptions is the root of serious difficulties and contradictions in quantum theory. At the same time he realized that it would be impossible to construct an independent appropriate conceptual apparatus (in the sense of a descriptive language) which would give an adequate intuitive interpretation of the abstract formalism, Heisenberg saw no other alternative than to retain classical intuitive notions and to restrict their applicability [24, p. 325].

The possibility of formulating physical theories by using classical mathematics shows (at least) the strong capacity of standard mathematical language in expressing science. In fact, even in the case of QM (which seems to require the consideration of a class of entities that have a completely different behaviour from that of the objects of classical physics) the theory can be completely described in the usual mathematical framework, as the Copenhagen school has shown.

According to Heisenberg, the basic 'paradox' of the Copenhagen interpreta-

M.L. Dalla Chiara et al. (eds.), Language, Quantum, Music, 171–181.

tion consists precisely in the fact that every physical experiment must be described in terms of classical physics (hence, they can be described by usual mathematics) [22, Chap. 3], [23, Chap. 46]. In fact, even in the theory of microscopic objects "... we need not speak of particles at all" [23, p. 49]. Schrödinger's approach by means of the wave function is also a description of quantum systems within the scope of usual mathematics. Other approaches could be mentioned, but we will not discuss them here.

Notwithstanding the above apology of classical mathematics, a lot of fundamental questions in this field seems to motivate philosophical discussions, from the 'very nature' of the basic entities of matter [37], [38], [39] (see [5])[3] to the validity of the principle of the identity of indicernibles, which according to some authors is violated in the quantum mechanical realm.[4] Schrödinger, for instance, in questioning the individuation of elementary particles, has suggested that in order to surmont atomism, quantized waves should be considered (perhaps in the sense of second quantization); at the same time, he felt that such an approach could hardly satisfy his own demands for an intuitive picture of quantum objects (cf. [13, p. 258]). As is well known, Schrödinger's position is more concerned with particles as that in a sense may be regarded as 'non-individuals' [39].[5] We will refer to this topic in the final section.

It is obvious that these philosophical questions motivate strict mathematical problems, as it was noted by M.L. Dalla Chiara [10]. Furthermore, if elementary particles are to be considered as some kind of non-individuals,[6] (taking into account the fact that permutations of 'identical' particles should not be regarded as observables) it might be 'natural' to ask for a mathematical device for treating such collections, which, due to the indistinguishability of their elements, could not obey the axioms of (say) Zermelo-Fraenkel set theory.[7] In fact, the search for axioms which could be useful in describing such collections was suggested by Yu. I. Manin (in his contribution to the list of Problems of Present Day Mathematics, presented at the 1974 AMS Congress on the Hilbert Problems) as one of the most relevant topics to be investigated by researchers interested in the foundations of mathematics [31] (see [7], [12]).

All these topics suggest that *quasi-set theories*, which have been pursued by M.L. Dalla Chiara, G. Toraldo di Francia, S. French, R. Giuntini and by ourselves, are mathematically, philosophically and physically relevant.

In this paper, we show how the quasi-set theory S^{**} ([12]) can be translated into the usual Zermelo-Fraenkel system with the Axiom of Choice –ZFC. This (obviously) shows that if ZFC is consistent, so is S^{**}.[8] This result has interesting philosophical consequences, since it is connected with the above discussion on the possibility of expressing QM within the scope of classical mathematics. Does this mean that quasi-set theories can be dispensed with or that they have no significance, since (roughly speaking) all that can be described in such a theory can also be described in classical set theory? The answer is of course negative: quasi-sets have significance in formalizing much of the underlying presuppositions of quantum physics, as we suggested above. Anyway, from the strict mathematical point of view our result only shows that these theories are formally equivalent. In practice, at least in principle, we may proceed as the mathematician, when she uses geometrical as well as algebrical tools in some areas of mathematics. A similar example is represented by the possibility of translating classical logic into intuitionistic logic; and of course it would be suspicious to claim that classical logic is not important.

2. A 'CLASSICAL' INTERPRETATION FOR S^{**}

The quasi-set theory S^{**} considered here is presented with more details in [12]. The theory S^{**} allows for the existence of two kinds of *Urelemente*: *m*-atoms and *M*-atoms.The usual concept of (self-) identity does not apply to *m*-atoms, while the other *Urelemente* (the *M*-atoms) act as usual atoms of the theory ZFU (Zermelo-Fraenkel with *Urelemente*). The underlying logic of S^{**} is classical quantificational logic without identity; the specific symbols are three unary predicate letters m, M and Z; $m(x)$ is read 'x is an *m*-atom', $M(x)$ says that x is an standard *Urelement*, while $Z(x)$ says that x is a set –in the sense that x is a 'copy' of a set in ZFU. Furthermore, the language has two binary predicates $\in$ (membership) and $\equiv$ (indistinguishability) and one unary functional symbol *qcard* ($qcard(x)$ stands for the 'quasi-cardinality' of x).

The terms and formulas are defined in the usual way. A quasi-set x ($Q(x)$) is defined as an entity that is not an *Urelement*; the concept of quasi-cardinal generalizes the concept of cardinal for arbitrary quasi-sets. 'Sets' are particular quasi-sets. The axioms of indistinguishability state that $\equiv$ has all the properties of an equivalence relation. The concept of 'extensional equality' $=_E$ is defined in the following way:

$$x =_E y := (Q(x) \wedge Q(y) \wedge \forall z(z \in x \leftrightarrow z \in y)) \vee (M(x) \wedge M(y) \wedge x \equiv y) \quad (1)$$

In other words, extensionally identical entities are indistinguishable standard *Urelemente* or quasi-sets that have exactly the same elements. It is then postulated that the substitutivity principle is valid only in connection with extensionally identical entities:

$$\forall x \forall y(x =_E y \rightarrow (A(x,x) \rightarrow A(x,y))) \quad (2)$$

with the usual syntactical restrictions. It is easily verified that the extensional equality has all the properties of classical equality. As a consequence of the above definitions, it results that identity, as it is usually understood, lacks sense for *m*-atoms; that is, the *m*-atoms are linked only by the weaker relation of indistinguishability which mirrors an equivalence relation. At the same time, the language does not allow us to talk about either about the identity or about the diversity of m-atoms.

In what follows, we will show how quasi-set theory can be translated into ZFC. The remaining axioms of quasi-set theory will be referred below. All that follows is performed in ZFC.

Let m be a non empty set and let R be an equivalence relation on m. The equivalence classes of the quotient set m/R are denoted by $C_1, C_2 \ldots$. If $x \in m$, define $\widehat{x} = \langle x, C_x \rangle$, where C_x is the equivalence class to which x belongs and call $\widehat{m}$ the set of all $\widehat{x}$ with $x \in m$.

Let X be the set $X = \widehat{m} \cup M$, where $\widehat{m}$ is as above and M is a set such that $\widehat{m} \cap M = \emptyset$.

Then we define a superstructure $\mathcal{Q}$ over the set X, called *the Q-set universe.* As we will see, $\mathcal{Q}$ is a 'model' for the quasi-set theory S^{**}. The definition is as follows:

$$\mathcal{Q}_0 = X$$
$$\mathcal{Q}_1 = X \cup \mathcal{P}(X)$$
$$\vdots$$
$$\mathcal{Q}_\lambda = \bigcup_{\beta<\lambda} \mathcal{Q}_\beta \text{ if } \lambda \text{ is a limit ordinal}$$
$$\mathcal{Q} = \bigcup_{\alpha \in On} \mathcal{Q}_\alpha$$

In accordance with the terminology of S^{**}, the elements of M are called M-atoms, M-elements or M-objects, while the elements of $\widehat{m}$ are called m-atoms, m-elements or m-objects. The final goal is to interpret the basic elements of S^{**} as the corresponding objects in $\mathcal{Q}$.

For the sake of simplicity, we introduce another superstructure which we will call $\mathcal{Q}^{\int}$.The construction of $\mathcal{Q}^{\int}$ is similar to that of $\mathcal{Q}$. However, differently from $\mathcal{Q}$, the only 'ground' basis of $\mathcal{Q}^{\int}$ is the set M. The sets of S^{**} (that is, those x that satisfy the predicate $Z(x)$) will be the elements of $\mathcal{Q}^{\int}$.

Now we define a translation from the language of S^{**} into the the language of ZFC. Let us first define the following relation on the set m/R:

$$\widehat{x} \sim \widehat{y} \textit{ iff } C_x = C_y$$

If $\widehat{x} \sim \widehat{y}$, we say that x and y are *indistinguishable.* It is easy to see that $\sim$ is an equivalence relation. Notice that in this way we are identifying x and y by means of the class (or 'state', or 'sort') they are in, represented by the equivalence class which they belong to, and this is done without direct reference to the objects themselves. This approach is essentially H. Weyl's idea of 'aggregates' of individuals, a concept he showed to be important in quantum mechanics as well as in chemistry and biology [44, App. B] (see also [25]).

Let us turn to the translation. Suppose that A is an atomic formula of S^{**}; let us call A' its translation into the language of ZFC.[9] We suppose that all the sets (of ZFC) involved in the definition below belong to $\mathcal{Q}$ and that the quantifiers are restricted to this class. Then,

1. If A is $m(x)$, then A' is $x \in \widehat{m}$.

2. If A is $M(x)$, then A' is $x \in M$

3. If A is $Z(x)$, then A' is $x \in \mathcal{Q}^{\int} \wedge x \notin M$

4. The translation of the term $qcard(x)$, is $card(x)$, the cardinal of the set x.

5. If A is $x \equiv y$, then A' is $(x \in \widehat{m} \wedge y \in \widehat{m} \wedge x \sim y) \vee x = y$

6. If A is $x \in y$, then A' is $x \in y$

The other formulas are translated in the expected way. By means of the above procedure, the definitions of S^{**} can be translated into ZFC. In particular, we have:

1. In S^{**}, a quasi-set is an object which is neither an m-atom nor an M-atom. The formal definition is $Q(x) := \neg(m(x) \vee M(x))$. Due to the translation, in ZFC this simply means that $x \in \mathcal{Q}$ but neither $x \in \widehat{m}$ nor $x \in M$. In other words, a set, which in ZFC 'represents' a quasi-set, is a set of the class $\mathcal{Q}$ that neither belongs to M nor is an ordered pair of the form $\langle x, C_x \rangle$.

2. In S^{**}, the 'pure' quasi-sets are those quasi-sets whose elements are only m-atoms. In the present case, they are interpreted (in ZFC) as subsets of $\widehat{m}$. Furthemore, in S^{**} we define a *classical object* as an x which obeys the predicate C defined by $C(x) := M(x) \vee Z(x)$. This simply means that x is either an element of M or of $\mathcal{Q}^{\int}$.

3. The quasi-set inclusion is defined as in ZFC; so, its translation coincides with the standard inclusion of sets in such a theory.

4. The Extensional Equality (1) expresses (in ZFC) the usual identity governed by the axiom of extensionality (in the case of sets) or the identity of the elements of the set M.

Now let us turn to the translations of the axioms of S^{**}. We will state only informally how the translations of the 'very peculiar' axioms of S^{**} can be proven as theorems of ZFC. Those axioms of S^{**} which are adaptations of the axioms of ZFU (see [12]) will be mentioned only in brief. The details can be performed without difficulty.[10]

In S^{**}, the Axioms of Indistinguishability state that $\equiv$ has the properties of an equivalence relation (axiom (A1)), while the substitutivity law (axiom (A2)) holds for 'extensionally identical' objects which are not m-objects. If we consider the above translation, it is easy to see that the images of the pairs $\langle x, y\rangle$ such that $x \equiv y$ define an equivalence relation in ZFC. Concerning the extensional identity, since this means nothing more than the usual identity for certain sets of ZFC, it results that the substitutivity law (2) is also valid. Hence, the translations of the axioms (A1) and (A2) are true in ZFC.

We consider now the other axioms of S^{**} and their informal translations. (A3) is $\forall x \forall y(y \in x \to Q(x))$. Since $Q(x)$ means (after the translation) that x is neither an element of $\widehat{m}$ nor an element of M, then the informal translation of (A3) says that if y has an element, then $rank(y) > rank(X)$, and this conforms with the translation of the definition of quasi-sets (cf. item 1 above).[11] Axiom (A4) states that every set is a quasi-set. The translations of the formulas $Z(x)$ and $Q(x)$ entail that in both cases x is an element of the class $\mathcal{Q}$; furthermore, it is neither an element of $\widehat{m}$ nor of M; since that sets are particular quasi-sets (due to the fact that the elements of $\mathcal{Q}^{\int}$ are also elements of $\mathcal{Q}$) it results that the translation of the axiom is of course true. Axiom (A5) says that no set has m-atoms as elements. In symbols, $\forall_Q x(\exists_m y(y \in x) \to \neg Z(x))$; the translation, informally stated, is true due to the meaning of the word 'set' given by the definition: a 'set' is an element of $\mathcal{Q}^{\int}$, and the m-atoms are ruled out of such 'sets' by definition.

Axiom (A6) of S^{**} is: $\forall_Q x(\forall y(y \in x \to C(y)) \to Z(x))$. In words, every quasi-set whose elements are either sets or 'classical' *Urelemente* is a set. It is an immediate consequence of the above definitions that the translation of this axiom is true in ZFC. (A7) is $\forall x(m(x) \wedge x \equiv y \to m(y))$. In words, the translation says that only elements of $\widehat{m}$ must be in the relation $\sim$ with elements of $\widehat{m}$, which is a consequence of the definition of $\sim$. Axiom (A8) states that indistinguishable sets are extensionally identical: $\forall_Z x \forall_Z y(x \equiv y \to x =_E y)$. According to the translation, *sets* are elements of the structure $\mathcal{Q}^{\int}$, and for these entities the indistinguishability relation is the ZFC equality, which coincides of course with the translation of the extensional equality relation as we have seen above.

Axioms (A9)–(A15) are the existence axioms. They are very similar to the existential axioms of ZFC (null set, union, separation, pair, power set, infinity, replacement). (A16) is the regularity axiom postulated for *sets* (it says that no infinite chains $\ldots x_2 \in x_1 \in x$ are admissible for sets x). We do not provide the details of their translations into the language of ZFC since they were adapted from the standard ZFC axioms, but it is easy to see that their translations, in the sense of the above definition, can be proved in ZFC. We only remark that S^{**} has a *pseudo-pair* axiom instead of the pair axiom. This is obtained as in the usual formulation of the pair axiom, but by using $\equiv$ instead of $=$. In other words, for any two objects x and y there exists a quasi-set, denoted by $[x, y]$, which contains as elements all the objects indistinguishable either from x or y and nothing more. When $x \equiv y$, the quasi-set $[x, x]$ is written simply as $[x]$, which is the *pseudo singleton* of x. In quasi-set theory, it results that $[x, y]$, as well as $[x]$, may have more than two (respect., one) elements. After the translation, this of course does not happen, since (A10) turns to be a particular case of the pair axiom of ZFC.

Based on the axioms of S^{**} just mentioned, it is easy to see that in this theory there exists a 'copy' of ZFC, since the language of such a theory can be translated into the language of S^{**} in an obvious way (see [12], [26]). In particular, all the mathematical constructions performed in ZFC can be done also in S^{**}, as for instance the theory of cardinals as well as the theory of ordinal numbers, which result to be *sets* in S^{**}. So, we may consider the following concepts as definable in this theory: $ord(x)$ means the ordinal of x, $Card(x)$ says that x is a cardinal, $card(x)$ is the cardinal of x, while $\leq_E$ denotes the order among cardinals, defined as usual.

A natural question is to ask for the possibility of associating cardinals even to quasi-sets that are not sets. The problem is that since the concept of identity lacks sense for the m-atoms, the quasi-sets that contain such objects as elements cannot be (well-)ordered. As a consequence, cardinals for general quasi-sets cannot be defined in the standard way. Even so, we suppose that every quasi-set has a cardinal, which is called its *quasi-cardinal*; in order to formalize this idea, we have assumed that the language of S^{**} contains a primitive functional symbol of arity one, *qcard*, so that the term $qcard(x)$ intuitively denotes the quasi-cardinal of (the quasi-set) x. The axioms for *qcard* simply generalize the concept of cardinal for arbitrary quasi-sets, and for the present paper it is sufficient to remark that the translations of the axioms (A17)–(A21) for the concept of quasi-cardinal are theorems of ZFC (for details concerning axioms for *qcard*, see [26], [12]).

Due to the fact that m-objects cannot be labeled or named, it results that a quasi-set cannot be determined by its elements. Quasi-set theory is a kind of 'semi-extensional' theory in this sense, since what holds is a weaker form of the extensionality axiom which is stated as follows:

(A23) [Weak Extensionality]
$\forall_Q x \forall_Q y [\forall u (u \in x \rightarrow \exists v (v \in y \ \wedge \ u \equiv v)) \wedge$
$\forall u (u \in y \rightarrow \exists v (v \in x \ \wedge \ u \equiv v)) \wedge qcard(x) =_E qcard(y) \rightarrow x \equiv y]$.

This axiom entails interesting consequences. First of all, it can be proven that if an element z of a quasi-set x is 'exchanged' by a certain object w such that $w \equiv z$, then 'nothing happens with x at all' [26], [30]. That is, the original x and the 'new' quasi-set turn out to be indistinguishable by (A23). So, 'permutations' of indistinguishable objects are not 'observable', a fact which reflects in (quasi-) set-theoretical terms one of the most basic supositions of quantum theory. In order to say more, let us reason informally supposing m-objects denote elementary particles like electrons, protons, etc. Then, according to (A23), only those quasi-sets whose elements are indistinguishable may be in the relation $\equiv$. So, apparently we are unable to say, for instance, that a 'quasi-set' x which intuitively has two electrons and three protons as elements may be indistinguishable from 'another' quasi-set y also with two electrons and three protons as elements. But in this case we can use the quotient quasi-set $x/_{\equiv}$, and obtain $x/_{\equiv} \equiv y/_{\equiv}$; because the elements of $x/_{\equiv}$ and $y/_{\equiv}$ are indistinguishable in the sense of (A23). In spite of this, $\bigcup x/_{\equiv}$ is not indistinguishable from $\bigcup y/_{\equiv}$ (that is, x and y are not indistinguishable).

Quasi-set theory encompasses also versions for the axioms of choice and replacement, but these versions, when translated to the language of ZFC, coincide with their analogues in this theory.

3. RELATED QUESTIONS AND FURTHER PROBLEMS

Max Jammer calls our attention to the fact that according to Heisenberg the uncertainties concerning whether a particle does have a well determined position and velocity has an ontological nature, rather than an epistemic one ([24, p. 332]). Schrödinger's intuitions about the very basic components of matter are also closely connected with the idea of certain entities which he tryed to characterize (ontologically) as 'non-individuals' [39], that is, as entities to which the usual concept of identity cannot be applied [37, pp. 16–18] (see also [2], [13], [1]). More recent authors have also stressed that there are important problems of a logico-mathematical nature regarding the philosophical foundations of quantum mechanics, raised by the way elementary particles are considered. Among other examples, we could mention those concerning the use of names and descriptions in the microworld [9], [43], [33], so as those concerning semantics for the languages of microphysics [11], [7], and even the suggestion that quantum particles offer examples of vague objects (see [17], [21]). In essence, all these problems apparently arise from the supposition of the existence of a kind of entity which is quite distinct from the 'classical' physical object, to which even the term 'particle' seens inadequate, since this term resembles us an 'individuatable' thing.

Taking into account that the language of quasi-set theory is stronger than the language of ZFC, we might expect that in that theory we are able to express some of the fundamental facts about the quantum world. For instance, in S^{**} we can derive things like the existence of a subquasi-set of $[x]$ (the pseudo-singleton of x) that has quasi-cardinal 1. Such a quasi-set, which was termed the *strong singleton* of x in [26], stands intuitively for a term which designates some entity, not further specified, which has the property of being indistinguishable from x. The element of such a quasi-set may perhaps be regarded as a non-individual in the above sense; on these grounds, if the *quanta* are considered as entities such as quantum mechanics describes them, mainly in the second quantized approach, then it seems natural to use a theory like S^{**} as an alternative mathematical device for discussing the just mentioned philosophical intuitions on the 'real' quantum object (see [19]).

ACKNOWLEDGMENTS

The authors would like very much to thank M.L. Dalla Chiara and R. Giuntini for helpful suggestions.

Department of Philosophy
University of São Paulo
and
Research Group in the Mathematics Methods of Science
Federal University of Paraná

NOTES

[1] It has been claimed that 'quantum mechanics does not really have its own language' [32, p. 84].
[2] Jammer made reference to space-time descriptions only, but of course what he said may be generalized.
[3] For a more recent discussion on the 'ontic' characteristics of quantum objects, see [17], inclusive for references.
[4] The literature on this particular topic is wide; some papers are [14], [15], [16], [35], [36], [34].
[5] Schrödinger's position regarding the concept of 'object' is treated in [2].
[6] For some explanations on the main features of what could be called a 'logic of non-individuality', see [30], [17].
[7] This question is subtle and demands careful analysis, which is outside the scope of this paper. See the references, in particular [11], [26], [12].
[8] A distinct proof of this fact was presented for some quasi-set theories in [29]. The converse of this result is the 'trivial' side of the proof of the equiconsistency between S^{**} and ZFC and was presented also in [26].
[9] We use $x, y, \ldots$ as individual variables in both theories.
[10] In the present case, the elements of X play the rôle of the *Urelemente* in $\mathcal{Q}$: the elements of $\widehat{m}$ act as the m-atoms while those of M act as the atoms in ZFU.
[11] The concept of *rank* is standard.

REFERENCES

[1] Ben-Menahem, Y.: 1992, 'Struggling with Realism: Schrödinger's Case', in Bitbol, M., Darrigol, O., eds., *Erwin Schrödinger: Philosophy and the Birth of Quantum Mechanics.* Frontières, Paris; 25–40.
[2] Bitbol, M.: 1992, 'Esquisses, Forme et Totalite (Schrödinger et le Concept d'Objet)', in Bitbol, M., Darrigol, O., eds., *Erwin Schrödinger: Philosophy and the Birth of Quantum Mechanics.* Frontières, Paris; 41–79.
[3] Bohr, N.: 1937, 'Causality and Complementarity', *Philosophy of Science* **4(3)**, 289–298.
[4] da Costa, N.C.A., French, S., Krause, D.: 1996, 'Some Remarks on Sortal Logics and Physics', in Etcheverría, J., de Lorenzo, J. y Peña, L., eds., *Calculemos ... Matemáticas y libertad.*, Madrid, Trotta; 159–172.

[5] da Costa, N.C.A., Krause, D., French, S.: 1992, 'The Schrödinger Problem', in Bitbol, M., Darrigol, O., eds., *Erwin Schrödinger: Philosophy and the Birth of Quantum Mechanics*. Paris, Frontières; 445–460.

[6] da Costa, N.C.A., Doria, F.A.: 1992, 'Suppes Predicates for Classical Physics', in Echeverria, J., Ibarra, A., Mormann, T., eds., *The Space of Mathematics*. Berlin and New York, Walter de Gruyter; 168–191.

[7] da Costa, N.C.A., Krause, D.: 1994, 'Schrödinger Logics', *Studia Logica*, **53(4)**, 533–550.

[8] da Costa, N.C.A., Krause, D.: 1997, 'An Intensional Schrödinger Logic', *Notre Dame J of Formal Logic*, **38(2)**.

[9] Dalla Chiara, M.L.: 1985, 'Some Foundational Problems in Mathematics Suggested by Physics', *Synthese*, **62**, 303–315.

[10] Dalla Chiara, M.L.: 1987, 'An Approach to Intensional Semantics', *Synthese* **73**, 479–496.

[11] Dalla Chiara, M.L., Toraldo di Francia, G.: 1993, 'Individuals, Kinds and Names in Physics', in Corsi, G. et al., eds., *Bridging The Gap: Philosophy, Mathematics, Physics*. Kluwer, Dordrecht; 261–283.

[12] Dalla Chiara, M.L., Giuntini, R., Krause, D.: 1998, 'Quasiset Theories for Microobjects: A Comparision', in Castellani, E., ed., *Interpreting Bodies: Classical and Quantum Objects in Modern Physics*. Princeton University Press, Princeton, 142–152.

[13] Darrigol, O.: 1992, 'Schrödinger's Statistical Physics and Some Related Themes', in Bitbol, M., Darrigol, O., eds. *Erwin Schrödinger: Philosophy and the Birth of Quantum Mechanics*, Frontières, Paris 237–276.

[14] French, S.: 1989, 'Identity and Individuality in Classical and Quantum Physics', *Australasian Journal of Philosophy*, **67**, 432–446.

[15] French, S.: 1989, 'Why the Principle of the Identity of Indiscernibles is not Contingently True Either', *Synthese*, **78**, 141–166.

[16] French, S., Redhead, M.L.G.: 1988, 'Quantum Physics and the Identity of Indiscernibles', *British Journal for the Philosophy of Science*, **39**, 233–246.

[17] French, S., Krause, D.: 1995, 'Vague Identity and Quantum Non-Individuality', *Analysis*, **55**(1), 20–26.

[18] French, S., Krause, D.: 'Opaque Predicates, Veiled Sets and their Logic', forthcoming, in Carnielli, W., D'Ottavianno. I, eds, *Proceedings of the XIth Brazilian Conference on Mathematical Logic*, Salvador, May 1996.

[19] French, S., Krause, D.: 1999, 'The Logic of Quanta', in Cao, T.L., ed., *Conceptual Foundations of Quantum Field Theory*. Cambridge University Press, 324–342.

[20] French, S., Krause, D.: 1996, 'Quantum Objects are Vague Objects', *Sorites*, **6**, 21–23.

[21] French, S., Krause, D., Maidens, A.: 'Quantum Vagueness', preprint.

[22] Heisenberg, W.: 1949, *The Physical Principles of the Quantum Theory*. New York, Dover, 1949. (First published by the University of Chicago Press, 1930).

[23] Heisenberg. W.: 1958, *Physics and Philosophy*. London, Allen & Unwin.

[24] Jammer, M.: 1966, *The Conceptual Development of Quantum Mechanics* New York, McGraw Hill.

[25] Krause, D.: 1991, 'Multisets, Quasi-sets and Weyl's Aggregates', *The Journal of Non–Classical Logic*, **8(2)**, 9–39.

[26] Krause, D.: 1992, 'On A Quasi-set Theory', *Notre Dame Journal of Formal Logic*, **33(3)**, 402–411.

[27] Krause, D.: 1993, 'The Theories of Quasi-sets and ZFC are Equiconsistent', in Carnielli, W.A., Pereira, L.C., eds., *Logic, Sets and Information*, Proceedings of the Xth Brazilian Conference on Mathematical Logic, Unicamp, Col. CLE, Vol. 14, 1995, 145–155.

[28] Krause, D.: 1994, 'Non-reflexive Logics and the Foundations of Physics', in Cellucci, C., Di Maio, M.C., Roncaglia, G., eds., *Logica e Filosofia della Scienza: Problemi e Prospettive*, Atti del Congresso Triennale della Societa Italiana di Logica e Filosofia della Science, Lucca, 7–10 Gennaio 1993. Pisa, Edizioni ETS, 1994, 393–405.

[29] Krause, D., 1996, 'Axioms for Collections of Indistinguishable Objects', *Logique et Analyse*, 153–154, 69–93

[30] Krause, D., French, S.: 1995, 'A Formal Framework for Quantum Non-Individuality', *Synthese*, **102**, 195–214.

[31] Manin, Yu.I.: 1976, 'Problems of Present Day Mathematics: I (Foundations)', in Browder, F.E., ed., *Proceedings of Symposia in Pure Mathematics*, **28** American Mathematical Society, Providence, p.36.

[32] Manin, Yu.I.: 1977, *A Course in Mathematical Logic*. New York, Springer-Verlag.

[33] Mittelstead, P.: 1985, 'Constituting, Naming and Identity in Quantum Logic', in Mittelstaed, P., Stachow, E.W., eds. *Recent Developments in Quantum Logics*, Mannheim, 215–234.

[34] Post, H.: 1963, 'Individuality and Physics', *The Listener*, **70**, 534–537. Reprinted in *Vedanta for East and West*, **32**, 1973, 14–22.

[35] Redhead, M., Teller, P.: 1991, 'Particles, Particle Labels and Quanta: The Toll of Unacknowledged Metaphysics', *Foundations of Physics*, **21(1)**, 43–62.

[36] Redhead, M., Teller, P.: 1992, 'Particle Labels and the Theory of Indistinguishable Particles in Quantum Mechanics', *Br J Phil Sc*, **43**, 201–218.

[37] Schrödinger, E.: 1952, *Science and Humanism*. Cambridge, Cambridge University Press.

[38] Schrödinger, E.: 1953, 'What is Matter?', *Scientific American*, 52-57.

[39] Schrödinger, E.: 1957, *Science Theory and Man*. London, Allen and Unwin.

[40] Schrödinger, E.: 1992, *Physique Quantique et Représentation du Monde*. Paris, Seuil.

[41] Toraldo di Francia, G.: 1978, 'What is a Physical Object?', *Scientia*, **113**, 57–65

[42] Toraldo di Francia, G.: 1981, *The Investigation of the Physical World*. Cambridge, Cambridge University Press.

[43] Toraldo di Francia, G.: 1985, 'Connotation and Denotation in Microphysics', in Mittelstaed, P., Stachow. E.W., eds., *Recent Developments in Quantum Logics*, Mannheim, 203–214.

[44] Weyl, H.: 1949, *Philosophy of Mathematics and Natural Science*. Princeton, Princeton University Press.

MARCO DEL SETA AND MAURICIO SUÁREZ

NON-IDEAL MEASUREMENTS AND PHYSICAL POSSIBILITY IN QUANTUM MECHANICS

Albert and Loewer [1,2,3] have posed an important difficulty for the modal interpretations advocated by Kochen [12], Healey [10] and Dieks [20] (KHD interpretations). They have argued that KHD interpretations cannot cope with physically realistic measurement situations. One of us (M.S. in [19]) has contested some of the assumptions underlying Albert and Loewer's argument. In this paper we review this and other criticisms of Albert and Loewer's assumptions, and we argue that a sound argument for Albert and Loewer's conclusion against KHD modal interpretations *can* be made, without relying on all of Albert and Loewer's assumptions.

The debate arises out of the need to provide a reliable solution to the problem of quantum measurement. We are looking for a way to model the process of measuring properties of physical quantum systems and a good starting point seems to be to consider two quantum systems, representing the physical entity on which the measurement is to be made and the measurement apparatus, modelled by two Hilbert spaces $\mathcal{H}_{\mathbb{S}}$ and $\mathcal{H}_{\mathbb{M}}$, respectively. These spaces will be tensored together as a standard quantum mechanical composite system, and the initial state of the object-apparatus system will be evolved to a final state after the measurement. What satisfactory conditions will we then impose on the operator which evolves the state and mathematically characterises the measurement?

The most uncontroversial necessary conditions for such an operator $W^{\mathbb{S},\mathbb{M}}$ are the so-called *probability reproducibility condition* and the *calibration condition* respectively (see, for instance, Busch, Lahti and Mittelstaedt [4]):

(1) $\langle\varphi|P_{[\varphi_i]}\varphi\rangle = \langle W^{\mathbb{S},\mathbb{M}}(\varphi\otimes\psi_0)|(I_{\mathbb{S}}\otimes P_{[\psi_i]})(W^{\mathbb{S},\mathbb{M}}(\varphi\otimes\psi_0))\rangle$

(2) $W^{\mathbb{S},\mathbb{M}}(\varphi_i\otimes\psi_0) = \tilde{\varphi}_1\otimes\psi_1$

for any vector $\varphi\in\mathcal{H}_{\mathbb{S}}$, observable specified by the complete orthonormal set $\{\varphi_i\}$ on $\mathcal{H}_{\mathbb{S}}$ to be measured, initial state $\psi_0\in\mathcal{H}_{\mathbb{M}}$ of the apparatus, and apparatus observable specified by the complete orthonormal set $\{\psi_i\}$ on $\mathcal{H}_{\mathbb{M}}$. The most general linear operator on $\mathcal{H}_{\mathbb{S}}\otimes\mathcal{H}_{\mathbb{M}}$ satisfying such condition has the form

M.L. Dalla Chiara et al. (eds.), Language, Quantum, Music, 183–195.

$$W(\cdot) := \sum_{i=1}^{n} \langle \varphi_i \otimes \psi_0 | (\cdot) \rangle (\tilde{\varphi}_i \otimes \psi_i)$$

This operator is a contractive partial isometry on the subspace $\mathcal{H}_S \otimes [\psi_0] := \{\varphi \otimes \alpha\psi_0 : \varphi \in \mathcal{H}_{\mathbb{S}}, \alpha \in \mathbb{C}\}$ of $\mathcal{H}_{\mathbb{S}} \otimes \mathcal{H}_{\mathbb{M}}$ whose null space is $\mathcal{H}_{\mathbb{S}} \otimes [\psi_0^{\perp}] := \{\varphi \otimes \alpha\psi_0^{\perp} : \varphi \in \mathcal{H}_{\mathbb{S}}, \alpha \in \mathbb{C}\}$.

The next usual condition is that the operator should be unitary. This leads us to look for unitary completions of the partial isometry W, of which there are, for $\dim \mathcal{H}_{\mathbb{M}} \geq \dim \mathcal{H}_{\mathbb{S}} > 2$, infinitely many. From here onwards we denote by W any such completion of the partial isometry.

At this point the measurement problem can be stated. The operator W maps an initial state of the object-apparatus system to a final state which shows interference between pointer position eigenstates of the apparatus observable. *If* we assume the eigenvalue-eigenstate link for value attribution (EEL), we cannot attribute a definite value to the pointer position of our measuring apparatus and hence have failed to model the real measurement processes of quantum physics, where values are actually measured.

(EEL) states that a system in a pure state φ has a definite value for an observable A if and only if $A(\varphi) = \alpha\varphi$, that is if φ is an eigenstate of the observable A. For an ideal measurement interaction we have that $W(\varphi \otimes \psi_0) = \sum_i \alpha_i \varphi_i \otimes \psi_i$, with φ_i eigenstates of the observable A one is trying to measure on the object system and ψ_i eigenstates of the pointer observable for the apparatus system. Given this overall state for the combined system, the reduced states for the component subsystems are given by two mixtures: $\sum_i |\alpha_i|^2 P_{[\varphi_i]}$ for the object system, and $\sum_i |\alpha_i|^2 P_{[\psi_i]}$ for the measuring apparatus. We could try to give the ignorance interpretation to either mixture, and assume that the state of the system really is one of $P_{[\varphi_i]}$, or the state of the apparatus really one of $P_{[\psi_i]}$. However, this move is well known to contradict the original supposition that the state of the combined system is a superposition, and hence a pure state. For if the reduced states are really pure (and this is what the ignorance interpretation entails) then the state of the combined system is a mixed (non-pure) state, which contradicts our original assumption. So, by reductio, even though the reduced state for the apparatus system is $\sum_i |\alpha_i|^2 P_{[\psi_i]}$, a mixture over eigenstates of the pointer observable, we are not warranted in giving an ignorance interpretation to this mixture and say that the apparatus is definitely in one of the states $P_{[\psi_i]}$. But then we can't appeal to (EEL) and say that the pointer definitely has as value one of the eigenvalues associated with one of the eigenstates ψ_i.

1. THE MODAL INTERPRETATION AND ITS CRITICISM

A host of interpretations try to avoid the problem of measurement by renouncing (EEL). Thus, even though the combined system is in a superposition over pointer eigenstates (more precisely eigenstates of the observable

$I \otimes A$), we assume it is possible to assign a definite value to the pointer observable A. These interpretations originate from Everett's work in the 1950's and include many-world interpretations and modal interpretations.

KHD modal interpretations claim that, given a state T, a density operator in the combined system $\mathcal{T}^1_+(\mathcal{H}_{\mathbb{S}} \otimes \mathcal{H}_{\mathbb{M}})$, the possible values for the individual systems are selected by the properties picked by the projection operators which diagonalise the reduced density operators in the spaces $\mathcal{T}^1_+(\mathcal{H}_{\mathbb{S}})$ and $\mathcal{T}^1_+(\mathcal{H}_{\mathbb{M}})$, obtained by partial trace operations on the state of the combined system.

In particular for every pure state in the composite space $\mathcal{H}_{\mathbb{S}} \otimes \mathcal{H}_{\mathbb{M}}$ there is a unique polar decomposition of such state (up to multiplicities), so the basis of this decomposition is formed of vectors from the two subsystems which are themselves pairwise orthogonal. Calculating such decomposition yields the very properties which diagonalise the reduced density operators and thus enables one to 'read off' the properties of the two subsystems and the associated values with ease.

Albert and Loewer's argument against modal interpretations has two parts. Firstly, they argue that physically possible measurement interactions are non-ideal. An ideal measurement interaction maps a vector $\varphi_i \otimes \psi_0$ to a special case of the vector $\tilde{\varphi}_i \otimes \psi_i$, namely $\varphi_i \otimes \psi_i$. However, argue Albert and Loewer, it is impossible for measurement devices to record outcomes always in such a perfect way: $\varphi_i \otimes \psi_0$ will be mapped to $\varphi_i \otimes \psi_i$ most of the time, but sometimes our apparatus will misfire, and map the initial state to $\varphi_j \otimes \psi_i$, $j \neq i$. Or more generally the interaction will be such as to yield a final state that is a superposition of pure states of the form $\tilde{\varphi}_i \otimes \psi_i$ such that the $\tilde{\varphi}_i$'s are not pairwise orthogonal.

Secondly, Albert and Loewer argue that modal interpretations cannot cope with non-ideal measurement interactions. If the biorthogonal decomposition algorithm is applied to such final state then the states of the apparatus system will not pick out the right pointer observable. Hence KHD modal interpretations will fail to give values to the pointer position observable in any real measurement.

2. OBJECTIONS TO THE ARGUMENT

There have been roughly speaking two kinds of responses to Albert and Loewer's objection to the modal interpretation. The overwhelmingly most common response has been to accept the first part of Albert and Loewer's argument, to the effect that ideal interactions are physically impossible, or at least that they are very rare, and to reject the second part of the argument, namely that the modal interpretation cannot cope with non-ideal interactions. Albert and Loewer's latter claim seems to rely upon the assumption that the correct model for imperfect real laboratory measurements is the generalisation of interactions of object system plus measuring apparatus from the ideal to the non-ideal case. However, alternative models of real measurement interactions have been proposed which are not prone to Albert and Loewer's criticisms. These models take the physical environment into account, besides the object system and the apparatus, and consequently expand the tensor product space. The application of the biorthogonal decomposition then becomes a substantive issue, for there are different ways to partition the tensor product space, and the success of the modal interpretation will now depend on the partition that is chosen. As long as these models can be shown to yield empirically indistinguishable predictions as regards measurements of the pointer position observable (and the jury is, arguably, still out on this) the modal interpretation can have a way out. Healey puts the question thus: "What determines whether a particular model of a physical phenomenon within a particular physical theory is appropriate?" [11, p. 51]

Another possible response to Albert and Loewer is to object directly to the first part of their argument. In a previous paper ([19]) one of us has contested the claim that ideal measurements are physically impossible. Why should we think that ideal measurements are physically impossible? We know this, according to Albert and Loewer, because the set of all ideal measurements has measure zero in the set of all possible measurement operators for the measured observable: "on natural measures the measure of the set of Hamiltonians which correspond to ideal measurements is 0" [3, p. 301]. Hence the probability of an ideal measurement taking place is zero.

In particular Albert and Loewer can reason as follows. It turns out that the set I of all unitary measurement operators that yield 'good' modal interactions, namely those for which $\tilde{\varphi}_i \otimes \psi_i$ are such that the $\tilde{\varphi}_i$'s are pairwise orthogonal, forms a set of co-dimension strictly greater than zero in the set of all possible measurement operators. It is then possible to parametrise this set in such a way that the set will need less variable parameters than the whole set of measurement operators. This we can think of in a similar way to the parametrisation of a sphere in $\mathbb{R}^3$ by two angular variables, with the radius variable being held constant. We can then treat the sphere as a set in a cartesian product space of the spaces of all possible angular variables for 'latitude' Θ and 'longitude' Φ, and of all possible radius lengths R, the sphere being a cartesian product $\{r_0\} \times \Phi \times \Theta$ of a singleton set in radius space by Θ and Φ. Provided singleton sets are given measure zero, in the theory of measures on product spaces such sets are known to have measure zero, as the measure of the product set is equal to the (real number) product of the measures on the factor sets: $\mu_{\mathbb{R}}{}^3 = \mu_R \cdot \mu_\Theta \cdot \mu_\Phi$. We assume, for the sake of argument, that the set I of all such unitary measurement operators can be parametrised so that there will be a section of the set consisting of a singleton set. Therefore the set I has measure zero.

However, to argue along this line Albert and Loewer need to make two substantive assumptions, namely that it is *physically appropriate* to treat the set of all possible operators as a product space, with measures defined on all the individual spaces, and that singleton sets on the individual spaces have measure zero. These assumptions are necessary for the measure zero claim to be provably true, rather than true by definition. Without the assumption that the set of all measurement operators is a product space we are not warranted in claiming that the set of ideal measurements has measure zero, as this set, far from being a singleton set, contains infinitely many operators all realising, under current theories of quantum measurement, the same measurement interaction and attribution of measure 0 is a matter of definition. On the other hand we also need the assumption that singleton sets have measure zero in the factor spaces, otherwise our product factorisation of the set will clearly not be sufficient to prove that the measure of the set I of operators is 0.

Now, as Halmos has put it (p 152)

> A measure space might appear as three-dimensional in one context and two-dimensional in another; if for instance, $n = 3$ [the 'dimension' of a product space], then we may view X [the whole space on which we define the measure] as $X_1 \times X_2 \times X_3$ or as $X_0 \times X_3$, (where $X_0 = X_1 \times X_2$). [9, p. 152]

Halmos is stressing that the dimensionality of a measure space is not taken to be an intrinsic structural property of the space, but merely serves as a reminder of how we have built the space, and the measure on this space as a product of measures on the factors. In other words, dimensionality is conventional from the measure theoretic point of view.

A probability measure is a totally finite real valued set function μ on a set X such that $\mu(X) = 1$. We can construct a measure on a set X by taking a function $f : \mathcal{P}(X) \to \mathbb{R}$ which is real valued and non-negative, for instance in the following way. Take $X = \mathbb{R}^2$ and define a measure μ on it such that $\mu\{x\} = 12\chi_{\{0,1\}} + k2e^{-x^2} \cdot \chi_{\{x,0\}}$. Here x is a subset of $\mathbb{R}^2$ (of the appropriate kind), $\chi_{\{0,1\}}$ is the characteristic function over the singleton set $\{0, 1\}$, $\chi_{\{x,0\}}$ is the characteristic function over the real line obtained by setting the y coordinate to be 0 and $k2e^{-x^2}$ is normalised so that $\mu\{\mathbb{R}^2\} = 1$. This measure is not expressible as a product measure, it might seem very pathological, but it is nevertheless a probability measure on $\mathbb{R}^2$. Presumably whether this is an appropriate measure for $\mathbb{R}^2$ in a physical context, or whether so-called natural measures are physically appropriate, is a matter for empirical investigation.

The topic here has some similarity with discussions of measure theoretic problems in classical and statistical mechanics. Earman, for example, when discussing the problem of singularities of solutions of the equation of motion in Newtonian gravitational theory of point mass particles, remarks that, if all singularities are due to collisions of point particles, a qualified doctrine of determinism might hold. This might seem reinforced by the fact that this is almost always true in the sense that the set of initial conditions which lead in a finite time to collisions is of (Lebesgue) measure zero. But

> There are, however, some caveats about [this]. [. . .] measure zero need not imply either insignificant or ignorable. We would not judge the set of initial conditions giving rise to collisions to be insignificant if, for example, it proved to be dense within the set of states that eventuate in strong interactions (in some appropriate sense) among the particles. Nor would we regard the measure

zero set as ignorable if it loomed large within the range of cases we regard, for whatever reason, to be physically interesting [7].

Or as Sklar has put it,

any proposal to utilise some mathematical concept, whether measure theoretic or topological, to express the fundamental posits of one's statistical mechanical theory is going to require backing up in the form of a careful exploration of what it is about the physical world, and possibly what it is about the manipulability of that world by nature including the natural things called experimentalists, that makes the mathematical formalism the appropriate one to capture our physical idea [18].

There are, arguably, some differences. Sklar also points out that, in an attempt to fill gaps in measure theoretic arguments for the foundations of statistical mechanics with appeals to topological notions, we use a topology linked to standard distance measures. There is an intuition that this distance measure is 'intrinsic' to the basic phase-space representation, while the choice of some measure or other over an algebra of subsets of phase space points is an 'imposition'. This is presumably because we retain the intuition that phase space points are but a short step beyond the 3-dimensional space in which we live.

The physical intuition about what the appropriate probability measure might be for the set of all possible unitary measurement operators is much more unclear than in discussions of these issues in statistical mechanics, and the connection with realistic geometric representations is harder to establish. The Hilbert space is of course a far more abstract structure, but giving a strongly 'visual' interpretation to the Hilbert space and the operators acting on it yields no automatic justification for adopting a probability measure on the space of operators which will assign measure zero to ideal measurement operators. The justification for the relevant probability measure is problematic, as Sklar's discussions amply exemplify, even in the context of measures over sets of particle trajectories in physical space.

Consider the case of a space representing all the possible orientations of an unbiased die with respect to an observer perpendicularly looking at the die on a plane. We can specify the possible orientations of the die through two angular parameters and identify them with points on a sphere, providing an intuitive picture of this space. What probability measure shall we define for the question as to what orientation will arise when a die is thrown on the plane? I think most of us will agree that on the basis of past physical experience it would make sense to adopt as measure the characteristic function over a set of the obvious six possible orientations, divided by $^1\!/_6$. This does not tally up with 'natural' distance measures, but it is still the probability measure that is physically most warranted in this case.

We see then that both assumptions behind Albert and Loewer's appeal to measure theory are questionable. This further strengthens the analysis that Suárez has given of the question [19]. He there claimed that Albert and Loewer's argument, if supported by an appeal to measure-theoretic properties, requires that the Hilbert space structure be interpreted realistically. Indeed it is now clear that even this is not sufficient. For only one particular measure makes sense of Albert and Loewer's claim, and that measure requires justification on its own right.

The second response to Albert and Loewer is then the following: the claim that the set of ideal measurements of an observable has measure zero is an empty claim unless some good *physical warrant* is given for the measure in question. In the absence of such good warrant one can infer measure-zero from probability zero, and one can infer probability zero from physical impossibility, but not vice versa: the inference of physical impossibility from measure-zero is flawed.

To sum up, there are two very different objections to Albert and Loewer's argument, and hence there are two different strategies to restore confidence in modal interpretations. The response that has become standard is to argue that the probability distribution over the measurement observable will typically underdetermine the correct model for an imperfect real measurement interaction; so, different models of any real measurement will be possible; and which model is chosen will be relevant to the success of modal interpretations. Another response points out to the prior fact that we need to back up the claim that real measurement interactions are non-ideal with physics, not measure theory.

3. PHYSICALLY POSSIBLE NON-IDEAL MEASUREMENTS

The point that Albert and Loewer make stands, though, as follows: if we can find an example of a real measurement situation which we can all agree is necessarily modelled by a non-ideal measurement operator of a certain kind, then this will pose a problem for the modal interpretation of quantum mechanics. Such an example would constitute a genuinely empirical challenge to modal interpretations. Modal interpretations will not fail *tout court* of course. The example cannot show that ideal measurements are physically impossible, nor even that they are measure-zero. It can only show that KHD modal interpretations fail to ascribe values to the correct observable in the particular physical situation. This section will show that there is at least one such example, namely destructive photon measurements.

These measurements are part of the folklore of philosophy of quantum theory. Henry Margenau evoked them in the context of a famous argument against the projection postulate [15, p. 377]

> ...perfectly good measurements in atomic physics ...may 'kill' or annihilate a system, and we may be pardoned if we refuse to discuss the eigenstates of nonentities.

A very good example of a destructive measurement was given by Margenau himself, namely the recording of a photon by a photographic plate as a means of measuring its position. Thus, Margenau colourfully claimed [14, p. 30] a blackened grain is at once position record and tomb stone of the photon, and [...] projection into an eigenstate has certainly not taken place.

In general the detection of a photon's polarisation (which is what must happen if a measurement is to have taken place according to Margenau), by light detectors, photographic plates and the like, result in the absorption of the photon and thus the effective disappearance of it. It will not do to claim that we measure the polarisation of a photon by placing the polariser in the path of a photon. For Margenau this is what we should call preparation; measurement must surely involve detection and, however the two processes of preparation and measurement might be similar, indeed virtually the same in classical mechanics, they certainly aren't the same in quantum physics. The sorts of measurements that Margenau discusses are indeed routine laboratory measurements. On '50's technology they would have worked by having a photomultiplier absorbing the photon, and the resultant energy increase would have triggered the release of an electron. A 'chain reaction' mechanism would have led to a signal sufficiently strong to be recorded by a mechanical device increasing the value for number of photon detected by one, on dials pretty similar to a car's mileage counter. By this time the photon would be well and truly dead.

Margenau's objective in discussing destructive measurements was, first of all, to criticise the projection postulate, which he described as follows[1]:

It is often said that, when a single measurement of an observable Q is made and the result is found to be q, the state is then *known*, and known to be ψ_q, the eigenstate belonging to the measured value.

Margenau noticed that, in the case of destructive measurements, the projection postulate entails the existence of eigenstates of non-entities. If a successful measurement is carried out of an observable Q on a quantum system (perhaps by means of a measurement of the correlated pointer position observable on an interacting measuring device) and a value q of that observable is found, then the projection postulate entails that the state of the quantum system is the corresponding eigenstate. However, if the quantum entity has been destroyed by the act of measurement, this turns out to be the eigenstate of a nonentity; and, as Margenau points out, we should "refuse to discuss the eigenstates of nonentities".

Margenau's argument would seem to cut against us too. We want to argue that destructive measurements are an instance of typical non-ideal laboratory measurements. Thus we would like to show that there is a final state of the composite (and hence an associated reduced state of the system) that exhibits the result of a non-ideal interaction. Consider the case where a photon interacts with an apparatus in the ready state. A measurement interaction is set up – one that obeys the probability reproducibility condition (1). The outcome of the measurement is a faithful reading by the apparatus of the values of the observable to be measured on the photon; but it also results in the destruction of the photon. We may characterise a destructive measurement interaction D as the unitary completion of an operator like

$$W(\varphi \otimes \psi_0) = \sum_{i=1}^{n} \alpha_i \varphi_0 \otimes \psi_i = \varphi_0 \otimes \sum_{i=1}^{n} \alpha_i \psi_i \tag{1}$$

where the initial state of the photon is $\varphi = \sum_{i=1}^{n} \alpha_i \varphi_i$, a superposition over the eigenstates of the required observable Q. (The interaction operator is characterised by the calibration condition mapping every element of the original superposed state of the combined system, $\varphi_i \otimes \psi_0$, into the element of the final state of the combined system $\varphi_0 \otimes \psi_i$.)

Margenau would argue that the final state in equation (1) fails to describe the physical situation, because it does not account for the fact that the measurement has destroyed the photon. The projection postulate would map the above final state to the density operator:

$$\sum_{i=1}^{n} |\alpha_i|^2 P_{[\varphi_0 \otimes \psi_i]}.$$

which enables us to say that 'the photon is in state φ_0'. The photon however is in no such state, or in any other state, for the photon *exists no more.*

So if Margenau is right, destructive measurements should not really be modelled by (1); and hence they need not constitute an example of measurement interactions that everyone will agree are best modelled as non-ideal. Therefore we must argue, first, that *pace* Margenau, equation (1) correctly models a destructive measurement; and we need to show, second, that the interaction described by (1) is indeed a non-ideal interaction.

Fred Kronz [13] has made clear that the supporter of the projection postulate can contest Margenau's claim. Kronz points out two important things. First, we need not restrict ourselves to think of the projection postulate as a claim that is made of a single system, like a photon or an electron. It is quite possible to state a form of the projection postulate for composite systems: this claims that the non-linear 'projective' transition does not happen at the level of a single system in a measurement situation, but rather at the level of the combined system $\mathcal{H}_S \otimes \mathcal{H}_M$. And what will this 'projective' transition do? It will take the final state of the system as it has evolved through the measurement interaction into a density operator which will precisely satisfy the objectification requirement for the pointer observable, thus allowing an ignorance interpretation of the reduced density operator characterising the pointer system *regardless of what the state of the measured system is at this stage.* So projection here becomes a global phenomenon rather than one characteristic of individual systems.

Such a generalisation of the projection postulate is then coupled by Kronz with the observation that in fact, in the case of destructive measurements, there is no substantive argument to stop us from actually having a no-photon state in the Hilbert space representing an instance of, say, photon absorption by an atom. In other words φ_0 need not represent a state of the photon. But isn't it the case that, in classical mechanics, a state stands for the complete set of values, at a given time, of the dynamic properties of an entity? The entity referred to (usually a point particle) is identifiable independently of the description of the physical situation surrounding it, and of the forces acting upon it.

There is no reason to suppose that exactly the same notion figures in quantum mechanics. A state may represent a contextualised property of the overall physical situation. Perhaps this is a more instrumentalistic rendition of what a 'state' is, in terms of its role in the theory. The state is a formal tool in the prediction of values of certain quantities to be measured in some physical set-up; it is not to be thought of as 'belonging' to a particular entity.

Or, alternatively, one could insist that the same classical notion of state applies in quantum mechanics too, but expand the concept of an 'entity' to cover all kinds of physical set-ups. Such quantum mechanical 'entities' would not be identifiable independently of the description of the overall physical situation; instead, they would be *constituted* by the physical situation. In discussions of holism and non-supervenience of EPR systems, the entangled composite system is often presented as an instance of such quantum mechanical 'entity'.

Margenau must be assuming that the quantum state is, as its classical counterpart, a catalogue of the physical values of properties of an existing entity. It need not be – quantum state can be said to describe a set of contextual properties of the problem-situation. Indeed in the branch of quantum physics where the photon is treated, namely QED, not only is the absence-of-photons state, the 'vacuum' state, explicitly postulated, it generates some of the more interesting and unusual features of the theory. It appears that we are entitled to discuss the eigenstates of nonentities after all.

For our purposes we are interested in the operator D. This operator most definitely represents a kind of non-ideal measurement as there is only one feasible state that will denote absence-of-photons. The biorthonormal decomposition theorem fails to yield the desired results: if the initial state is $\varphi = \sum_i \alpha_i \varphi_i$, the final state will in fact be $\sum_i \alpha_i \varphi_0 \otimes \psi_i$, which is a factorised state, and the biorthogonal decomposition will pick out properties φ_0 and $\sum_i \alpha_i \psi_i$ as the possessed properties for system and apparatus respectively.

The operator we discuss is formally a special case of a class of operators considered in the literature as possible candidates for measurement operators by Fine amongst others[2] [8].

Contrary to the Albert and Loewer example, this operator does not model a measurement process which is 'inefficient' in its recordings, but rather a measurement which is intrinsically non-ideal. And we can 'piggy-back' on Kronz's work: his arguments show that if we want to respond to Margenau's attack against the whole of the quantum theory of measurement we better accept that (1) correctly models destructive measurements. So we have now landed a genuine empirical difficulty for modal interpretations, one which cannot be solved by claims that W models inefficiency of the measurement in the wrong way; such considerations are besides the point here.

Our example of a non-ideal interaction has real force against the first set of responses to Albert and Loewer's criticism. In particular it is worth mentioning that Bub's response to the Albert and Loewer criticism, a variant of the 'wrong model' response, can not work against this example. Bub [5] claims that Albert and Loewer's model of inefficient measurements can be interpreted as such a model only by assuming that the probabilities that the states of the composite system codify are in fact classical probabilities over actual but unknown values for the instrument readings. But this is precisely what we can't assume at the outset when interpreting quantum states in the context of the measurement problem.

However, our example of destructive measurement interactions has no such problem: that the non-ideal destructive measurement is a genuine model of an actual physical phenomenon does not depend on assigning classical probabilities of an instrument making mistakes on the basis of the state's encodings. Furthermore, the final state in the evolution carried through by the destructive measurement is not an entangled state: (1) can be rewritten as $\omega = \varphi_0 \otimes \sum_{i=1}^{n} \alpha_i \psi_i$. Bub's construction of the observables which have actual values for the system in consideration includes the apparatus observable R, characterised by an orthonormal set $\{\psi_i\}$, times the identity operator on the measured system, namely the observable $\mathbb{I} \otimes R$, as well as observables compatible with the projection $P_{[\omega]} \in \mathcal{T}_{+}^{1}(\mathcal{H}_{\mathbb{S}} \otimes \mathcal{H}_{\mathbb{M}})$. The latter, though, can be written as

$$P_{[\varphi_0]} \otimes P_{[\sum_{i=1}^{n} \alpha_i \psi_i]},$$

with

$$P_{[\varphi_0]} \in \mathcal{T}_{+}^{1}(\mathcal{H}_{\mathbb{S}})$$

and

$$P_{[\sum_{i=1}^{n} \alpha_i \psi_i]} \in \mathcal{T}_{+}^{1}(\mathcal{H}_{\mathbb{M}}).$$

This implies that the projector $P_{[\sum_{i=1}^{n} \alpha_i \psi_i]}$ will guarantee the presence in the partial algebra of observables which have actual values of observables of the form $\mathbb{I} \otimes O$, M-observables in Bub's own terminology, which will be incompatible with the observable $\mathbb{I} \otimes R$, given that the projection $P_{[\sum_{i=1}^{n} \alpha_i \psi_i]}$ is not orthogonal to the ψ_i's, the eigenstates of the observable R.

Summing up, destructive measurements seem to satisfy the following desiderata:

- They are actual, fairly common measurements in quantum physics;
- They can be modelled within the traditional framework for quantum measurements;
- They are essentially non-ideal, in that the final states of the measured system are not pairwise orthogonal – in fact, if the process is to be modelled in a satisfactory manner, they must be the same state.

Of course, it is of no use to reappropriate Albert and Loewer's measure-theoretic arguments to deny plausibility for this kind of measurements. It certainly is the case that the co-dimensionality of unitary operators satisfying a condition such as 1, on page , is larger than that of ideal operators, so that there are even 'fewer' of them. But this, as explained before, is besides the point. Just as measure theoretic arguments cannot be used to substantiate Albert and Loewer's critique, nor can they be used in the opposite direction, to undermine Albert and Loewer's critique. Destructive measurements *happen*, and hence they are physically possible; whether these measurements are assigned probability zero or non-zero by some particular measure is simply irrelevant to the fact that they are real, physical, laboratory measurement interactions.

4. CONCLUSIONS

Destructive measurements must then fall outside the domain of modal interpretations of quantum theory, which indicates that these interpretations do not possess full generality. All this is of course not a claim for ruling out modal interpretations in general. We already established that no general refutation can be forthcoming. Indeed, in reviewing the recent literature springing from Albert and Loewer's original proposal, one may be forgiven for wondering what the fuss has been all about. In the absence of a detailed discussion of concrete examples of real, physical, non-ideal interactions, Albert and Loewer's criticism is rather empty, and can have no bite against any interpretation of quantum theory.

We have tried to redress the balance by pointing out a family of real laboratory interactions that seem to be necessarily non-ideal in the required sense. We don't claim to have the definitive analysis of destructive measurement interactions, but we hope to be focusing attention on what really matters. Albert and Loewer's argument against the modal interpretation is an empirical, not a mathematical, argument—and hence it must be discussed at the right level of empirical detail.

Department of Philosophy, Logic and Scientific Method
The London School of Economics and Political Science

NOTES

[1] This argument has been subsequently taken up by Park [16] and Chang [6].
[2] Laura Reutsche [17] includes it among her General Unitary Measurement (GUM) operators and uses it to criticise van Fraassen's modal interpretation. These operators fall under the characterisation of measurement evolutions given by Beltrametti, Cassinelli and Lahti [4]. Fine rules out operators such as this as possible measurement operators in the context of measurement with objectification because they do not yield objectification; if objectification is not an issue, however, there seems to be little ground for excluding them.

REFERENCES

[1] Albert, D.: 1992, *Quantum Mechanics and Experience*. Harvard University Press, Cambridge, Mass.

[2] Albert, D., Loewer, B.: 1990, Wanted Dead or Alive: Two Attempts to Solve Schrödinger's Paradox, in Fine, A., Forbes, M., Wessels, L., eds., *Proceedings of the 1990 Biennial Meeting of the Philosophy of Science Association, Volume 1*, Philosophy of Science Association, East Lansing, Michigan.

[3] — 1993, 'Non-Ideal Measurements', *Foundations of Physics Letters*, **6**, 297–303.

[4] Beltrametti, E. Cassinelli, G., Lahti, P.: 1990, 'Unitary Measurements of Discrete Quantities in Quantum Mechanics', *Journal of Mathematical Physics*, **31**, 91–98.

[5] Bub, J.: 1993, 'Measurement: It Ain't Over Till It's Over', *Foundations of Physics Letters*, **6**, 21–35.

[6] Chang, H.: 1997, 'On the Applicability of the Quantum Measurement Formalism, forthcoming, *Erkenntnis*.

[7] Earman, J.: 1986, *A Primer on Determinism*. D. Reidel Publishing Company, Dordrecht.

[8] Fine, A.: 1970, 'Insolubility of the Quantum Measurement Problem', *Physical Review*, **2**, 2783–2787.

[9] Halmos, P.R.: 1974, *Measure Theory*. Springer-Verlag, New York.

[10] Healey, R.: 1989, *The Philosophy of Quantum Mechanics: An Interactive Interpretation*. Cambridge University Press, Cambridge, UK.

[11] — 1993, 'Why Error-Prone Measurements Have Outcomes', *Foundations of Physics Letters*, **6**, 37–54.

[12] Kochen, S.: 1987, *A New Interpretation of Quantum Mechanics*, in Lahti, P.J., Mittelstaedt, P., eds., Symposium on the Foundations of Modern Physics. World Scientific, Singapore; 151–69.

[13] Kronz, F.: 1991, 'Quantum Entanglement and Nonideal Measurements: A Critique of Margenau's Objections to the Projection Postulate', *Synthese*, **89**, 229–51.

[14] Margenau, H.: 1958, 'Philosophical Problems Concerning the Meaning of Measurement in Physics', *Philosophy of Science*, **25**, 23–34.

[15] — 1977, *The Nature of Physical Reality*. Ox Bow Press, Woodbridge, Conn.

[16] Park, J.L.: 1968, 'Quantum Theoretical Concepts of Measurement, Parts 1 and 2', *Philosophy of Science*, **19**, 205–231.

[17] Reutsche, L.: 'Measurement Error and the Albert-Loewer Problem', *Foundations of Physics Letters*, **8**, 327–44.

[18] Sklar, L.: 1996, *Topology Versus Measure in Statistical Mechanics*. Talk given at LSE in May, 1996.

[19] Suárez, M.: 1996, 'On the Physical Impossibility of Ideal Quantum Measurements', *Foundations of Physics Letters*, **9**, 425–36.

[20] Vermaas, P., Dieks, D.: 1995, 'The Modal Interpretation of Quantum Mechanics and its Generalisation to Density Operators', *Foundations of Physics*, **25**, 145–58.

MAURO DORATO

TIME, RELATIVITY, AND THE SPATIALITY OF MENTAL EVENTS

Henceforth, space by itself, and time by itself, are doomed to fade away into mere shadows, and only a kind of union between the two will preserve an independent reality.
H. Minkowski

1. INTRODUCTION: MENTAL TIME AND PHYSICAL TIME

Sellars once wrote that " 'the problem of time' is rivaled only by the 'mind-body problem' in the extent to which it inexorably brings into play all the major concerns of philosophy" (1971, 527). Considering that time plays a major role both in our inner life and in the description of the outer world, one could suggest that two problems are deeply related: our progress in understanding bits of the problem of time might shed light into the mind-body problem and *viceversa*.[1] In this paper, I will test the plausibility of this suggestion, by focusing on a fundamental aspect of the relationship between the 'time of physics' and the 'time of mind', namely the problem of their compatibility.[2]

As a first approximation, the 'time of mind' is here synonymous with 'time as it appears to our conscious experience'. The time of physics refers instead to time as it is presupposed by our best spacetime theories. If, as many philosophers and physicists of our century have claimed, physical and mental time possess incompatible properties, the possibility of modelling our phenomenological experience within physics would become more remote, by having to depend on future, radical revisions of current physical theories, perhaps of the type speculatively envisaged by Penrose (1989, 1994). If, on the contrary, our experience of the passage of time, which is essential to mental time, can be regarded as a merely intensified expression of a becoming that we share with the physical universe, both a materialistic theory of mind, and a naturalistic approach to a theory of human knowledge (Quine's naturalized epistemology), would be entitled to rosier prospects. In a word, the reducibility of mental time to physical time seems to be a presupposition not just of physicalism, but of any version of reductionism of the mental to the physical within the so-called 'mind-body problem'. It is therefore rather curious to note that philosophers of mind have, with very few exceptions, paid very little attention to this aspect of the problem, and it is one of the purposes of this paper to stimulate them to do so.

M.L. Dalla Chiara et al. (eds.), Language, Quantum, Music, 197–207.

The distinction between what I called 'physical' and 'mental' time dates back at least to Russell (1915), and can be made more precise in terms of our two fundamental ways of distinguishing events in time, the 'static', *tenseless* way of 'before' and 'after', and the 'dynamical', *tensed* way of 'past', 'present' and 'future'. The 'static-ness' in question refers to the fact that the truth-value of sentences containing the tenseless relation of temporal precedence is independent of the time of assertion. By relying on declensions of verbs and on temporal indicators like 'yesterday', 'tomorrow' and 'next year', the truth-value of tensed sentences is instead a function of time, and is therefore time-dependent: different tokens of the same type of tensed sentences have different truth-values as a function of time.[3] As a matter of fact, while our description of the physical world seems to require just the tenseless, binary relation of 'beforeness', the tensed distinction between past, present and future is presupposed by any adequate phenomenological account of our temporal awareness (Husserl, 1964), by a theory of human action (Mellor, 1981; Brockelman, 1985), and, *a fortiori*, by any historical or social science (Denbigh, 1981).

In order both to rehearse some arguments for the claim that 'tenses'[4] are just in the mind and to introduce the specific problem to be discussed, it is appropriate to recall that the special theory of relativity has often been invoked to claim that the distinction between past, present and future depends on how mind-possessing organisms perceive a tenseless reality (Gödel, 1949). Since in the theory of relativity a world-wide separation between past and future must depend on a conventional choice of an inertial reference frame, it is urged that within that theory an ontological, *global* distinction between a determinate past and an indeterminate future cannot be made, no matter how we define the relation of 'being determinate'. On the other hand, if defining such a relation in terms of the relation of past causal connectibility might reinstate the above distinction *locally* – so that, for each point in Minkowski spacetime, only the closure of the past light-cone could count as determinate with respect to that point[5] – it can be argued that nonlocal correlations of measurement outcomes in Bell-type experiments destroy even this possibility, forcing any event to be definite with respect to any other event.[6] It seems that we must conclude that events in Minkowski spacetime are given once and for all, in the sense that they don't become, that is, they neither come into being nor cease to exist

How can we square this claim with our experience of time, which presupposes, intuitively, that events *come into being* in the present and *are real only when they are present or past*? If tenses were essential to the description of the physical world, how could physicists neglect them as they do without impairing the explanatory power of their theories?[7] Considering that physics does not need the notion of being present (except in the trivial or pragmatic sense in which each point in Minkowski spacetime is 'here-now'),[8] and even treats it as relative or frame-dependent, *the tensed distinction between past and future*, which is essential to mental time, *could be regarded as a genuine watershed between mind and matter*. Failure of inter-translatability of tensed and tenseless discourse, recognized even by tenseless theorists of time like Mellor (1981) or Faye (1989), could then be used to lend further support to the claim that *any attempt to reduce the features of mental time to those of physical time is bound to fail.*

Even though such an argument for a more general *dualism of properties* of mental and physical events cannot be taken to imply by itself a *dualism of substances* of a Cartesian type, it certainly seems to speak against the theory of identity of the mental and the physical in its two familiar brands. That it, it speaks *both* against the claim that *types* of mental events are identical with *types* of physical events (type-type identity theory), *and* against the claim that *tokens* of mental events instantiating (or corresponding to) a mental property are identical with *tokens* of physical events instantiating (or corresponding to) some physical property (token-token identity theory). In fact, since we have seen that there are at least two families of mental and physical properties that are mutually irreducible – namely, those given by the tensed and tenseless determinations – it is hard to see how a particular token of a mental event instantiating the 'property' of being *tensed* (by being present, for instance) can be identical with a token of a physical event exemplifying a tenseless relation.[9]

Although the argument sketched above cannot be considered conclusive – one could try to avoid its conclusion by denying that the tensed determinations of past, present and future are genuine properties of mental events – it should sound in any case surprising that the special theory of relativity has recently been used to argue in favor of the token-identity theory between mental and physical events (Weingard, 1977; Lockwood, 1984a, 1984b, 1985; Lockwood, 1989, ch. 5).[10] In order to discuss the relevance of these papers to the problem of the relationship between mental and physical time, in what follows I will attempt a critical review of such relativistic arguments.

2. WEINGARD'S AND LOCKWOOD'S ARGUMENT IN FAVOR OF THE IDENTITY THEORY

In the discussion between rival conceptions of mind, an important role is usually played by the problem of the spatial localizability of mental events, already raised by Descartes. Can we endorse Wittgenstein's and Ayer's point that mental events cannot be assigned a spatial location, at least independently of a prior assumption that they are identical with some physical event or other?[11]

Weingard (1977) and Lockwood (1989) have recently argued that mental events must be in space, provided (i) that the special theory of relativity is correct, and (ii) that mental events are located in time in the 'same sense' in which physical events are. Since (i) is usually taken to imply the inseparability of time from space (whatever is in time must also in space and *vice versa*), (ii) guarantees the conclusion that any mental event is in space-time *if* it belongs to the same temporal order of physical time. On the additional, reasonable hypothesis that whatever is located in space-time is physical (material) in principle, it follows that each token of a mental event must be identical to a token of a physical (brain) event. If the two premises were true, this valid argument would score an important point in favor of the token-identity theory, of the kind argued on a different basis by Davidson in his 'Mental Events' (1980, ch.11). More dramatically, it would provide an interesting argument against the Cartesian model of the mind as a non-extended point in time, since mental events, by occupying a location in space-time, would be by necessity *spatio-temporally extended.* How about the truth of the two premises? The real question raised by (i), of course, is not the correctness of special relativity, but the particular arguments used by Weingard and Lockwood to maintain that any event that is in time is also in space-time. Unless it is shown that the special theory of relativity 'applies to mental events', as Weingard put it, the argument in (i) will have no force, but we will see that to argue in favor of (i) implies the use of hypotheses which make (ii) sound very dubious. Since Lockwood and Weingard offer slightly different arguments in favor of these two premises, it is appropriate to discuss their version of the argument in turn.

3. LOCKWOOD'S ARGUMENT FOR THE SPATIALITY OF MENTAL EVENTS

To begin with, a certain imprecision in the terminology of Lockwood's argument is worth-noting. I will first point out such a difficulty in the formulation of his argument, and I will then illustrate the consequences that such a formulation has *vis à vis* the connection between the first and the second premise. He starts by assuming the correctness of the special theory of relativity, which he takes to be meaning, in our case, that 'any two events that are temporally separated with respect to one frame of reference must be spatially separated with respect to some other frame' (1989, 72).

What does it mean exactly to be 'temporally' and 'spatially' separated? Despite the fact that Lockwood curiously affirms that 'being spatially separated' has *not* the same meaning as the technically more legitimate notion of 'being spacelike-related' (1989, 73),[12] we must take him to refer to the well-known fact that two events that can be temporally ordered by 'earlier than' (temporally separated) according to the coordinate time associated to one inertial frame, *can* be simultaneous (spatially separated) in another frame.

However, it is essential to note against Lockwood's distinction between 'spatial separation' and 'spacelike separation' that *the above fact holds only for spacelike-related events*, since it is only for such events that the temporal order is non-invariant across different frames and the two events in question can be simultaneous for at least one inertial observer. If 'temporally separated' meant in fact *timelike separated*, there would be no frame in which the two events could be simultaneous, although we could find a frame in which they would be 'spatially separated' in a different sense, given by the fact that *the spatial component of the Minkowski metric* (giving us the purely spatial distance of the events) would be non-zero. In other words, in this second sense, a projection parallel to the direction of the time axis of one of the two events on the surface of simultaneity containing the other would yield a positive number.

The problem with this second sense of spatial separateness is that it not appropriate for Lockwood's purposes, since *one can properly talk about spatial separation only when the events lie on the same surface of simultaneity*, so that they share the same time. Against his intentions, Lockwood must then be identifying 'spatial separateness' and 'spacelike-relatedness', an identification that makes the adoption of his terminology potentially misleading, given that his term 'spatial separateness' does not belong to the standard language of the theory of relativity, and he never explicitly tells us what it means.

This remark, which uncovers what should be regarded only as a minor lapse, has unfortunately consequences which could be fatal to Lockwood's own project. In fact, as we are about to see, if one assumes that mental events belong to the same temporal order of physical events, which is the second premise (ii), it becomes implausible to use (i) as formulated by Lockwood to reach his wished-for conclusion. First of all, it seems non-controversial that 'being in a relation of simultaneity with some event *F*' is to be regarded as a *necessary* condition for a mental event *E* to be in space with respect to *F* in Lockwood's intended sense. As we have shown, from this claim it follows that the only pairs of events that could be spatially separated are those that are causally non-connectible, or spacelike-related. Despite the possibility of a 'dissociated' consciousness – compare Lockwood's 'compound I' model (1989, ch. 6) with Dennett's 'multiple drafts' model of consciousness (1991) – this is certainly not what Lockwood can have in mind!

Suppose Lockwood is right in assuming the truth of (ii), that is, that mental events really are in time in the same sense in which physical events are. Given the small size of the brain compared to the speed of light, it is empirically plausible to assume that *most, if not all*, of the conscious mental events of a single person's life, call them *E*, must be pairwise timelike-related, or causally connectible. *At least to our direct introspection*, the mental events of which we are conscious appear to us to be timelike-related exactly in the special relativistic sense of being *one after the other*. Relativity then implies that there is no observer for whom *any two such events can be spatially separated*, which simply means that there is no observer for whom the mental events in anyone's mind can be in space with respect to each other. Since Lockwood is not an eliminativist about mental events, he must take introspection seriously. To this remark, we can add that even though introspection is not infallible in all cases, in judgements of temporal priority it cannot be doubted, since any other attribution of temporal relations depend on it. When I perceive the hand of a chronometer striking one before it strikes two, the order of the physical events is the same as the order of my perceptions, and the latter are temporally ordered in an invariant way.

Granted the timelike-relatedness of mental events as witnessed by introspection, we have seen that 'the temporally separated' events of Lockwood's argument must instead be referring to a mental event and a physical event which are *spacelike separated*, in such a way that there exists a frame in which they can be simultaneous. If each of the members *e* of the set of mental events *E* of an observer *O* can be in space with respect to the members of a set of physical event *F* if and only if *e* in *E* and *f* in *F* are spacelike-related, then *e* and *f* are temporally related in a sense that is *fundamentally different* from the sense in which mental events are in time by being *one after the other*. To such sets of causally non-connectible events, premise (ii) cannot be applied: the mental events *e* in *E* and the physical events *f* cannot share the same temporal order, because they don't bear any causal relation, and in the special theory of relativity the causal order fixes the temporal order.

To this, it could be replied that all Lockwood needs to show is that there are some physical events *f* with which the mental events *e* are simultaneous: this fact suffices to grant spatiality to each mental event *e* in the set *E*. It is then possible to apply (ii) not to the temporal relations holding between *E* and *F*, but to the set of mental events *E*, ordered by 'earlier than', and the set *C*, whose members are physical events with which the *e*'s in *E* are causally interacting. Since the mental events in the set *E* are temporally related in an invariant way, now we can prove the conclusion. Assume that two events can be causally related in Minkowski spacetime only if they are temporally related in an invariant way. Given that it is plausible to accept that mental and physical events are causally related – as witnessed by the pain I feel after I have been stung by a bee and by my attempt to run away from the bee because of the pain – they must share the same temporal order. If they share the same temporal order, they are in time in the same sense in which physical events are in time, so that mental events are also in space. In a word, this reply to the objection above points out that we first use the spacelike-relatedness of the set *E* with *F* to prove the spatiality of *e*, and then use (ii) on the partially ordered set of events *E* to prove that they must share the temporal order of the physical events with which they interact.

4. THE PROBLEM OF THE EXISTENCE OF EVENTS IN SPACE

Unfortunately, this reply, which seems so natural, presupposes an additional, subtle premise that in these circumstances cannot be granted: that 'being simultaneous with' be also *sufficient*, and not just necessary, to confer *mutual reality in space* to spacelike related events. This difficulty is linked also to an implication that is given above but that is invalid: from the fact that mental and physical event share the same temporal order we *cannot* deduce that they are in time in the same sense. On the one hand, being able to regard simultaneity as sufficient to confer reciprocal reality to events in space is absolutely central to argue in favor of the spatiality of mental event *e* with respect to *f*. On the other hand, as we are about to see, such a condition is incompatible with the requirement that mental and physical events 'are in time in the same sense', as Lockwood has it. The problem is that if such a requirement is to be fulfilled, as (ii) imposes, *we must assume that there is becoming also between physical events, that is, that events in the future light-cone are not as real as those in the past light-cone.*[13] In this way, the compatibility of the properties of mental and physical time would be ensured from the start, and the order of time would be given, locally, by the becoming real of future events in the 'here-now' (the order of becoming). The mere tenseless order imposed by 'earlier than' in fact does not suffice to claim that mental and physical events are in time in the same sense, since from §1 we know that one could always object that the two sets of events are characterized by irreducible properties. On the contrary, by adding the hypothesis of an ontological asymmetry between the past and the future light-cone of any event of Minkowski spacetime, the implication which in Lockwood's text was invalid becomes acceptable: from 'mental and physical event share the same temporal order', *we can now infer* that they are in time in the same sense, because the order of time is the order of becoming, common to the mental and the physical domain.

The only conceptual link that is now left to explain is the following: why is it that granting becoming to physical events makes it impossible to attribute reciprocal reality to events that are simultaneous in the coordinate time relative to an inertial reference system? This apparently counterintuitive claim follows from a formal result according to which for each event p in Minkowski spacetime, the only events that are real with respect to p are those in the causal past of p.[14] Assume that for any event p in time-oriented Minkowski spacetime there are events that are unreal as of p, the essential hypothesis that is needed to fulfil (ii); assume furthermore that the relation of reality is transitive, reflexive and invariant for automorphisms preserving the time order. Then it can be proved that events that are spacelike-related with respect to p are not real at all as of p.

In this case, the restriction of Lockwood's criterion to events that are 'temporally separated' but spacelike-related is fatal to his own argument: if we can't prove that the mental event e is real with respect to the physical event f by being simultaneous with it, *then the concept of being in space invoked by Lockwood and Weingard won't serve their purpose.* In fact, without some additional premise and discussion, relativistic space is not amenable to the kind of argument that Lockwood and Weingard have put forward.[15] Since relativity makes time more spacelike only by making space more timelike, one can gain the spatiality of mental events only at the cost of making space intrinsically dynamical in the sense that will be clarified in what follows. The point deserves some more labour, since generally it is not appreciated that in relativity not just the notion of time, but also that of space, is radically different from its Newtonian or Galileian counterpart.

5. THE DILEMMA OF LOCKWOOD AND WEINGARD'S ARGUMENT

In order both to summarize the essential claims of this paper, and to show how Weingard's version of the argument stresses even more the need of an additional premise concerning the determinate existence of spacelike related events, we can present the following dilemma. Either Lockwood's 'temporally separated events' refers to causally connectible events or it doesn't: by choosing the first horn, (i) comes out false, and by choosing the second, it is possible to argue that spacelike related events are not real as of each other. Since two events are either causally connectible or they are not (*tertium non datur*), Lockwood's conclusion must be rejected.

To see why this follows, let us start by taking the first horn: it is natural and empirically plausible to assume that the mental events in the life of any one person are ordered by the relation of 'being before' or, at most, are simultaneous with one another in a way that clearly does *not* require a comparison of different times at a distance. That is, they are timelike related. Since it is an elementary truth of special relativity that any two events that are timelike-related in any one frame are timelike-related in any other frame, there are no frames in which the events *e*'s can be simultaneous. According to Lockwood's own argument, most if not all of the events in a person's life *E* could not be 'located in space' with respect to each other, provided that simultaneity in at least one frame is a necessary condition of whatever Lockwood means for 'spatial separateness'. The events *e*'s can be in space only if each of them is simultaneous with some other physical event with respect to which it is spacelike related, which brings us to the second horn of the dilemma.

In this case, since the pairs of physical events and mental events picked out by Lockwood's 'temporally separated' are spacelike related, they are temporally incomparable, and so they don't belong to a unique, intrinsic (coordinate-free) temporal order. It is at this point that Weingard's argument for (i) can be injected into the discussion. Weingard correctly argues that if an event *E* occurs at coordinate time *t* in frame *F*, it will lie in 'slice' *F*(*t*) by definition of coordinate time. 'Slice' *F*(*t*), however, just *is* a three-dimensional spatial hypersurface simultaneous with *E* in frame *F*. *For Weingard, this automatically implies that any datable event to which special relativity applies has a certain spatial location relative to F at time t.* If mental events are datable in the same sense in which physical events are, it would follow that mental events are in space in the same sense in which physical events are.

This revised argument is correct to the extent that it expresses the well-known inseparability of time from space (and *vice versa*) codified in the Minkowski space-time metric. However, also this version of the argument has the shortcoming that we have already pointed out, because the concept of time that is more appropriate to express the kind of temporal succession typical of mental events is *proper* time measured along the worldlines representing events in our brains, and not *coordinate* time referred to global surfaces of simultaneity. This point shows in a different way how reliance on coordinate time can create the illusion that material, physical events exist simultaneously in 'space' in the pre-relativistic sense provided by Newton's absolute space.

It is of capital importance to point out that the existence of 'space' as a set of simultaneous and determinate events at a distance cannot be taken to derive from the axioms of the theory of relativity, but is an additional metaphysical assumption. Of course, such an assumption can be based on another physical theory that can be given a spacetime formulation. For instance, if quantum mechanics requires action at a distance, one must assume the reciprocal determinateness for spacelike-related event. The important point is to realize that in order to use the claim that the spatially distant is real with respect to here-now we need some further assumption, which Lockwood and Weingard did not provide. We must not forget the empiricist, operational critique that lies at the foundations of the special theory. Influenced by Hume's criticism of an ontologically based necessity of the causal links between events, Einstein was led to postulate the relativity of simultaneity because there is no way to ascertain operationally which temporal relation really holds for two events that are far away from each other. The only information or signal that can reach an event comes from its causal past. Since there is no identity of instants at different places, in relativity a present instant is here and nowhere else (Robb, 1921, 12): "'There', in respect to 'here', is always *in the past* or (potentially) *in the future*" (Čapek, 1961, 216). The main point is that since physical interactions take time, and light is 'a limit signal', there are no *instantaneous* connections of distant parts in space, but only *successive*, dynamic connections. Such a 'dynamization' of space, considered together with the geometrization of time, are two faces of the same coin.

It should be clear how this observation creates problems for the sense in which physical events, let alone mental events, can be *in space*. In fact, Lockwood and Weingard must assume that whatever can be located in space (because is simultaneous with my here-now) is real with respect to my here-now. *The claim that mental events can be located in space won't help very much if the event in question does not exist with respect to my here-now.* Although the thesis of the reality of spatially distant events seems to be taken for granted by many philosophers of relativity, it is far from obvious. If we want the theory of relativity to 'apply to mental events', as Weingard has it, it'd better meet the essential requirement of mental time, concerning an ontological difference between a determinate past and an indeterminate future. However, relativity teaches us that any reason we might have to deny reality to events in the future light cone must be extended also to the spatially distant. If we assume that events in the future light-cone are unreal with respect to the here-now, as a consequence of a series of theorems proved by Stein (1991) and Clifton and Hogarth (1995), *we also must assume that events that are spacelike related with respect to each other are mutually unreal.* The set of events that are real with respect to my 'here-now' and are in space is restricted to a single member, which is the 'here-now' itself. In short, *if* Minkowski spacetime must reflect the essential property of mental time as (ii) requires, *then* we can prove that any two spacelike-related events are unreal each from the perspective of the other. Not only are mental events not in space in the sense intended by Lockwood and Weingard, but not even physical events are. 'Space' as referred to by Lockwood and Weingard does not exist in the classical sense invoked by them.

As a concluding remark, I would like to make clear that I have not tried to defend a particular form of the so-called 'dualism of properties' based on the philosophy of time. What I have tried to show is rather that the theory of special relativity cannot be used to argue in favor of the token-identity theory without further additional premises which cannot be easily granted. In order to establish whether such a dualism of properties would be justified with respect to time, an inquiry into the question whether the kinds of features that we intuitively attribute to experienced time – its flow, or the actualization of future events – can be modelled in principle within physical time does certainly seem more pressing.[16]

Department of Philosphy
University of Rome

NOTES

[1] This view has been forcefully expressed by Pauri (1997).
[2] For a clear statement of the problem, see van Fraassen (1978).
[3] For the distinction between these two ways of describing events in time, see McTaggart (1908) and Mellor (1981).
[4] The word 'tenses' is here used in Mellor's sense (1981) to refer to temporal determinations which make the truth value of a sentence containing them depend on the time of assertion.
[5] See Stein (1991), Clifton, Hogarth (1995).
[6] For this argument, see Dorato (1996).
[7] Says Reichenbach: " ... if there is Becoming, the physicist must know it; but if time is merely subjective and Being is timeless, the physicist must have been able to ignore time in his construction of reality" (1956, 16).
[8] The same conclusion carries over to general relativistic spacetime, despite Smith's contentions (1993, 21–23), since cosmic time is available only in some models of Einstein's field equations.
[9] For the claim that dualism of properties is incompatible with a reductionism approach, see J. Kim (1993).
[10] Gordon (1984) and Gibbins (1985) offer some criticism of Lockwood (1984a) and Weingard (1977).
[11] See Lockwood (1989, 72) for sources.
[12] This notion is invariant across reference frames, since it is equivalent to 'non-connectible by a light signal'.
[13] Without considering non-local, quantum correlations, there is no difficulty in defining a binary relation of 'being ontologically definite' in terms of the structure of Minkowski spacetime.
[14] See Stein (1991) and Clifton, Hogarth (1995).
[15] A possible argument to regard events in a surface of simultaneity as reciprocally determinate is offered in Dorato (1996). The problem is that if this argument is valid, then objective becoming and special relativity are not compatible, and there is no way in which one can argue that mental events and physical events can be 'in time in the same sense'.
[16] For tentative answers to such a question, I refer the reader to Dorato (1995).

REFERENCES

Brockelman, P.: 1985, *Time and Self*. New York, The Crossroad Publishing Company & Scholars Press.
Čapek, M.: 1961, *The Philosophical Impact of Contemporary Physics*. Princeton, van Nostrand.
Clifton, R., Hogarth, M.: 1995, 'The Definability of Objective Becoming in Minkowski Spacetime', *Synthese*, **103**, 355–387.
Davidson, D.: 1980, *Essays on Action and Events*. Oxford, Clarendon Press.
Denbigh, K.: 1981, *Three Concepts of Time*. Berlin, Springer Verlag,
Dennett, D.: 1991, *Consciousness Explained*. London, Penguin Books.
Dorato, M.: 1995, *Time and Reality*. Bologna, Clueb.
Dorato, M.: 1996, 'On Becoming, Relativity and Nonseparability', *Philosophy of Science*, **63**, forthcoming.
Faye, J.: 1989, *The Reality of the Future*. Odense, Odense University Press.
Gödel, K.: 1949, 'A Remark about the Relationship Between Relativity Theory and Idealistic Philosophy', in Schilpp, P., ed. *Albert Einstein: Philosopher-Scientist* , La Salle IL, Open Court, 557–562.
Gibbins, P.F.: 1985, 'Are Mental Events in Space and Time', *Analysis*, **45**, 145–147.
Gordon, D.: 1984, 'Special Relativity and the Location of Mental Events', *Analysis*, **44**, 126–127.
Husserl, E.: 1964, *The Phenomenology of Internal Time Consciousness*. Translated by J. Churchill, Indiana University Press, Bloomington IN.
Kim, J.: 1993, *Supervenience and the Mind*. Cambridge University Press, Cambridge.
Lockwood, M.: 1984a, 'Einstein and the Identity Theory', *Analysis*, **44**, 22–25.
Lockwood, M.: 1984b, 'Reply to Gordon", *Analysis*, **44**, 127–128.
Lockwood, M.: 1985, 'Einstein, Gibbins, and the Unity of Time', *Analysis*, **45**, 148–150.
Lockwood, M.: 1989, *Mind, Brain and the Quantum. The Compound I*. Oxford, Basil Blackwell.
Malament, D.: 1977, 'Causal Theories of Time and the Conventionality of Simultaneity', *Nôus*, **11**, 293–300.
McTaggart, J.: 1908, 'The Unreality of Time', *Mind*, **68**, 337–362.
Mellor, D.: 1981, *Real Time*. Cambridge, Cambridge University Press.
Pauri, M.: 1997, 'The Physical Worldview and the Reality of Becoming', in Faye, J., Scheffler, U., Urchs, M., eds., *Perspectives on Time*. Dordrecht, Reidel. pp. 267–297.
Penrose, R.: 1989, *The Emporer's New Mind. Concerning Computers, Mind, and the Laws of Physics*. Oxford, Oxford University Press.
Penrose, R.: 1994, *Shadows of the Mind*. Oxford, Oxford University Press.
Reichenbach, H.: 1956, *The Direction of Time*. Berkeley, University of California Press.
Robb, A.: 1921, *The Absolute Theory of Time and Space*. Cambridge, Cambridge University Press.
Russell, B.: 1915, 'On the Experience of Time', *Monist*, **25**, 212–233.
Russell, B.: 1927, *The Analysis of Matter*. London, Allen and Unwin.
Sellars, W.: 1971, 'Time and World-Order', in Feigl, H., Maxwell, G., eds., *Scientific Explanation, Space and Time*, Vol. 3 of the Minnesota Studies in the Philosophy of Science, Minneapolis, University of Minnesota Press, 527–616.
Smith, Q.: 1993, *Language and Time*. Oxford, Oxford University Press.
Stein, H.: 1991, 'On Relativity Theory and the Openness of the Future', *Philosophy of Science*, **58**, 147–167.
van Fraassen, B.: 1978, 'Time: Physical and Experiential', *Epistemologia*, **1**, 323–336.
Weingard, R.: 1977, 'Relativity and the Spatiality of Mental Events', *Philosophical Studies*, **31**, 279–84. Princeton, Princeton University Press.

ANATOLIJ DVUREČENSKIJ

BELL-TYPE INEQUALITIES AND ORTHOMODULAR LATTICES[1]

1. INTRODUCTION

Today one of acceptable mathematical models of quantum mechanics is an *orthomodular lattice* (OML for short) (Beltrametti and Cassinelli [3], Varadarajan [19]), which generalizes models of Kolmogorovian measurable spaces [10] or Boolean algebras which describe classical mechanics. The main difference between OMLs and Boolean algebras is the absence of the distributive law in OMLs, in general. However, in particular cases, it can happen that the local character of events in an OML has a classical form.

J.S. Bell [1] studied some probability inequalities holding in any Boolean algebra, and a violation of such an inequality entails that the measured data come from a classical observation (see also [5]). Inspiring by the papers of Beltrametti and Maczyński [4], Länger and Maczyński [11], and Dvurečenskij and Läanger [8], [9], we say that an inequality of the type

$$\sum_{I\subseteq N} f(I)p(\bigwedge_{i\in I} a_i) \in [0,\ 1], \tag{1}$$

which holds for a state p in an orthomodular lattice L (and for all $a_1, \dots, a_n \in L$) is said to be a *Bell type inequality of order n*. The inequality (1) gives a linear combinations of correlations (or joint distributions) $p(\bigwedge_{i\in I} a_i)$ of the set of events $\{a_i : i \in I\}$ in the state p, $a_1, \dots, a_n \in L$, with real coefficients $f(I), I \subseteq N := \{1, \dots, n\}$.

A very special and important form of (1) is the inequality of type 2

$$p(a) + p(b) - p(a \wedge b)$$

when in (1) we put $N = \{1,2\}, f(\emptyset) = 0, f(\{1\}) = f(\{2\}) = -f(\{1,2\}) = 1$.

Proposition 1.1 *Let $f : 2^N \to \mathbf{R}$ and let p be a state on an OML L. Then the property*

$$\sum_{I\subseteq N} f(I)p(\bigwedge_{i\in I} a_i) \in [0,\ 1] \quad \forall\, a_1, , a_n \in L, \tag{2}$$

implies

M.L. Dalla Chiara et al. (eds.), Language, Quantum, Music, 209–218.

$$\sum_{I \subseteq K} f(I) \in [0,1] \quad \forall\, K \subseteq N. \tag{3}$$

Proposition 1.2 *Let $f : 2^N \to \mathbf{R}$ and let p be a state on an OML L. Then the statement*

$$\sum_{I \subseteq N} f(I) p(\bigwedge_{i \in I} a_i) \in [0,\ 1] \quad \forall\, a_1, , a_n \in \{0,\ 1\} \tag{4}$$

is equivalent to (3).

Theorem 1.3 *(Beltrametti-Maczyński* [4]*) The statements* (2) *and* (3) *are equivalent for any state p on a Boolean algebra L.*

Theorem 1.3 provides a very simple method of verification of Bell-type inequalities Beltrametti-Maczyński [4]. We take a function $f : 2^N \to \mathbf{R}$ and we investigate whether does it satisfy (3). Because N has n elements, this means we examine 2^n inequalities: If all sums occurring in (3) lie in the interval $[0,\ 1]$, then the Bell inequality (4) holds in any classical model; in the opposite case we reject the inequality as not valid in any Kolmogorov probability model, and, in addition, in view of Propositions 1.1 and 1.2 also not valid in any OML. In the first case, this Bell-type inequality can be used as a test for a given system of events in order to see if it comes from a classical or non-classical physical system. In physical praxis, for f we take a function with $f(I) = 0$ whenever $\bigwedge_{i \in I} a_i$ is not physically measurable. The verification procedure simplifies when we consider a function f with integer values, in particular, when $f : 2^N \to \{-1, 0, 1\}$. In general, for large values of n, some computational problems can appear, Pitowski [14], but as we shall see, the most important cases are $n = 2$ and $n = 3$. We recall that today Bell-type inequalities and Bell phenomena are intensively studied in various frames: using (1) orthomodular lattices, see e.g. [16], [8]–[7], [6], [15], (2) convex approach [2] which allows not only orthomodular lattice approach, (3) tensor products [17], etc.

2. BELL-TYPE INEQUALITIES OF ORDER 2

Santos [18] introduced the *separation*, or the *distance*, $S_p(a, b)$, between two propositions a, b in the state p via

$$S_p(a, b) = p(a) + p(b) - 2p(a \wedge b), \qquad a, b, \in L. \tag{5}$$

He proved that if L is a Boolean algebra, then S_p is a pseudometric on L. For any two elements a and b in L, we define

$$a\triangle b := (a \vee b) \wedge (a \wedge b)^{\perp},$$

and if p is a state on L we put

$$\rho_p(a,b) := p(a\triangle b), \quad a,b \in L. \tag{6}$$

In the paper [16], the connection between Bell-type inequalities of order 2 and states which are subadditive was shown. This result can be generalized as follows.

Theorem 2.1 *Let p be a state on an OML L. Then the following statements are equivalent:*

1. $p(a) + p(b) - p(a \wedge b) \leq 1$ *for all $a, b \in L$ (Bell inequality).*
2. *p is subadditive.*
3. *p is a valuation.*
4. *p is 0-additive.*
5. *ρ_p is a pseudometric on L.*
6. *S_p is a pseudometric on L.*
7. $S_p = \rho_p$.
8. *For every $f : 2^{\{1,2\}} \to \mathbf{R}$ ($f : 2^{\{1,2\}} \to \mathbf{Z}$) with*

$$\sum_{I\subseteq\{1,2\}} f(I)p(\bigwedge_{i\in I} a_i) \in [0,\ 1] \quad \forall\ a_1, a_2 \in \{0,\ 1\},$$

we have

$$\sum_{I\subseteq\{1,2\}} f(I)p(\bigwedge_{i\in I} a_i) \in [0,\ 1] \quad \forall\ a_1, a_2 \in L.$$

9. *For every $f : 2^{\{1,2\}} \to \mathbf{R}$ ($f : 2^{\{1,2\}} \to \mathbf{Z}$) with*

$$\sum_{I\subseteq K} f(I) \in [0,\ 1] \quad \forall\ K \subseteq \{1,2\}$$

we have

$$\sum_{I\subseteq\{1,2\}} f(I)p(\bigwedge_{i\in I} a_i) \in [0,\ 1] \quad \forall\ a_1, a_2 \in L.$$

10. *There are a modular OML M, a homomorphism h from L onto M and a (positive) subadditive state P on M such that $P(h(a)) = p(a)$ for any $a \in L$.*
11. *There exists an $\alpha > 0$ such that $1 - \alpha + \alpha p(a) + \alpha p(b) - \alpha p(a \wedge b) \leq 1$ for all $a, b \in L$.*

12. *There exists an* $\alpha < 0$ *such that* $0 \leq -\alpha + \alpha p(a) + \alpha p(b) - \alpha p(a \wedge b)$ *for all* $a, b \in L$.
13. $0 \leq p(b) - p(c) - p(a \wedge b) - p(b \wedge c) - p(c \wedge d) + p(a \wedge d),\ a, b, c, d \in L$. *(Clauser-Horne-type inequality.)*

Example 2.2 *There exists a non-Boolean OML L with the non-empty system of all states such that any Bell-type inequality of order 2 satisfying (3) holds for any state on L.*

3. BELL-TYPE INEQUALITIES OF ORDER 2 IN HILBERT SPACES

Hilbert space formalism is of big importance for quantum mechanics, where $L(H)$ is the system of all closed subspaces of a real, complex or quaternionic Hilbert space H. Here we have only particular cases when Bell-type inequalities of order 2 are valid for finite-dimensional Hilbert spaces:

Proposition 3.1 *Let* $1 \leq \dim H = n < \infty$. *Then we have:*

1. *Any subadditive state on* $L(H)$ *has a support.*
2. *The support of a subadditive state on* $L(H)$ *is equal to* H.
3. *Any subadditive state p on* $L(H)$ *is of the form*

$$p(M) := \dim M/n, \quad M \in L(H), \tag{7}$$

Proposition 3.2 *Let* $\dim H = \infty$. *Then we have:*

1. *If p is a subadditive state on* $L(H)$, *then* $p(M) = 0$ *for any finite-dimensional subspace M of H.*
2. *There is no subadditive, completely additive state on* $L(H)$.

The system $L(H)$ is isomorphic with the system $\mathcal{P}(H)$ of all orthogonal projectors on H. If we now take into account not only orthogonal projectors, but also skew projectors ([13]), we have the following situation. Let $S(H)$ denote the set of all skew operators on H, i.e., of all linear operators $P: H \to H$ such that $P^2 = P$. Then any idempotent operator is continuous and $S(H)$ contains as a subset the set $\mathcal{P}(H)$ of all orthogonal projections on H, i.e., of all Hermitian idempotents on H. Put $E := \mathrm{Ran}P = \{Px : x \in H\} = \{x \in H : x = Px\} \in L(H)$ and $F := \mathrm{Ker}P = \{x \in H : Px = 0\} \in L(H)$. Then $E \cap F = \{0\}$ and $E + F = H$, and P projects any vector $x \in H$ onto E parallel with F. This relationship among P, E, and F will be written as $P = \pi(E, F)$. If, for $E, F \in L(H)$ we have $E \cap F = \{0\}$ and $E + F = H$, then E, F determine a unique skew operator $P = \pi(E, F) \in S(H)$. Indeed, we put $Px = x_1, x \in H$, whenever $x = x_1 + x_2$, $x_1 \in E, x_2 \in F$. We have $I - \pi(E,F) = \pi(F,E)$ and $\pi^*(E,F) = \pi(F^\perp, E^\perp)$, where I is the identity operator on H. For $P, Q \in S(H)$, we write $P \leq Q$ iff $PQ = QP = P$, and $P^\perp := I - P$. Then $S(H)$ is an OMP.

Choose a skew operator $P \in S(H)$ and define

$$A = P^*P + (I-P)^*(I-P). \tag{8}$$

Then A is a Hermitian, positive invertible operator on H ($A^{-1} = P_{\mathrm{Ran}P}P^* + P_{\mathrm{Ker}P}(I - P^*)$) and it defines on H a new inner product $(\cdot,\cdot)_A$ via

$$(x,y)_A := (Ax, y), \quad x, y \in H.$$

Then H with respect to $(\cdot,\cdot)_A$ is again a Hilbert space, and the topologies induced by $||\cdot||$ and $||\cdot||_A$ are the same. Let $\mathcal{P}_A(H)$ denote the set of all orthogonal projections on H with respect to $(\cdot,\cdot)_A$. We note that $\mathcal{P}(H) = \mathcal{P}_I(H)$. Then

$$S(H) = \bigcup_A \mathcal{P}_A(H),$$

where A is defined via (5.4), or $S(H) = \bigcup_A \mathcal{P}_A(H)$, where A is any positive invertible operator on H. In addition, $P \in \mathcal{P}_A(H) \Leftrightarrow AP = P^*A \Leftrightarrow P^* \in \mathcal{P}_{A^{-1}}(H)$.

Proposition 3.3 *Let* $1 \leq \dim H = n < \infty$. *Then on* $S(H)$ *there is a unique state, which is subadditive on any* $\mathcal{P}_A(H)$, *namely* $p : S(H) \to [0,1]$, *defined via*

$$p(\pi(E,F)) := \dim E/n, \quad P = \pi(E,F) \in S(H). \tag{9}$$

Another important structure connected with Hilbert spaces are **Krein spaces** (see also [12]). Let H be a Hilbert space and P any orthogonal projection on H. We put $J = P - (I - P)$ and define a new inner product $[\cdot,\cdot]$ on $H \times H$ via

$$[x,y] := (Jx, y),\ x, y \in H.$$

Then $[\cdot,\cdot]$ is not necessarily a positive inner product. Put $\Gamma^+ := \{x \in H : [x,x] = 1\}$, $\Gamma^- := \{x \in H : [x,x] = -1\}$, $\Gamma^0 := \{x \in H : [x,x] = 0\}$, and $H^+ := PH$, $H^- := (I-P)H$. A vector $x \in H$ is said to be *isotropic* if $x \in \Gamma^0$. For example, if $H = \mathbf{R}^2$ and $\dim H^+ = 1$, then $\Gamma^+ \cup \Gamma^-$ consists of two hyperbolas $x^2 - y^2 = \pm 1$, and Γ^0 consists of two lines $y = \pm x$. If $H = \mathbf{R}^3$ and $\dim H^+ = 2$, then two rotational hyperboloids $x^2 + y^2 - z^2 = \pm 1$ form $\Gamma^+ \cup \Gamma^-$. We recall that the case $H = \mathbf{R}^4$ and $\dim H^+ = 1$ is used as a Minkowski space in special relativity theory.

Denote by $K_J(H)$ the set of all idempotent linear operators P on H such that $[Px, y] = [x, Py]$ for all $x, y \in H$, or equivalently $JP = P^*J$. For two elements $P, Q \in K_J(H)$ we write $P \leq Q$ iff $PQ = QP = P$, and put $P^{\perp_J} := I - P$. Then $K_J(H)$ with respect to $\leq$ and $\perp_J$ is an OMP with the least and greatest elements O and I, respectively. We recall that $P \perp_J Q$ iff $PQ = QP = O$, and then $P \vee Q$ exists in $K_J(H)$ and $P \vee Q = P + Q$. If $P = I$, then $J = I$ and $K_J(H) = \mathcal{P}(\mathcal{H})$. We note that M is a one-dimensional projectively complete subspace of H iff $P = \pi(\mathrm{sp}(x)) = [x, x][\cdot, x]x$ for some $x \in \Gamma^+ \cup \Gamma^-$. In addition, for any subspace M of H we have $M^{\perp_J} = JM^{\perp}$, and it is a mirror image of M with respect to the axis H^+.

Example 3.4 *Assume* $\dim H = 2$. *Then* $K_J(H)$ *is an OML, and any subadditive state* p *on* $K_J(H)$ *has the form*

$$p(\pi(M)) = \dim M/2, \quad \pi(M) \in K_J(H).$$

An element $P \in K_J(H)$ is said to be *positive* (*negative*) if $[Px, x] > 0$ ($[Px, x] < 0$) for any non-zero $x \in PH$. It is well-known that every element $P \in K_J(H)$ can be expressed as

$$P = P^+ + P^-,$$

where P^+ and P^- are positive and negative elements of $K_J(H)$, respectively. Equivalently, if $P = \pi(M)$, then

$$M = M^+ + M^-,$$

where $P^+ = \pi(M^+)$ and $P^- = \pi(M^-)$. According to the inertia law, if $P = P_1 + P_2$, where P_1 and P_2 are positive and negative elements of $K_J(H)$, respectively, then

$$\dim M^+ = \dim M_1 \quad \text{and} \quad \dim M^- = \dim M_2,$$

whenever $M = M_1 + M_2$ and M_1 and M_2 are positive and negative subspaces of H.

Proposition 3.5 *Let* $2 \leq \dim H = n < \infty$, *and* $\kappa := \min\{\dim H^+, \dim H^-\} \geq 1$. *Then any of the following functions*

$$p_1(\pi(M)) = \dim M/n, \quad \pi(M) \in K_J(H), \tag{10}$$

$$p_2(\pi(M)) = \dim M^+/\dim H^+, \quad \pi(M) \in K_J(H), \tag{11}$$

$$p_3(\pi(M)) = \dim M^-/\dim H^-, \quad \pi(M) \in K_J(H), \tag{12}$$

defines a state on $K_J(H)$. Consequently, any convex linear combination $p = \lambda_1 p_1 + \lambda_2 p_2 + \lambda_3 p_3$, $\lambda_1, \lambda_2, \lambda_3 \geq 0, \lambda_1 + \lambda_2 + \lambda_3 = 1$, of p_1, p_2, p_3 *is a state on* $K_J(H)$, *too.*

We recall that (10) can be rewritten equivalently as

$$p_1(P) = \mathrm{tr}(P)/n, \qquad P \in K_J(H).$$

4. BELL–TYPE INEQUALITIES OF ORDER 3

For any pair $a, b \in L$, we define the *commutator*, $\mathrm{com}(a, b)$, of a, b via

$$\mathrm{com}(a, b) = a \wedge b \vee a \wedge b^{\perp} \vee a^{\perp} \wedge b \vee a^{\perp} \wedge b^{\perp}. \tag{13}$$

Let p be a state on an OML L. If there are a non-void set Ω, an algebra $\mathcal{S} \subseteq 2^{\Omega}$, a finitely additive probability measure P on $\mathcal{S}$, and a mapping $z : L \to L_2(\Omega, \mathcal{S}, P)$ such that, for all $a, b \in L$, (i) $E(z(a)z(b)) = p(a \wedge b)$, and (ii) $z(a \vee b) = z(a) + z(b)$ whenever $a \perp b$, then z is said to be a *random measure of the first kind* related to p. The following result says, roughly speaking, that for Bell-type inequalities of order 3, we have practically a classical system, and for Bell-type inequalities of higher order the case n = 3, is a limiting case (for more details on Bell-type inequalities of higher order see [9]).

Theorem 4.1 *Let p be a state on an OML L, and let n be a positive integer. Then the following statements are equivalent:*

1. $p(\mathrm{com}(a, b)) = 1$ *for all* $a, b \in L$.
2. *For p and all* $a, b, c \in L$
$$p(a) + p(b) + p(c) - p(a \wedge b) - p(a \wedge c) - p(b \wedge c) \leq 1 \tag{14}$$
holds.
3. *For all* $a, b, c \in L$
$$p(a) + p(b) + p(c) - p(a \wedge b) - p(a \wedge c) - p(b \wedge c) + p(a \wedge b \wedge c) \leq 1. \tag{15}$$
4. *p is distributive.*
5. *There are a Boolean algebra B, a homomorphism h from L onto B and a (positive) state P on B such that* $P(h(a)) = p(a)$ *for any* $a \in L$.
6. *There is a random measure of the first kind related to p.*
7. *There exist real numbers* $\alpha, \beta, \gamma, \delta$ *with* $0 < \beta \leq \alpha \leq 1$, $-\alpha - 1 \leq \gamma \leq -\alpha$ *and* $-1 - \alpha + \beta - \gamma \leq \delta \leq -\alpha + \beta - \gamma$ *such that, for all* $a, b \in L$, *it holds that*

$$1-\alpha+\alpha p(a)+\alpha p(b)+\beta p(c)+\gamma p(a\wedge b) \\ -\ \beta p(a\wedge c)-\beta p(b\wedge c)+\delta p(a\wedge b\wedge c)\leq 1.$$

8. *There exist real numbers* $\alpha, \beta, \gamma, \delta$ *with* $-1 \leq \alpha \leq \beta < 0$, $-\alpha \leq \gamma \leq 1-\alpha$ *and* $-\alpha+\beta-\gamma \leq \delta \leq 1-\alpha+\beta-\gamma$ *such that, for all* $a, b \in L$, *it holds that*

$$0 \leq -\ \alpha+\alpha p(a)+\alpha p(b)+\beta p(c)+\gamma p(a\wedge b) \\ -\ \beta p(a\wedge c)-\beta p(b\wedge c)+\delta p(a\wedge b\wedge c).$$

9. $p(t_1(a_1,\dots,a_n)) = p(t_2(a_1,\dots,a_n))$ *holds* $\forall\ a_1,\dots,a_n \in L$ *and for every positive integer* n, *if* $t_1(x_1,\dots,x_n) = t_2(x_1,\dots,x_n)$ *is a law holding in any Boolean algebra.*

10. *For every* $n \geq 1$ *and for every* $f : 2^{\{1,\dots,n\}} \to \mathbf{R}$ $(f : 2^{\{1,\dots,n\}} \to \mathbf{Z})$ *with*

$$\sum_{I\subseteq\{1,\dots,n\}} f(I)p(\bigwedge_{i\in I} a_i) \in [0,\ 1] \quad \text{for any } a_1,,a_n \in \{0,1\}$$

we have

$$\sum_{I\subseteq\{1,\dots,n\}} f(I)p(\bigwedge_{i\in I} a_i) \in [0,\ 1] \quad \text{for any } a_1,,a_n \in L.$$

11. *For every* $n \geq 1$ *and for every* $f : 2^{\{1,\dots,n\}} \to \mathbf{R}$ $(f : 2^{\{1,\dots,n\}} \to \mathbf{Z})$ *with*

$$\sum_{I\subseteq K} f(I) \in [0,1] \quad \text{for any } K \subseteq \{1,\dots,n\}$$

we have

$$\sum_{I\subseteq\{1,\dots,n\}} f(I)p(\bigwedge_{i\in I} a_i) \in [0,\ 1] \quad \text{for any } a_1,,a_n \in L.$$

12. $p(a) \leq p(a\wedge b) + p(a\wedge b^\perp)$ *for all* $a, b \in L$.

13. $p(b)+p(c)-p(a\wedge b)-p(b\wedge c)-p(c\wedge d)+p(a\wedge d)\leq 1$ *for all* $a, b, c, d \in L$. *An inequality of Clauser-Horne-type.*

14. p *is Jauch-Piron and if* $a, b \in L$ *and* $p(a\wedge b) = p(a\wedge b^\perp) = 0$, *then* $p(a) = 0$.

Theorem 4.2 *Let* p *be a state on an OML* $L, f : 2^N \to \mathbf{R}$, *and assume that* (1) *or* (2) *holds:*

1.

$$\sum_{I \subseteq N} f(I)p(\bigwedge_{i \in I} a_i) \leq 1 \text{ for all } a_1, \ldots, a_n \in L,$$

and there exist $j, k, m \in N, j \neq k \neq m \neq j$, *with* $f(\{j\}) = f(\{k\}) = 1 - f(\emptyset)$ *and* $f(\{m\}) = -f(\{j, m\}) = -f(\{k, m\}) > \gtrless$.

2.

$$0 \leq \sum_{I \subseteq N} f(I)p(\bigwedge_{i \in I} a_i) \text{ for all } a_1, \ldots, a_n \in L,$$

and there exist $j, k, m \in N, j \neq k \neq m \neq j$ *with* $f(\{j\}) = f(\{k\}) = -f(\emptyset)$ *and* $f(\{m\}) = -f(\{j, m\}) = -f(\{k, m\}) < 0$.

Then p is distributive.

5. CONCLUDING REMARKS

1. For every Boolean algebra and for every state on it all Bell-type inequalities are valid.
2. This property does not characterize the class of Boolean algebras. This means that there exist OMLs L with a non-empty state space which have the property that all Bell-type inequalities hold for all states on L, but which are not Boolean algebras. All Bell-type inequalities are valid for a state p on an OML L iff L is distributive with respect to p (i.e., if p is distributive) (Theorem 4.1).
3. The original Bell inequality implies all possible Bell-type inequalities of order 2 (Theorem 2.1).
4. There exist OMLs L and states p on L such that all Bell-type inequalities of order 2 are valid, but not all Bell-type inequalities of order 3 hold (Proposition 3.1).
5. There exist single Bell-type inequalities of order 3 (Theorem 4.1) and also of higher order (Theorem 4.2) which imply all Bell-type inequalities.

Mathematical Institute
Slovak Academy of Sciences

NOTES

[1] Research supported by grant G–4033 of the Slovak Academy of Sciences.

REFERENCES

[1] Bell, J.S.: 1964, 'On the Einstein Podolsky Rosen Paradox', *Physics*, **1**, 195–200.

[2] Beltrametti, E.G., Bugajski, S.: 1996, 'The Bell Phenomenon in Classical Frameworks', *J Phys A Math Gen*, **29**, 247–261.

[3] Beltrametti, E.G., Cassinelli, G.: 1981, *The Logic of Quantum Mechanics.* Addison-Wesley, Reading.

[4] Beltrametti, E.G., Maczyński, M.J.: 1994, 'On Bell-type Inequalities', *Found Phys*, **24**, 1153–1159.

[5] Clauser, J.F., Horne, M.A., Shimony, A., Holt, R.A.: 1969, 'Proposed Experiment to Test Local Hidden-Variable Theories', *Phys Rev Lett*, **23**, 880–884.

[6] D'Andrea, A.B., Pulmannová, S.: 1996, 'Quasivarieties of Orthomodular Lattices and Bell Inequalities', *Rep Math Phys*, **37**, 261–266.

[7] Dvurečenskij, A., Länger, H.: 1994, 'Bell-type Inequalities in Horizontal Sums of Boolean Algebras', *Found Phys*, **24**, 1195–1202.

[8] Dvurečenskij, A., Länger, H.: 1995, 'Bell-type Inequalities in Orthomodular Lattices. I. Inequalities of Order 2', *Int J Theoret Phys*, **34**, 995–1024.

[9] Dvurečenskij, A., Länger, H.: 1995, 'Bell-type Inequalities in Orthomodular Lattices. II. Inequalities of Order *n*'. Int J Theoret Phys, **34**, 1025–1036.

[10] Kolmogorov, A.N.: 1933, *Grundbegriffe der Wahrscheinlichkeitsrechnung*, Berlin.

[11] Länger, H., Maczyński, M.J.: 1995, 'On a Characterization of Probability Measures on Boolean Algebras and Some Orthomodular Lattices', *Math Slovaca*, **45**, 455–468.

[12] Matvejchuk, M.S.: 1991, 'Measures on Quantum Logics of Subspaces of a J-Space', *Sibir Mat Zurnal*, (in Russian) **32**, 104–112.

[13] Mushtari, D.Kh., Matvejchuk, M.S.: 1985, 'Charges on the Logic of Skew Projections', *Soviet Math Doklady*, **32**, 35–39.

[14] Pitowsky, I.: 1989, *Quantum Probability–Quantum Logic*, Lecture Notes in Physics. Vol. **321**, Springer-Verlag, New York, Berlin.

[15] Pulmannová, S.: 1994, 'Bell Inequalities and Quantum Logics', in Accaradi, L., ed., *The Interpretation of Quantum Theory: Where Do We Stand?*, Enciclopedia Italiana, 295–302.

[16] Pulmannová, S., Majerník, V.: 1992, 'Bell Inequalities on Quantum Logics', *J Math Phys*, **33**, 2173–2178.

[17] Pykacz, J., Santos, E.: 'Bell-type Inequalities on Tensor Products of Quantum Logics, (to appear).

[18] Santos, E.: 1986, 'The Bell Inequalities as Tests of Classical Logic', *Phys Letters A*, **115**, 363–365.

[19] Varadarajan, V.S.: 1968, *Geometry of Quantum Theory*, Vol. 1. van Nostrand, Princeton.

CLAUDIO GAROLA

AN INFORMAL PRESENTATION OF SEMANTIC REALISM AND ITS SOLUTION OF QUANTUM PARADOXES

1. INTRODUCTION

It is well known that the fathers of Quantum Physics (QP) maintained that a new epistemological attitude, not only a new physical theory, was needed in order to get over the crisis of Classical Physics (CP). But it is also well known that a universal agreement on the epistemological conception that must support QP has not yet been attained: this has deep consequences on the proposed interpretations of the theory, which have multiplied beyond any reasonable limit.

The main evidence that a satisfactory quantum epistemology has not yet been conceived is provided by the so called *quantum paradoxes*, the most important of which ensue from the thought experiment proposed by Einstein, Podolski and Rosen (EPR) in 1935 with the aim of proving that QP is not complete. Indeed, these paradoxes show that there are predictions of the theory which conflict with some basic conditions that one expects to be fulfilled by any physical theory: for instance, *locality* (see Sect. 5). Of course, one can give up all requirements that are not satisfied by QP, and this actually is the solution of the paradoxes proposed by many physicists; but this procedure is not unique, since there are various interpretations of QP, and opens the way to a number of serious philosophical problems on the very nature of physics and science.

In my opinion, the unsettled quarrel on the interpretation of QP and the paradoxes follows from some difficulties underlying the positions of the discussants. Indeed, on one side, a number of attempts have been made at providing a 'realistic' interpretation, or modification, of QP (for instance, by introducing hidden variables). Most of these attempts are characterized by the attribution, explicit or not, of an ontological status to physical entities that appear in the theory, as 'wave function', 'electron', 'quark', etc. (this occurs because of the widespread tendency to ontologize the models of physical theories, especially whenever they are intuitive models) and I will call this position *ontological realism*. On the other side, most interpretations of QP that refuse such kind of 'metaphysical' assumptions, like those based on the Copenhagen interpretation of QP, adopt more or less explicitly a theory of truth and meaning (*empirical verificationism* in the following) which implies, first, that no meaning can be attributed to physical statements that cannot be

M.L. Dalla Chiara et al. (eds.), Language, Quantum, Music, 219–230.

tested according to QP, even if they are syntactically correct, and, second, that the truth value of a testable statement may depend on the physical context in which the test occurs. Now, it seems to me that the premise on which ontological realism is based is untenable; indeed it underestimates, in particular, the fundamental distinction between theoretical and observative terms in a theory, thus attributing 'reality' to entities that are refined constructions of our mind, which usually change with the evolution of research and are different even in different but empirically equivalent theories (e.g. Braithwaite, 1953). But the widespread adoption of empirical verificationism (although verificationism had been severely criticized by scholars like Russell, 1940, and Popper, 1969) collapses the *pragmatic* notion of *testability* and the *semantic* notion of *truth*. This can be seen as irrelevant, or even desirable, from a physicist's viewpoint. Yet, it actually constitutes the deep root of a number of difficulties in QP. For instance, it engenders the problem of *nonobjectivity* of quantum measurements (e.g. Busch *et al.*, 1991). Furthermore, it may lead one to ignore (or to deny explicitly) the basic difference between logical laws (that should hold independently of a specific interpretation, rule our inference processes and provide *a priori* rationality criteria for physical theories) and physical laws (the validity of which depends on the interpretation); this occurs, for instance, in some quantum logical approaches to QP, as witnessed by the paper that started this kind of approaches (Birkhoff and von Neumann, 1936).

The above arguments obviously raise a number of questions. In particular, can we escape the dichotomy between ontological realism and empirical verificationism, so that we can avoid the difficulties that affect both positions? Moreover, can this be done while preserving some fundamental features of these positions that one cannot give up without encountering deep philosophical difficulties (e.g. the difficulties occurring in the existing alternatives to QP that introduce some kind of contextual hidden variables, as Bohm's theory)? I maintain that an affirmative answer exists to these questions, as I tried to show in a number of recent papers, where I proposed a new approach to QP (Semantic Realism, or, briefly, SR) and proved that paradoxes disappear in the new framework (the reader interested in placing SR in the context of recent research on the foundations of QP can locate it, with some cautions due in particular to the fact that SR is not a conventional hidden variables approach, among the theories that maintain that QP is incomplete, e.g. the book by Busch *et al.* quoted above). Therefore I provide an informal picture of some aspects of this approach to QP that are epistemologically relevant in the next section, without affording a deepened discussion that would be long and difficult.

I must, however, warn the reader: I think that the studies on the foundations of physics require an interdisciplinary approach and that, in particular, they can highly profit of the tools of modern logic. This is not a very popular viewpoint between physicists, who usually prefer elaborating new models and odd interpretations rather than criticizing the semantics of their language, and consider logic as a somewhat empty formal machinery. Of course, a number of arguments can be raised against this position. But the above remarks on the necessity of distinguishing between logical and physical laws, and the possible role of empirical verificationism in creating paradoxes, should be sufficient to demonstrate that a deeper analysis of the truth criteria underlying the standard interpretation of QP is desirable, and this is a typical logical issue. As a further argument, I would like to mention the distinction between predicates of different orders, that is standard in formal logic but only implicit or ignored in the natural language used by physicists in order to set up their theories. This has led some scholars to confuse first order properties, that can be tested directly on individual samples of physical systems, and second order properties, that can be tested only by taking into account ensembles of samples and measuring correlations between first order properties (this distinction is crucial when dealing with compound physical systems in QP, as I will briefly show in Sect. 3). In any case, I will refer throughout this paper to basic logical concepts (even if I will not undertake here any explicit formalization of the language).

2. SEMANTIC REALISM

First of all, SR avoids attributing an ontological status to theoretical entities but maintains a minimal realistic attitude. Indeed, it assumes that in every *laboratory* (that is, a space-time domain in the actual world) one can perform physical procedures, called *preparations*. These procedures prepare *physical objects*, which could be identified with the acts of preparation themselves, but are intuitively interpreted as individual samples of physical systems. Then, SR requires that every elementary (first order) property of a physical object can be attributed or not to the object independently of the observations that one may decide to perform on it. Consistently with this perspective, a physical system, or entity, can be defined in terms of a suitable set of characterizing properties, and one can say that a weaker 'realism of properties' is substituted to the 'realism of entities' that characterizes ontological realism.

It is apparent that SR is not compatible with the standard interpretation of QP, and its compatibility with the mathematical apparatus and the observative content of QP (that I do not intend to question at all) must be carefully checked. However, SR embodies some operational requirements inherent in the standard interpretation of QP. To be precise, let us consider the general language of a given physical theory: this necessarily contains theoretical terms, and its formalization requires a predicate calculus L^* with quantification on predicative variables. Then, SR assumes that L^* contains a sublanguage L that is observative in the sense that all *atomic* (that is, noncompound) statements of L either regard preparation procedures or measuring procedures that can be directly performed on physical objects (intuitively, this means that an atomic statement of L either takes the form 'the physical object x has been prepared according to the procedure π' or 'the physical object x would yield answer yes if a measurement on it should be done by means of the dichotomic registering device r'). Furthermore, SR requires that all theoretical terms in L* be connected, via *correspondence rules*, to the terms appearing in the observative language L.

The above operational features of SR are obtained within the approach by constructing L as a first order predicate calculus which formalizes adequately the observative languages of CP and QP (the formalization of the general language L* is not required for coping with the problems pointed out in the Introduction). The language L has two sets, $\mathcal{F}$ and $\mathcal{S}$, of (monadic) predicates. The set $\mathcal{F}$ contains only predicates, called *effects* (e.g. Ludwig, 1983), that are interpreted on classes of physically equivalent dichotomic registering devices; a subset $\mathcal{F}_0$ of $\mathcal{F}$, the set of *exact effects*, contains predicates interpreted on idealized devices and can be bijectively mapped on the set of all elementary properties of a given physical system that are *testable* (I say here that a physical property is testable if one can test whether it can be attributed or not to a given physical object: but it is important to note explicitly that testable properties not necessarily are *conjointly*, or simultaneously, testable). The set $\mathcal{S}$ contains only predicates, called *states*, that can be interpreted on classes of physically equivalent preparations of a physical system (this differentiates SR from some modal approaches to QP where states appear as *possible worlds* and belong to the semantics rather than to the syntax of the language of QP). One can then fulfil the minimal realistic requirements underlined at the beginning of this section by endowing L with a correspondence theory of truth, that is, a theory according to which any atomic or compound interpreted sentence has a truth value (true/false) independently of the existence of an experimental apparatus that allows one to test it (in technical terms, L is endowed with a Kripkean semantics, constituted by a family of Tarskian interpretations labeled by laboratories rather than by possible worlds).

The language L has a number of features that are introduced in order to make it adequate to express formally some nontrivial concepts that are usually stated by means of the natural language of physics; in particular, it contains a set of *statistical quantifiers* that are introduced in order to express statistical statements. More important, it provides a suitable environment in which *truth* and *testability* can be distinguished, so that SR avoids the criticism that applies to the standard interpretation of QP. Indeed, while all syntactically correct formulas (*well formed formulas*, or, briefly, wffs) of L have, when interpreted, a truth value, it may occur that this truth value cannot be known by means of suitable experiments. Thus, one can distinguish between *testable* and *non-testable* wffs of L, which constitutes perhaps the main innovation introduced by SR.

However, the sets of testable and nontestable wffs depend on the theory one is dealing with (one can express this more properly by saying that the metalinguistic concept of testability is theory-dependent). Thus, the distinction does not become effective if one does not introduce suitable axioms that specify the physical theory. Some years ago I made a first attempt of stating by means of L a set of axioms that hold both in CP and in QP (Garola, 1991): the two theories are then distinguished by some further axioms on the Lindenbaum-Tarski algebra of L (it is interesting to observe that the set of non-testable wffs is void in CP, nonvoid in QP, as it can be expected from an intuitive viewpoint). Within this framework one can find an answer to some canonical questions regarding QP (Garola, 1992a,b,c; 1993). In particular, the controversial role of Quantum Logic (QL) can be clearly specified. Indeed, QL can be obtained by selecting the subset of all atomic wffs of L of the form F(x), with $F \in \mathcal{F}_0$, that are certainly testable, and restricting the logical order to this subset; hence QL can be seen as a mathematical structure that is picked out within a classical logic (the language L) by using the theory dependent pragmatic concept of testability in QP (from a technical viewpoint, the elements of QL are equivalence classes of wffs of L, and the nondistributive lattice of QL is embedded, in the sense of order, in the Boolean Lindenbaum-Tarski algebra of L, but obviously it is not a subalgebra of it). Furthermore, QP proves to be an incomplete theory (in a well defined technical sense), which helps, in particular, to avoid some old paradoxes following from a quantum description of the EPR experiment.

3. THE MGP PRINCIPLE

In spite of the achievements mentioned at the end of the last Section, SR is seriously challenged by two results that are well known and debated between scholars concerned with the foundations of QP: the Bell-Kochen-Specker (Bell-KS) and Bell theorems (e.g. Mermin, 1993). Indeed, let us recall that all statements in L have a truth value. This contradicts a well known assumption in the standard interpretation of QP, according to which a statement attributing two or more non-compatible properties to a sample of a given physical system is meaningless, but it is not sufficient to invalidate SR, which could simply be offered as an alternative interpretation of QP. Yet the Bell-KS theorem asserts that the truth value of a statement attributing a property belonging to a set of properties that are measured on a physical object may depend on the set, hence it cannot be thought of as prefixed (one usually says that QP is a *contextual theory*). Furthermore, the Bell theorem states that contextuality occurs even if properties are measured that belong to different and spatially separated subsystems of the given physical system (*nonlocality* of QP). Thus, these results seem to show that the adoption of empirical verificationism, or, at least, of a contextual theory of truth, is unavoidable, since it is imposed by results that are internal to QP itself. Should this be true, the fundamental assumptions of SR, that is, the adoption of a correspondence theory of truth, would be invalidated.

In order to cope with this challenge, the original apparatus of SR must be refined and improved. Indeed, the set of axioms mentioned in Sect. 2 applies to noncompound physical systems only, while the Bell and Bell-KS theorems are proved by considering compound physical systems.

A detailed presentation of a generalized version of SR which takes into account compound systems has been provided in some recent papers (Garola, 1995; Garola and Solombrino, 1996a). I will limit myself here to point out some features of the generalized approach that are epistemologically relevant and to illustrate from an intuitive viewpoint how they lead one to circumvent the Bell-KS and Bell theorems.

The first feature follows from a critical analysis of the technical procedure adopted in Hilbert space quantum physics in order to describe compound physical systems. Indeed, this procedure introduces, via linear combination of tensor products of vectors, a class of pure states that are called *second type*, or *entangled*, states. The one-dimensional projections associated to these vectors are usually interpreted, as any other projection, as representing properties of the compound physical system. But if one tries to devise a procedure for measuring a property of this kind, one sees that no direct measurement on a given physical object can allow him to establish whether the property is true or false for the object itself, and that only correlation measurements can be done, which means measuring some elementary properties of the component sub-systems on a great number of samples (ensembles). Therefore, the properties associated to entangled states cannot be considered testable according to the meaning of the word 'testable' specified in Sect. 2. Now, our logical viewpoint comes to our aid, since it suggests that what one actually measures with a correlation measurement is a second order property (that is, a property of first order properties), which cannot be expressed by an atomic wff of the first order language L (though they can be expressed by quantified wffs of L). The projections characterized by entangled states should then be considered as representing merely *theoretical* first order properties, to be considered as labels for the higher order properties mentioned above. It is then remarkable that, from a technical viewpoint, the introduction of these properties within the formal apparatus of SR is mathematically quite natural (though not immediate), since theoretical properties appear as a kind of mathematical limit of suitable sets of standard testable properties whenever compound physical systems are considered. Thus, the subset $\mathcal{F}_0$ of predicates of L that are interpreted on testable properties can be enlarged by means of a set $\mathcal{D}_0$ of predicates interpreted on theoretical properties. By using the set of predicates $\mathcal{E}_0 = \mathcal{F}_0 \cup \mathcal{D}_0$ one can construct a new language L_e, that cannot be classified as observative since not all its atomic wffs are testable, though it still has an observative part. The distinction between testable and nontestable wffs of L can be transferred to L_e: the set of nontestable wffs of L_e then contains molecular wffs where two or more predicates appear expressing testable properties that are not conjointly testable, and also wffs in which predicates appear expressing theoretical properties. This is relevant not only from an epistemological viewpoint, but even for technical reasons. Indeed, the ortho-modular lattice of QL for a compound physical system can be obtained by selecting the subset of all wffs of L_e of the form E(x), with $E \in \mathcal{E}_0$, which contains also nontestable wffs. Now, distinguished scholars, as Aerts (1982), have concluded that a quantum system resulting from the composition of different subsystems is nonseparable. But this result is reached assuming implicitly that all elements in the QL of a compound system are testable. Aerts' puzzling conclusion can then be avoided according to SR.

The second feature of SR to be taken into account here is the possibility of distinguishing in this approach between *semantic* and *pragmatic* compatibility of testable properties. Intuitively, one can say that two properties are semantically compatible (or *consistent*) whenever no contradiction occurs if they are simultaneously attributed to a given physical object, while they are pragmatically compatible whenever they are compatible in the standard sense in QP, that is, they can be tested conjointly in order to know for each of them whether it can be attributed or not to a given physical object (of course, the distinction between the two kinds of compatibility is irrelevant according to the standard interpretation of QP, because of the adoption of empirical verificationism).

I would finally illustrate a third feature of SR, that is, a new principle regarding physical laws (*Metatheoretical Generalized Principle*, or, briefly, MGP), the statement of which is possible since SR supplies a background for a deeper examination of the concept of physical law. Resting on the above distinction between testable and nontestable wffs of L_e, one can characterize *theoretical* and *empirical* physical laws whenever these laws can be stated by means of L_e: to be precise, a law will be said to be theoretical (empirical) if it is expressed by a non-testable (testable) wff of L_e (the possibility of stating theoretical laws in L_e is intuitively apparent because of the existence of predicates interpreted on theoretical properties in this language). Now, a theoretical law cannot be directly verified; one therefore concludes that its metatheoretical role consists in producing, via logical deduction, empirical physical laws that can be directly tested. Then, can we assert that a wff expressing an empirical physical law deduced from an accepted theoretical law must be true in every laboratory? Seemingly, the answer is yes. But a deeper insight into the matter leads one to note that the empirical laws are used in physics together with boundary conditions, or *premises*, in order to obtain physical predictions. In the formalized framework of SR the premises are expressed by wffs of L, or L_e, and it may occur that they attribute properties to a given physical object that are semantically but not pragmatically compatible: in this case, a physical situation is envisaged that is logically conceivable but not epistemically accessible. If one does not want to attribute an ontological status to the theoretical apparatus from which empirical laws are deduced, one cannot assert that these laws still hold in such hypothetical situations, where it is impossible in principle to test them (I stress that this does not mean that the wffs expressing empirical physical laws are meaningless in physical situations that are epistemically non-accessible, as Bohr would say: it only means that one cannot know whether their truth value is 'true' or 'false' in these situations).

The above remarks are formalized in SR by means of the MGP principle. I limit myself here to supply an informal and rather imprecise version of it.

MGP. *A statement expressing an empirical physical law deduced from a general theoretical law is true for every physical object in every laboratory where only pragmatically compatible properties are attributed to each physical object that is considered.*

The MGP principle does not introduce any change in the intuitive conception of empirical physical laws in CP, where all properties are pragmatically compatible, since it then states that these laws hold in every laboratory for all physical objects. But it is relevant in QP (or in every theory where semantic and pragmatic compatibility are distinguished and nontrivial) since it limits the validity of an empirical law to the set of laboratories in which it is not impossible, in principle, to test whether the law holds. This limitation has far-reaching consequences, and, in particular, it allows SR to face successfully the Bell and Bell-KS theorems.

4. THE SOLUTION OF THE PARADOXES

I have shown in some recent papers (Garola, 1994, 1995b,c) that all existing proofs of the Bell theorem rest on assuming the validity of some empirical correlation laws deduced from the general theoretical laws of QP outside the domain of validity established by MGP. Furthermore, Solombrino and myself (Garola and Solombrino, 1996b) have recently proven that this is also the case for the Bell-KS theorem.

Let me look into these subjects in more detail. The result called Bell theorem here was stated by Bell in 1964. Since then, the Bell argument has been refined by a number of authors (e.g. Selleri, 1988), who have supplied accurate analyses of the premises on which it is based, and a number of alternative statements and proofs. Nowadays most authors agree that the Bell theorem mainly shows that QP conflicts with a *locality* assumption, that can be briefly stated as follows.

LOC. *Whenever two subsystems, say 1 and 2, of a compound physical system are sufficiently far apart, a measurement on subsystem 1 (2) does not modify the values of the observables of subsystem 2 (1).*

The older proofs of this theorem essentially show that locality implies inequalities (Bell's inequalities) which are not consistent with QP (e.g. Bell, 1964; Wigner, 1970; Sakurai, 1985; Selleri, 1988). In all these proofs, empirical physical laws are assumed to hold for physical objects having noncompatible properties, which implies a violation of MGP. This is apparent, for instance, in the Wigner and Sakurai proofs, where many subsets of physical objects are considered, in each set being assigned the spin components along different directions, and then an inequality is obtained by applying a perfect correlation law to these objects; it is less apparent in other proofs, as the original Bell's one, where an inequality is obtained regarding sets of physical objects which have non-compatible properties without introducing quantum laws: however, these laws are then used when the inequality is compared with an analogous inequality predicted by QP. Thus, one can conclude that all these older proofs cannot be accepted according to SR.

More recent proofs do not resort to inequalities, but directly show that, by joining locality and QP, one obtains contradictory predictions (e.g. Greenberger *et al.*, 1990; Mermin, 1990, 1993; Clifton *et al.*, 1991). In these proofs, a number of empirical laws are deduced from a general theoretical quantum law, and it is implicitly assumed that they all hold simultaneously. But this contradicts MGP, since the properties predicted by some laws are not compatible with the properties predicted by other laws, hence the statements expressing these laws cannot be simultaneously true. Again, one concludes that none of these proofs can be accepted according to SR.

The Bell theorem is thus invalidated within SR. This implies that QP does not necessarily conflict with LOC. But, then, a crucial question arises. What would happen if one performs a suitable test of a given Bell's inequality? Would the inequality be violated or not? The answer of SR is simple but not trivial. In fact, SR asserts that a Bell inequality either is incorrect, since it is obtained by assuming the unrestricted validity of some quantum laws (e.g. the Wigner and Sakurai proofs) or it is a correct theoretical formula which is not epistemically accessible in QP (e.g. the Bell proof). Any physical experiment actually tests something else (correlations among properties of physical objects in accessible contexts) and obviously yields the results predicted by QP. No contradiction with QP can actually occur, since the inequalities that can be tested in QP cannot be identified with Bell's inequalities if one accepts MGP. Thus, a Bell inequality does not supply a method for testing experimentally whether either QP or LOC is correct, contrary to the belief of most physicists concerned with the foundations of QP. But one can say that the experimental confirmation of the predictions of QP indirectly shows that the empirical physical laws of QP fail to hold in physical contexts that are not directly accessible according to QP itself, otherwise the quantum inequalities would coincide with Bell's (this provides in particular new interesting constraints for hidden variables theories for QP).

Let us come to the Bell-KS theorem. The original proofs of this theorem (Bell, 1966; Kochen and Specker, 1967) were rather complicate, but there are some recent proofs (e.g. Mermin, 1993) that are quite simple and immediate. By considering these proofs, one can show (Garola and Solombrino, 1996b) that they are invalidated, according to SR, by the same argument used in order to reject the proofs of the Bell theorem that do not resort to inequalities. Let me discuss this argument briefly here with reference to the four-dimensional Mermin's proof, thus also providing a pattern for illustrating a typical intuitive application of MGP.

By using Mermin's symbols and procedures, one can introduce a set of nine dichotomic observables (eigenvalues ± 1) for a physical system made up of two subsystems (labeled 1 and 2; the observables are represented in terms of the Pauli matrices for two independent spin $\frac{1}{2}$ particles), arranged as in the following Table.

σ_x^1	σ_x^2	$\sigma_x^1\sigma_x^2$
σ_y^2	σ_y^1	$\sigma_y^1\sigma_y^2$
$\sigma_x^1\sigma_y^2$	$\sigma_x^2\sigma_y^1$	$\sigma_z^1\sigma_z^2$

By multiplying the elements in each row and column in the Table and using the commutation rules of spin-like observables, one gets the following equations.

(4.1) $(\sigma_x^1)(\sigma_x^2)(\sigma_x^1\sigma_x^2) = \mathrm{I}$

(4.2) $(\sigma_y^2)(\sigma_y^1)(\sigma_y^1\sigma_y^2) = \mathrm{I}$

(4.3) $(\sigma_x^1\sigma_y^2)(\sigma_x^2\sigma_y^1)(\sigma_z^1\sigma_z^2) = \mathrm{I}$

(4.4) $(\sigma_x^1)(\sigma_y^2)(\sigma_x^1\sigma_y^2) = \mathrm{I}$

(4.5) $(\sigma_x^2)(\sigma_y^1)(\sigma_x^2\sigma_y^1) = \mathrm{I}$

(4.6) $(\sigma_x^1\sigma_x^2)(\sigma_y^1\sigma_y^2)(\sigma_z^1\sigma_z^2) = -\mathrm{I}$

Each equation contains the product of three observables (the observables in parentheses, which can be further expressed as products of spin-like observables) that pairwise commute, but the observables that appear in one of the products do not necessarily commute with the observables that appear in another product. Now, Eqs. (4.1)–(4.6) cannot be simultaneously satisfied by values of the observables that appear in them, which is canonically interpreted as the proof that no value can be attributed to the observables themselves independently of the set of observations that one decides to perform (*contextuality*).

Let us argue against this conclusion in the framework of SR. The core of the argument consists in noticing that Eqs. (4.1)–(4.6) express general theoretical laws (I also note that they neither can be expressed by means of L nor by means of L_e, since they hold for every state of the physical system, hence stating them formally would require the general language L* mentioned at the beginning of Sect. 2, which admits quantification on state variables; these laws are however *derived* theoretical laws, since each of them is obtained by defining a new observable, that is the product on the left side, and by deducing that it must be equal to the identity because of *primitive* theoretical laws, that is the commutation rules of spin-like observables). Let us consider all the exact effects (properties) associated to the three observables in one of these theoretical laws (I remind that the standard procedure for associating a family of properties to a physical observable Ω consists in considering all pairs (Ω, Δ), with Δ a Borel subset of the real line, and interpreting (Ω, Δ) as the property 'the value of the observable Ω lies in the Borel set Δ'; of course, each property in the family is represented by a projection in Hilbert space quantum theory). Since the observables commute, we expect that all these properties be pragmatically compatible, and that all the empirical laws that can be deduced from the theoretical law (that, roughly speaking, take the form of statements attributing a property with a certain specified probability to all physical objects in a given state) can simultaneously hold in a laboratory *i*. But let us consider the empirical laws deduced from two or more of the laws (4.1)–(4.6) such that the products on the left member contain noncommuting observables (as $\sigma_x^1\sigma_x^2$ in Eq. (4.1) and $\sigma_x^1\sigma_y^2$ in Eq. (4.3). Then, one expects that there are properties associated to these observables that are not pragmatically compatible, so that, taking into account MGP, one cannot assert that all the aforesaid empirical laws can simultaneously hold in *i*: indeed, if some premises and laws are assumed to hold in *i*, and they lead to introduce properties that are not pragmatically compatible, one cannot assume that further empirical laws deduced from Eqs. (4.1)–(4.6) hold in *i*. This result is sufficient to invalidate Mermin's proof of the Bell-KS theorem. For, it means that in every empirical situation the truth values of all statements regarding physical properties of a given physical object (that are defined independently of any observative context) are not bound to respect simultaneously all the conditions imposed by Eqs. (4.1)–(4.6). Thus, the impossibility that the observables appearing in these equations take values that satisfy all them simultaneously does not imply contextuality as defined above, but rather that the values of the observables are not bound to satisfy in every laboratory and in every state Eqs. (4.1)–(4.6) simultaneously (it must however be stressed that no direct test on a physical object can reveal whether these equations are violated).

The results expounded above can be summarized by saying that contextuality and nonlocality of QP, that are taken for granted by most physicists concerned with foundational research, do not necessarily hold if SR is accepted. Since MGP is used, one can also say that this reassuring conclusion is obtained by enhancing the operational attitude of QP, and rejecting the unfair matching of empirical verificationism with a classical conception of physical laws that leads to prove locality and contextuality.

Finally, I would like to close this paper by noticing that SR also proves to avoid the old Furry (1936a,b) and Bohm-Aharonov (1957) quantum paradoxes (Garola and Solombrino, 1996b). This result is obtained by using the distinction between states and effects, and the distinction between semantic and pragmatic compatibility, respectively, and no use is required of the MGP principle.

Department of Physics
University of Lecce

REFERENCES

Aerts, D.: 1982, 'Description of Many Physical Entities without the Paradoxes Encountered in Quantum Mechanics', *Found Phys*, **12**, 1131.

Bell, J.S.: 1964, 'On the Einstein Podolski Rosen Paradox', *Physics*, **1**, 195.

Bell, J.S.: 1966, 'On the Problem of Hidden Variables in Quantum Mechanics', *Rev Mod Phys*, **38**, 447.

Birkhoff, G., von Neumann, J.: 1936, 'The Logic of Quantum Mechanics', *Ann Math*, **37**, 823.

Bohm, D.: 1951, *Quantum Theory*. Prentice-Hall, Englewood Cliffs, N.J.

Bohm, D., Aharonov, Y.: 1957, 'Discussion of Experimental Proofs for the Paradox of Einstein, Rosen and Podolski', *Phys Rev*, **108**, 1070.

Braithwaite, R.B.: 1953, *Scientific Explanation*. Cambridge University Press, Cambridge.

Busch, P., Lathi, P.J., Mittelstaedt, P.: 1991, *The Quantum Theory of Measurement*. Springer-Verlag, Berlin.

Clifton, R.K., Redhead, M.L.G., Butterfield, J.M.: 1991, 'Generalization of the Greenberger-Horne-Zeilinger Algebraic Proof of Nonlocality', *Found Phys*, **21**, 149.

Einstein, A., Podolski, B., Rosen, N.: 1953, 'Can Quantum Mechanical Description of Reality Be Considered Complete?', *Phys Rev*, **47**, 777.

Furry, W.H.: 1936a, 'Note on the Quantum-Mechanical Theory of Measurement', *Phys Rev*, **49**, 393.

Furry, W.H.: 1936b, 'Remarks on Measurements in Quantum Theory', *Phys Rev*, **49**, 476.

Garola, C.: 1991, 'Classical Foundations of Quantum Logic', *Int J Theor Phys*, **30**, 1.

Garola, C.: 1992a, 'Semantic Incompleteness of Quantum Physics', *Int J Theor Phys*, **31**, 809.

Garola, C.: 1992b, 'Quantum Logics Seen as Quantum Testability Theories', *Int J Theor Phys*, **31**, 1639.

Garola, C.: 1992c, 'Truth versus Testability in Quantum Logic', *Erkenntnis*, **37**, 197.

Garola, C.: 1993, 'Semantic Incompleteness of Quantum Physics and EPR-like Paradoxes', *Int J Theor Phys*, **32**, 1863.

Garola, C.: 1994, 'Reconciling Local Realism and Quantum Physics: a Critique to Bell', *Teoreticheskaya i Matematicheskaya Fizika*, **99**, 285.

Garola, C.: 1995a, 'Pragmatic versus Semantic Contextuality in Quantum Physics', *Int J Theor Phys*, **34**, 1383.

Garola, C.: 1995b, 'Criticizing Bell: Local Realism and Quantum Physics Reconciled', *Int J Theor Phys*, **34**, 269.

Garola, C.: 1995c, Questioning Nonlocality: an Operational Critique to Bell's Theorem, in Garola C, Rossi A, eds., *The Foundations of Quantum Mechanics. Historical Analysis and Open Questions*. Kluwer Academic Publishers, Dordrecht.

Garola, C., Solombrino, L.: 1996a, 'The Theoretical Apparatus of Semantic Realism: A New Language for Classical and Quantum Physics', *Found Phys*, **26**, 1121.

Garola, C., Solombrino, L.: 1996b, 'Semantic Realism versus EPR-like Paradoxes: The Furry, Bohm-Aharonov and Bell Paradoxes', *Found Phys*, **26**, 1329.

Greenberger, D.M., Horne, M.A., Shimony, A., Zeilinger, A.: 1990, 'Bell's Theorem without Inequalities', *Am J Phys*, **58**, 1131.

Kochen, S., Specker, E.P.: 1967, 'The Problem of Hidden Variables in Quantum Mechanics', *J Math Mech*, **17**, 59.

Ludwig, G.: 1983, *Foundations of Quantum Mechanics 1*. Springer-Verlag, New York.

Mermin, N.D.: 1990, 'Quantum Mysteries Revisited', *Am J Phys*, **58**, 731.

Mermin, N.D.: 1993, 'Hidden Variables and the Two Theorems of John Bell', *Rev Mod Phys*, **65**, 803.

Popper, K.R.: 1969, *Conjectures and Refutations*. Routledge and Kegan Paul, London.

Russell, B.: 1940, *An Inquiry into Meaning and Truth*. Allen & Unwin, London.

Sakurai, J.J.: 1985, *Modern Quantum Mechanics*. W.A. Benjamin, Reading, (Mass).

Selleri, F.: 1988, History of the Einstein-Podolski-Rosen Paradox, in Selleri, F., ed., *Quantum Mechanics versus Local Realism*. Plenum Press, New York.

Wigner, E.P.: 1970, 'On Hidden Variables and Quantum Mechanical Probabilities', *Am J Phys*, **38**, 1005.

PETER MITTELSTAEDT

INDIVIDUALISTIC VERSUS STATISTICAL INTERPRETATION OF QUANTUM MECHANICS

1. INTRODUCTION – HISTORICAL REMARKS

The quantum mechanical formalism which was discovered by Heisenberg [1] and Schrödinger [2] in 1925 was first interpreted by Born [3] in a statistical sense. The formal expressions $p(\varphi, a_i) = |(\varphi, \varphi^{a_i})|^2$, $i \in \mathbb{N}$ are interpreted as the probabilities to find the value a_i after the measurement of the observable A of the system S with preparation φ. In this improved version the statistical or 'Born interpretation' is used in the present day literature. Usually the statistical (Born) interpretation of quantum mechanics is taken for granted and the formalism of quantum mechanics is considered as a theory which provides statistical predictions. On the other hand, the meaning of the same formal terms $p(\varphi, a_i)$ for an individual system is highly problematic. Heisenberg [4] tried to understand the meaning of the terms $p(\varphi, a_i)$ for a single system by means of the Aristotelian concept of 'potentia'. Popper introduced for the same reason the new concept of 'propensity'. However, since both attempts to understand probability on the individual level don't give rise to any observable prediction, in the current literature the individualistic interpretation of quantum mechanics is not considered as an alternative which should be taken seriously.

In contrast to this well known situation recent results on quantum probability seem to be more in favour of the opposite way of reasoning. Quantum mechanics is then primarily a theory of individual systems and the statistical predictions of the same theory are merely resulting from the behaviour of a large number of individual systems. If this way of reasoning proves to be correct, the statistical (Born) interpretation of quantum mechanics would turn out to be merely a preliminary way of speaking about a domain of reality which can in principle be understood also on the individualistic level. In the present paper this way of reasoning will be elaborated more in detail.

2. INTERPRETATION OF QUANTUM MECHANICS

a. Minimal interpretation

Here we consider the minimal interpretation of quantum mechanics which

M.L. Dalla Chiara et al. (eds.), Language, Quantum, Music, 231–239.

connects the formal terms of the theory with experimental results. The minimal interpretation I_M is the weakest possible interpretation of the quantum mechanical formalism. It is in the spirit of David Hume and the positivism of Ernst Mach and was advocated in particular by Niels Bohr. It avoids any statements about properties of individual objects and instead refers only to observational data, i.e. to the results of a measuring process. Let S be an object quantum system prepared in a pure state φ and $A = \sum a_i P[\varphi^{a_i}]$ a discrete observable with eigenvalues a_i and eigenstates φ^{a_i}. We can then distinguish three postulates which constitute the 'minimal interpretation'.

The *calibration postulate* (C_M):
If φ is an eigenstate φ^{a_i} of A and the system possesses a value a_i of A, then a measurement of A leads to a pointer value Z_i of the measuring apparatus M indicating that the system S had the eigenvalue $a_i = f(Z_i)$ and the state φ^{a_i} before the measurement, where $f(Z)$ is the pointer function. – The postulate (C_M) represents the probability-free part I_M^0 of the minimal interpretation.

The *pointer objectification postulate* (PO):
If φ is not an eigenstate of A, then it is not possible to predict with certainty the pointer value after the measurement of A. However, the postulate (PO) requires that after an A-measurement the pointer possesses some objective value Z_i even if it cannot predicted with certainty.

The *probability reproducibility condition* (RP):
In addition to the pointer objectification the postulate (RP) requires that the formal probability $p(\varphi, a_i) = tr\{P[\varphi]P[\varphi^{a_i}]\}$ which is induced by φ and A, is reproduced in the statistics of the pointer values $Z_i^{(n)}$ after A-measurements on N equally prepared systems $S^n(\varphi)$ with $n = 1, 2, , N$.

b. Quantum theory of measurement

In order to illustrate the relevance of the measuring process for the minimal interpretation we recall some important features of the measuring process. The object system S and the measuring apparatus M are considerd as proper quantum systems the initial states of which are given by φ and Φ, respectivly. In the first step of the measuring process, the *preparation*, the system S and M are considered as a compound system $S + M$ in the state $\Psi(S + M) = \varphi(S) \otimes \Phi(M)$. In the second step, the *premeasurement*, the systems S and M are in interaction, which is described here by a unitary operator $U(t)$ acting on state $\Psi(S + M)$ within the time intervall $0 \leq t \leq t'$. Hence the compound state after the premeasurement reads

$$\Psi'(S + M) = U(t')(\varphi \otimes \Phi).$$

The state $\Psi'(S+M)$ and hence the interaction $U(t)$ is further specified by the calibration postulates (C_M) and (C_R). Indeed, for a system S with the preparation $\varphi = \varphi^{a_i}$ it should hold according to C_M (in the interpretation I_M)

$$U(\varphi^{a_i} \otimes \Phi) = \varphi_i' \otimes \Phi_i.$$

Here the states Φ_i are orthonormal eigenstates of the pointer observable $Z = \sum Z_i P[\Phi_i]$ and φ_i' are the correlated states of S which are not further specified. By means of the expansion $\varphi = \sum c_i \varphi^{a_i}$, with $c_i = (\varphi^{a_i}, \varphi)$, one obtains

$$\Psi' = U(\varphi \otimes \Phi) = \sum c_i U(\varphi^{a_i} \otimes \Phi) = \sum c_i \varphi_i' \otimes \Phi_i.$$

In the third step of the measuring process, *objectification and reading*, the two systems S and M are again dynamically independent but still correlated. As subsystems of the compound system $S+M$ in the state Ψ' the apparatus M can be described by the reduced (mixed) state

$$W_M' = \sum |c_i|^2 P[\Phi_i].$$

According to the *pointer objectification postulate* (*PO*), after the measurement the pointer should possess some objective value Z_i indicating the measuring result $a_i = f(Z_i)$ where f is again the pointer function.

3. THE PROBABILITY THEOREM

At first glance the postulates (C_M), (*PO*) and (*RP*) of the minimal interpretation I_M seem to be independent requirements which characterize this interpretation. However, it turns out, that the condition (*RP*), i.e. the probabilistic part of the interpretation I_M follows from the probability-free interpretation I_M^0. This important and somewhat surprising result is the content of the probability theorem discussed below.

a. The probability theorem in the interpretation I_M

Within the interpretation I_M the calibration postulate C_M corresponds to a probability-free minimal interpteration I_M^0 which merely states, that for a system which possesses a value a_i of A, a measurement of A leads with certainty to the pointer value $Z_i = f^{-1}(a_i)$ which is correlated with the value a_i. The quantum theory of measurement provides explicit exemples of unitary operators U_A which fulfill the requirement of the postulate C_M.

The calibration postulate C_M of the interpretation I_M implies that the post-measurement state of the apparatus system M reads $W_M' = \sum |(\varphi^{a_i}, \varphi)|^2 P[\Phi_i]$. However, the meaning of the coefficients $c_i = (\varphi^{a_i}, \varphi)$ is still open. It is easy to show, that the real positive numbers $p(\varphi, a_i) = |(\varphi^{a_i}, \varphi)|^2$ are probabilities in the formal sense. The interpretation of the probability distribution $p(\varphi, a_i)$ can then be obtained from the probability reproducibility condition (*RP*).

According to this condition (*RP*) the probability distribution $p(\varphi, a_i)$ induced by φ and A is reproduced in the statistics of the pointer values Z_i. In other words, the number $p(\varphi, a_i)$ is the probability to find after the measuring process the pointer value Z_i which indicates that the value $a_i = f(Z_i)$ was measured. This means, that if one would perform a series of N measurements of the observable A of equally prepared systems S_n $(n = 1, 2, , N)$, then the relative frequency $f^N(\varphi, Z_i)$ of pointer values Z_i would approach in some sense the probability $p(\varphi, a_i)$ for $N \to \infty$.

The probability reproducibility condition (*RP*), i.e. the probabilistic part of the minimal interpretation I_M follows from the calibration postulate C_M, i.e. from the probability-free interpretation I_M^0 by means of quantum mechanics, provided this theory is assumed to be universally valid. In particular, it must be presupposed that quantum mechanics can equally be applied to a single object system as well as to a compound system $S^{(N)}$ consisting of N equally prepared systems S_i.

In order to derive this interesting result, we consider N independent systems S_i with equal preparations $\varphi^{(i)}$ as a compound system $S^{(N)}$ in the tensor product state $(\varphi)^N$. For each system S_i a premeasurement of A transforms the initial state Φ of M into mixed state $W'_M = \sum p(\varphi, a_i)P[\Phi_i]$ with eigenstates Φ_i of the pointer observable Z corresponding to pointer values Z_i. If the observable A is measured of each system S_i then the full measuring outcome is given by a sequence of pointer values $\{Z_{l_i}\}$ and pointer states $\{\Phi_{l_i}\}$, respectively, with an index set $l = \{l_1, l_2, , l_N\}$. In the N-fold tensor product Hilbert space $H_M^{(N)}$ of the apparatus the states $\Phi_l^N = \Phi_{l_1} \otimes \Phi_{l_2} \otimes \cdots \otimes \Phi_{l_N}$ form a complete orthonormal basis. For any product state Φ_l^N the relative frequency $f^N(l, k)$ of pointer values Z_k is objectively determind and one can define an observable 'relative frequency of pointer value Z_k' by $F_k^N = \sum_l f^N(k, l)P[\Phi_l^N]$ where the sum runs over all sequences l.

After a premeasurement of A the apparatus is in the mixed state W'_M. The sequence of N measurements is then described by the N-fold product $(W'_M)^N$ of these mixed states. In general this product state is not an eigenstate of the relative frequency operator F_k^N which means that the relative frequency of pointer values Z_k is not an objective property in the mixed state $(W'_M)^N$. However, in the limit of large N the post-measurement product state $(W'_M)^N$ becomes an eigenstate of the relative frequency observable F_k^N and the relative frequency of the pointer value Z_k approaches the probability $p(\varphi, a_k)$. This is the content of the *probability theorem*, which can formally be expressed by the relation

$$\lim_{N\to\infty} tr\{(W'_M)^N (F_k^N - p(\varphi, a_k))^2\} = 0$$

It holds in the minimal interpretation I_M.

b. Discussion of the probability theorem

a. Calibration

For the derivation of the probability theorem the calibration postulate C_M plays a two-fold role. On the level of the single object system the calibration postulate requires, that for the objective properties of the system S the result of a measuring process can be predicted with certainty. In this case the property pertains to the system in an objective sense.[1] On the level of the many-body compound system $S^{(N)}$ the calibration postulate is needed again. Here it requires, that the property 'relative frequency of a value a_k' pertains to the system $S^{(N)}$ whenever the state of the compound system is an eigenstate of a conveniently defined relative frequency observable. The universal validity of quantum mechanics is used here in the sense, that the probability-free minimal interpretation, expressed by the calibration postulate, can equally be applied to the single system S as well as to the compound system $S^{(N)}$.

b. Objectification

It is obvious from the preceeding discussion, that the probability theorem has a physical meaning only, if in addition to the calibration postulate C_M also the pointer objectification (*PO*) is presupposed. The probability reproducibility condition (*RP*) requires that the initial probability $p(\varphi, a_i)$ is reproduced in the statistics of the pointer values. Clearly, this postulate can only be fulfilled if after the measurement the pointer possesses an objective value Z_i, i.e. if the pointer objectification (*PO*) is given. Otherwise one could not count the pointer values and compute the relative frequency $f^N(\varphi, Z_i)$.

4. INDIVIDUALISTIC INTERPRETATION OF QUANTUM MECHANICS

The probability theorem has important consequences for the problem of an individualistic interpretation of quantum mechanics. In order to demonstrate these implications let us briefly recall the usual probabilistic interpretation: For a system S with preparation φ the observables A with $[A, P[\varphi]] = 0$ possess objective values a_i of A. The observables $B = \sum b_k P(b_k)$ with $\forall_k [P(b_k), P[\varphi]] \neq 0$ have no objective value in the state φ. However, to the system S one can attribute probabilities $p(\varphi, b_k)$ which provide some additional information about the values b_k. Indeed, $p(\varphi, b_k)$ is the probability to obtain the result b_k after a measurement of the observable B. The experimental verification of these probability propositions is a statistical one: If one would perform a large number N of B-measurements on equally prepared systems $S_n(\varphi)$, then the relative frequency of outcomes b_k would approach in some sense the value $p(\varphi, b_k)$ in the limit $N \to \infty$.

This means that the probability proposition $P\{p(\varphi, b_k)\}$ is a proposition the referent of which is a compound quantum system $S^{(N)}$ composed of N equally prepared systems $S_n(\varphi)$. The proposition $P\{p(\varphi, b_k)\}$ states that the relative frequency for obtaining a result b_k in the ensemble $S^{(N)}$ approaches the value $p(\varphi, b_k)$ for $N \to \infty$. This is the content of the probability reproducibility postulate.

However, this explanation does not answer the question what the term $p(\varphi, a_i)$ could mean for an individual system S with preparation φ. This question is motivated by the fact, that quantum probabilities express some genuine indeterminism and cannot be reduced to the distribution of hidden inertial conditions. Clearly, a system $S(\varphi)$ possesses objectively values of all observables A with $[A, P[\varphi]] = 0$. On the other hand, probability statements concerning values of non-objective properties can neither be verified nor falsified by experimental means, since the single system can only be measured once.

In order to attach probabilities $p(\varphi, a_i)$ to a single physical system, two often discussed proposals have been made by Heisenberg and by Popper. In the first approach Heisenberg [4] refered to the philosophy of Aristotle. Beeing aware that quantum probabilities must not be considered as quantification of the observers subjective ignorance only, he tried to understand probabilities also as some kind of physical reality which pertains to the object system in question. In this sense he argued that quantum probabilities are a quantitative formulation of the concept of *potentia* in Aristotle's philosophy. Accordingly, in quantum mechanics one should interpret the 'possibility' or 'tendency' which is quantified by $p(\varphi, a_i)$ as some intermediate layer of reality, halfway between the objective reality of quantum systems and the subjective reality of the observers knowledge.

The second attempt to give some meaning to the probability of a single event was made by Popper [8]. He introduced the concept of *propensity* in order to provide a single case interpretation of probability. However, the propensity concept was not primarily conceived as a means for interpreting quantum mechanics. Instead it was understood by Popper as a new concept which can be used for all kinds of probabilistic problems. For this reason there is a subjectivistic interpretation which is without interest for our problems, and an objectivistic interpretation which might be of interest for physics. Within the objectivistic approach one has to distinguish the deterministic case, e.g. dice tossing experiments, and the indeterministic case, which presumably incorporates quantum mechanical situations. Whereas for deterministic situations the propensity concept is often considered to be void [9, 10, 11], some authors claim that propensities might be particulary useful in quantum mechanical situations which are considered to be indeterministic [12].

In spite of these intuitively rather convincing arguments in an actual experimental situation this way of reasoning amounts to saying almost nothing. Firstly, there is no measurable property of a single system which could be interpreted as a probability. Secondly, and this is more important, a pure state φ provides already a maximal information about the system $S(\varphi)$. Indeed, the values of all objective properties can be deduced from the knowledge of φ

and these values will be obtained with certainty by a measuring process which fulfills the calibration postulate.

Moreover, by means of the probability theorem it follows from the general validity of the calibration postulate, that the numbers $p(\varphi, a_i)$ are values of a relative frequency observable F_k^N which pertains to an ensemble $\{Z_{l_i}\}^N$ of N pointer values Z_{l_i} obtained by A-measurements on N equally prepared systems $S_n(\varphi)$ – in the limit $N \to \infty$. In other words, on the basis of the probability-free interpretation I_M^0 the values $p(\varphi, a_i)$ turn out to be objective properties of the ensemble $\{Z_{l_i}\}^N$. These arguments show, that for an individual system $S(\varphi)$ there is no need for an additional 'potentia' or 'propensity', since from the state φ one can deduce all objective properties, even the statistical ones which turn out to be yes–no properties of a conveniently defined ensemble. For a single system $S(\varphi)$ the totality of probabilities $p(\varphi, a_i)$, $p(\varphi, b_k)$ expresses nothing but the fact that φ is the state of S.

Summarizing the preceeding arguments we arrive at the following result. A single quantum system $S(\varphi)$ with preparation φ possesses values a_i of all objective observables $A = \sum a_i P(a_i)$. According to the calibration postulate these values can be measured with certainty. For nonobjective observables B nothing is known about its value b_k and no prediction can be made for the result of a B-measurement. The probabilities $p(\varphi, b_k)$ which are induced by φ and B do not refer to the system $S(\varphi)$ but to an ensemble $\{Z_{l_i}\}^N$ of pointer values or to an ensemble $\{S_i(\varphi^{b_{l_i}})\}^N$ of N equally prepared systems $S_i(\varphi)$ after measurements of B. According to the probability theorem the values $p(\varphi, b_k)$ are eigenvalues of the relative frequency observable of the ensemble $\{Z_{l_i}\}^N$ in the limit of infinite N.

Obviously the probability-free interpretation I_M^0 is completely sufficient for the interpretation of a quantum system. Any interpretation of an individual system, which goes beyond the calibration postulates making use of 'metaphysical' concepts like *potentiality* or *propensity* is purely hypothetical and no longer justified by quantum mechanics. There is no need and no room for further interpretations of the theoretical terms $p(\varphi, a_i)$. For an individual system S with preparation φ the values $p(\varphi, a_i)$ give no additional information except that S is in the state φ. Hence one gets the impression that the problem of a single case interpretation of quantum mechanical probabilities is merely a pseudo-problem.

5. CONCLUDING REMARKS

Since the discovery of quantum mechanics in the twenties this theory was considerd as intimately related to the concepts of uncertainty, indeterminacy and probability. Within the framework of different interpretations these concepts were understood as subjective ignorance, as deficiency of information, as objective indeterminacy, as potentiality or as propensity. Without refering to one of these meanings the statistical content of the concepts was incorporated into the 'minimal interpretation' of quantum mechanics.

In contrast to this well established understanding of quantum theory the probability theorem has led to the surprising result, that quantum mechanics is primarily not a probabilistic theory, but a theory which is concerned with yes–no propositions. A result of this kind was already conjectured by several authors, e.g. Everett [13], Finkelstein [14], Hartle [15], Graham [16] and DeWitt [17]. However, instead of referring to the reduced state of the object system or of the apparatus after the measurement, these authors prove an analog of the probability theorem for a pure state preparation prior to the measurement. The present investgations have shown that the original conjectures of the authors mentioned were nevertheless correct and can be justified within the quantum theory of measurement by means of the probability theorem mentioned.

Institute for Theoretical Physics
University of Cologne

NOTES

[1] In Einsteins terminology [7] this means, that an objective property is an 'element of the physical reality'.

REFERENCES

[1] Heisenberg, W.: 1925, 'Über quantentheoretische Umdeutung kinematischer und mechanischer Beziehungen', *Zeitschrift für Physik*, **33**, 879–893.

[2] Schrödinger, E.: 1926, 'Quantisierung als Eigenwertproblem', *Annalen der Physik*, **79**, 361–376.

[3] Born, M.: 1926) 'Zur Quantenmechanik der Stoßvorgänge', *Zeitschrift für Physik*, **37**, 863–867.

[4] Heisenberg, W.: 1959, 'Die Plancksche Entdeckung und die physikalischen Probleme der Atomphysik', *Universitas*, **14**, 135–148.

[5] Busch, P., Lahti, P., Mittelstaedt, P.: 1991, *The Quantum Theory of Measurement*. Springer, Heidelberg, 2nd ed.,1995.

[6] Mittelstaedt, P.: 1993, 'The Measuring Process and the Interpretation of Quantum Mechanics', *Int J Theor Phys*, **32**, 1763–1775.

[7] Einstein, A., Podolsky, B., Rosen, N.: 1935, 'Can Quantum-Mechanical Description of Physical Reality be Considered Complete?', *Phys Rev*, **32**, 777–780.

[8] Popper, R.K.: 1957, The Propensity Interpretation of the Calculus of Probability and the Quantum Theory, in Körner, S., ed., *Observation and Interpretation in the Philosophy of Physics*. Butterworth, London.

[9] Sklar, L.: 1970, 'Is Probability a Dispositional Property', *J Philosophy*, **67**, 355–366.

[10] Giere, R.: 1973, Objective Single-Case Probabilities, in Suppes, P. *et al.*, eds., *Logic, Methodology and Philosophy of Science*. 457-483.

[11] Schneider, C.: 1994, 'Two Interpretations of Objective Probabilities', *Philosophia Naturalis*, **31**, 107–131.

[12] Popper, R.K.: 1982, Quantum Theory and the Schism of Physics, Postscript III to the Logic of Scientific Discovery, Bartley, W.W. III, ed. Hutchinson, London, Melbourne 1988.

[13] Everett, H.: 1957, The Theory of the Universal Wave Function Theses. Princeton University. Reprinted in DeWitt, B.S., Graham, N., eds., *The Many Worlds Interpretation of Quantum Mechanics*. Princeton University Press, Princeton N.Y. 1973; 1–140.

[14] Finkelstein, D.: 1962, The Logic of Quantum Physics, *Transactions of the New York Academy of Science*, **25**, 621–637.

[15] Hartle, J.B.: 1968) Quantum Mechanics of Individual Systems, *Am J Phys*, **36**, 704–712.

[16] Graham, N.: 1973, The Measurement of Relative Frequency, in DeWitt, B.S., Graham, N., eds., *The Many–Worlds Interpretation of Quantum Mechanics*. Princeton University Press, Princeton N.Y; 229–252.

[17] DeWitt, B.S.: 1971, The Many-Universes Interpretation of Quantum Mechanics, in *Foundations of Quantum Mechanics*. IL Corso, Academic Press, New York; 167–218.

F.A. MULLER

THE LOCALITY SCANDAL OF QUANTUM MECHANICS

HISTORICAL INTRODUCTION

Einstein, Podolsky and Rosen (EPR) illustrated their celebrated 1935 argument against the completeness of quantum mechanics with a *Gedankenexperiment* [9]. They considered a 1-dimensional two-particle system in a peculiar improper quantum-mechanical state such that either the momentum or the position of one particle is predictable (i.e. probability equal to 1) as soon as the position or the momentum of the other particle is measured. Bohm's 1951 version of EPR's *Gedankenexperiment* in terms of the spin $\frac{1}{2}$-singlet state (EPRB experiment):

$$\frac{1}{\sqrt{2}}\begin{pmatrix}1\\0\end{pmatrix}\otimes\begin{pmatrix}0\\1\end{pmatrix}-\frac{1}{\sqrt{2}}\begin{pmatrix}0\\1\end{pmatrix}\otimes\begin{pmatrix}1\\0\end{pmatrix}=\frac{1}{\sqrt{2}}\begin{pmatrix}0\\1\\-1\\0\end{pmatrix} \tag{1}$$

yields a proper state in the 2×2-dimensional tensor-product Hilbert space $\mathbb{C}^2\otimes\mathbb{C}^2$ [5]. Although according to the standard interpretation of quantum mechanics [25] physical magnitudes do not possess values before a measurement is performed, the values appearing upon measurements of spin directed oppositely along *any*, at-the-last-minute chosen spatial axis will exhibit perfect correlation, even if the measurements are performed simultanously but zillion miles away from each other. Thus Einstein commented that the particles seem to communicate 'telepathically' [8]. Assuming instead that two spin magnitudes of each particle do possess values before measurement (Realism) that cannot be changed by measurements performed elsewhere (Locality), J.S. Bell derived in 1964 an inequality, pertaining to an ingenious combination of four EPRB experiments (the Bell-experiment), which contradicts quantum mechanics [4]. Hence Bell's Theorem: quantum mechanics is incompatible with Local Realism. In 1983 Aspect performed the Bell-experiment in the laboratory: the results confirmed quantum mechanics and refuted Bell's inequality [1]. Hence Bell's Fact: nature is incompatible with Local Realism.

Bell's Theorem and Fact have boosted foundational investigations of quantum mechanics as to what precisely formulated locality conditions are

M.L. Dalla Chiara et al. (eds.), Language, Quantum, Music, 241–248 .

violated by quantum mechanics (and by nature), and as to whether locality violations are in conflict with the special theory of relativity [28]. The reigning policy is that of Peaceful Coexistence between quantum mechanics and the special theory of relativity [28, Ch. 8]. The pillar of this policy is the acknowledgement of the existence of a benign kind of non-locality (outcome dependence) which however cannot be exploited to signal, superluminally or otherwise [7,13]. Thus a full scale war between quantum mechanics and the special theory of relativity seems avoided.

THE LOCALITY SCANDAL

All foundational-philosophical investigations contain however three lacunae which according to this author's loud mouth constitute The Locality Scandal of Quantum Mechanics.

Lacuna 1

Whereas the notions space and time figure prominently in the foundational-philosophical investigations of locality, and whereas the notions of space and time are necessarily presupposed in the very idea of *locality,* space and time are absent from the Bohm singlet (1). Moreover, space and time never enter the relevant formulae; they are served as ingredients of a verbal sauce on the mathematical pasta which do not have counterparts in that pasta. Not one member of the impressive parade of purported locality conditions formulated in quantum-mechanical language as for instance in Redhead [28], d'Espagnat [10], Van Fraassen [11], Eberhard [7] and Ghirardi [13], even *mentions* space or time. Locality conditions are always written as equations in which only various spin probability measures figure, e.g. Fine's factorisability and Shimony's outcome independence and parameter independence.

To fill Lacuna 1 means (i) to give a description of the Bell-experiment where Galilean space-time is part of the mathematical pasta, if only in order to express mathematically that the spin measurements are performed simultaneously in different regions of space, (ii) to formulate a locality condition of a manifestly spatio-temporal nature, and (iii) to prove Bell's Theorem in space-time.

Lacuna 2

All descriptions of the Bell-experiment are in the frame-work of orthodox quantum mechanics [25], which is in fact a *Galilean-covariant* theory [2] and therefore not supposed to satisfy a locality condition at all. On the contrary, it should *violate* locality generically just as the *actio in distans* potentials of Galilean-covariant classical mechanics violate locality generically. Strikingly this is hardly ever mentioned! Galilean space-time is the 'wrong' space-time to discuss locality. The sole two exceptions, i.e. who do leave the inappropriate Galilean-covariant frame-work and enter a Lorentz-covariant one for discussing locality, are Smith and Weingard [29], who provide a relativistic formulation of the EPR's original *Gedankenexperiment* with spinless particles in a Diracian 'constraint formalism', and Mittelstaedt and Stachow [21], who provide a description of the measurement outcomes of the EPRB experiment using *relativistic quantum logic*. Authors such as Clifton *et al.* [6], Ghirardi and Grassi [14], who claim to adress the EPR-paradox in a manner which meets the exigencies of the special theory of relativity, in actual fact fob us off with fair promises.

To fill Lacuna 2 is just like filling Lacuna 1 but now for Minkowski space-time. In particular, a Lorentz-covariant description of Aspect's experiments [1] is needed, which is not to be found in the extant literature. We emphasise that such a description is impossible in the frame-work of orthodox quantum mechanics because the particles in Aspect's experiments are *photons*. Photons are the relativistic particles *par excellence*; their behaviour is not governed by the Schrödinger equation.

Lacuna 3

The distinction between the concepts of *locality* and *separability* is usually emphasised, and traced back to Einstein by Howard [18], but their relation to Hegerfeldt's spatio-temporal concept of *localisability* is hardly ever discussed [16]. But separability, in so far as conceived as a spatio-temporal concept, suffers in the literature from the same defects as locality: neither space nor time is mentioned in the various separability conditions. To fill Lacuna 3 this should be accomplished and connections between all these spatio-temporal concepts should be established.

Needless to say that filling these three Lacunae in the extant literature in a manner which meets modest standards of informal rigour requires more space than we have at our disposal *hic et nunc*. We shall therefore only sketch *how* the first two Lacunae can be filled, and further fob the reader off with a fair promise: [22].

Mark the author's words: he is not saying that everybody should serve *these* particular dishes whenever discussing locality, separability, the EPR-paradox and the Bell-experiment; he is saying that these particular dishes should be on the menu for those who taste deliciously.

Ad. 1. Space and Time

In quantum mechanics both *space* and *time* have a peculiar status. As for example Michael Friedman [12] has demonstrated, classical mechanics, and the special and the general theory of relativity can all be re-formulated as so-called *space-time theories*. In such theories space and time merge into the prime ontological substance *space-time*, mathematically represented as a differentiable manifold. So far no such re-formulation has proved possible for quantum mechanics, i.e. quantum mechanics is not a space-time theory, not even a Galilean-covariant one. Within orthodox quantum mechanics *time* is not treated on a par with the other physical magnitudes because it reasonably cannot be mathematically represented by a self-adjoint operator, whereas all other physical magnitudes are; time is represented by a real parameter. In contrast to time, position is represented by a self-adjoint operator, *viz.* by the wave-mechanical multiplication operator. How to arrive from the spectrum of the multiplication operator, which is the real number set $\mathbb{R}^3 \equiv \mathbb{R}\times\mathbb{R}\times\mathbb{R}$, at Euclidean *space* (or a structure isomorphic to it) by employing the Galilean symmetry group of quantum mechanics, requires further analysis we set aside.

(i) The state space of two spin-12-particles wherein space, time and spin are all taken into account is the tensor-product Hilbert-space $\mathcal{H}_{\text{Gal}} \equiv L^2(\mathbb{R}^3) \otimes \mathbb{C}^2 \otimes L^2(\mathbb{R}^3) \otimes \mathbb{C}^2$. The elements of $\mathcal{H}_{\text{Gal}}$ are quadruples of two-particle wave functions, usually written as a four-component Pauli-bispinor $\Psi(t)$. Due to the Exclusion Principle the anti-symmetric spin singlet state (1) must be combined with a symmetric spatial state in $L^2(\mathbb{R}^6)$, say

$$\chi_{\text{sym}}(\mathbf{q}_1, \mathbf{q}_2, t) \equiv \big[\alpha(\mathbf{q}_1, t)\beta(\mathbf{q}_2, t) + \alpha(\mathbf{q}_2, t)\beta(\mathbf{q}_1, t)\big]/\sqrt{2}\,, \tag{2}$$

where $\alpha(\mathbf{q}_1, t)$ and $\beta(\mathbf{q}_2, t)$ are normalised members of $L^2(\mathbb{R}^3)$ and solutions of the one-body free Schrödinger equation in $L^2(\mathbb{R}^3)$ of particles 1 and 2, respectively. The solution of the free Schrödinger equation in $\mathcal{H}_{\text{Gal}}$, which is mathematically equivalent to $4 \times 2 \times 3 = 24$ independent free Schrödinger equations in $L^2(\mathbb{R})$, is just the product of χ_{sym} (2) and the Bohm singlet (1), because the free Hamiltonian is spin-independent.

Elements of Physical Reality. In stead of spin probabilities we should speak of *joint probabilities for spin and position at a time.* Only then can we legitimately say that the spin of particle 1 is measured *here* and the spin of particle 2 is measured *simultaneously over there.* If one considers spin probabilities only, one is simply not licensed to speak of measuring a property of a particle *here* rather than *elsewhere.* If and only if one considers joint probabilities of spin and position, one can legitimately claim to possess a description of the Bohm singlet *in Galilean space-time* $\mathcal{G}$. Luckily, we can consider such joint probabilities because the multiplication operator and the Pauli spin operators commute.

Hegerfeldt's theorem states that if a wave-function is prepared such that it has bounded support at $t = 0$ (localised state), then the support is unbounded at any later time $t > 0$ [16]. This means that joint probabilities for spin and position (any finite spatial region, such as occupied by a laboratory) equal to 1 are possible at the instant of preparation only. As a consequence, Redhead's modernisation of EPR's incompleteness argument [28, pp. 71–75], where the particles of a Bohm singlet are supposed to fly through space in opposite directions as time goes by, breaks down. This happens because EPR's sufficient condition of joint probability equal to 1 for inferring spin-position elements of physical reality is never satisfied as time goes by.

(ii) Since Galilean space-time does not have a light-cone structure such as Minkowski space-time, it is not possible to formulate a locality condition in a natural manner. For the sake of charity, let us ignore this *pro tem* and approach the issue in a disputable but rigorous manner. The definition of locality below is inspired by Hellman [15] and the author's criticism of Hellman [23].

Consider the theory of quantum mechanics as a class of set-theoretical structures ('models'). A quantum-mechanical model (restricted to pure states for expediency) is an ordered set $\langle \mathcal{H}, \psi, \mathcal{O}, \mathcal{G}, \mathcal{P} \rangle$, where $\psi : \mathbb{R} \rightarrow \mathcal{H}$, $t \mapsto \psi(t)$ is the path in Hilbert-space $\mathcal{H}$ of the physical system under consideration, $\mathcal{O}$ is a set of self-adjoint operators acting in $\mathcal{H}$, $\mathcal{G}$ is Galilean space-time (only included if $\mathcal{O}$ contains position operators), and P is a set of probability measures that ψ and members of $\mathcal{O}$ give rise to. In a coordinate frame of reference F we define a light-cone whose surface is determined by the speed of light in F. Since the speed of light is not a constant in $\mathcal{G}$, we are going to add in our definition of locality the odd proviso that models with Galilean space-times can only be compared in one and the same frame of reference – this will block spurious locality violations obtained by merely choosing another (moving) frame of reference. One should not be able to violate a locality condition by re-describing the same physical situation from a moving train! The *future domain of light-dependence* $D^{+}(R) \subset \mathcal{G}$ of spatio-temporal region $R \subset \mathcal{G}$ we define as the region that R slices of the top of a past light-cone. (The 'correct' definition of future domain of dependence of $R \subset \mathcal{G}$ yields the entire future because there is no upperbound of speed – but we have agreed to ignore this *pro tem.*) The notion of *agreement on spatio-temporal region R* between two models M and M' we define as coinciding joint probabilities for every common physical magnitude $\widehat{A} \in \mathcal{O} \cap \mathcal{O}'$ and $\widehat{E}^{Q}(\Delta)$, for all $t \in (t_1, t_2)$, where t_1 and t_2 are the temporal boundaries of R in frame F, and where $\widehat{E}^{Q}(\Delta)$ is the spectral position projector on an arbitrary spatial region Δ such that $(\Delta, t) \subseteq R$ – of course under the condition that for the states during time-interval (t_1, t_2) the joint probabilities are well-defined, i.e. the commutator $[\widehat{E}^{Q}(\Delta), \widehat{A}]$ acting on $\psi(t)$ of M and on $\psi'(t)$ of M' for all $t \in (t_1, t_2)$ must vanish.

Definition: quantum mechanics is *past stochastic local* iff for any two models M and M', having the same frame of reference F on their Galilean space-time manifolds $\mathcal{G}$, it holds for each bounded spatio-temporal region $R \subset \mathcal{G}$ that if M and M' agree on R, then they agree on its future domain of light-dependence $D^{+}(R) \subset \mathcal{G}$. The notions of *future stochastic locality* and *stochastic locality* are obtained by obvious modifications.

It is an elementary exercise to think of a quantum-mechanical *Gedankenexperiment* by choosing a free one-particle Hamiltonian in one model, say M, and the same Hamiltonian in the other model, say M', but now containing an additional potential term (graviational potential, Coulomb potential) which is smoothly switched on. Then one has a proof for the theorem that quantum mechanics is *not* a past stochastic local theory. We can also imagine the proverbial Bob and Alice signalling superluminally by means of an *actio in distans* potential. The validity of the No-Signalling Theorems in quantum mechanics [7,13], [28, p. 116] is restricted to situations where we have two particles *which do not interact*. Hence there is no *general* Peaceful Coexistence between quantum mechanics and the special theory of relativity – not even an Armed Truce, as Peacock would have it [27] – but a full scale war; there is only Peaceful Coexistence when we narrow our perspective to models pertaining to Bell-type experiments.

At the risk of belabouring the obvious, this is, of course, not to say that Bell's Theorem, stating the incompatibility between Galilean-covariant quantum mechanics and Local Realism, does not deserve our perplexity *despite* the fact that non-relativistic theories are not supposed to be local. In the case of the Bell-experiment there simply is no interaction, in particular no *actio in distans* potential, which is taken to carry over to relativistic descriptions. The fact that the quantum-mechanical states form a *vector space*, which gives rise to entangled states such as the singlet (1), is the first cause of our perplexity – and this has little to do with Locality [19] but everthing with Realism. A Bell theorem such as that some given *relativistic quantum theory is incompatible with Local Realism* deserves our perplexity however *at face value*: a theory designed in Minkowski space-time suggests to harbour locality necessarily because it rules out the very possibility of 'telepathical communication', whereas the theory still barks at Local Realism.

Ad. 2.

(i) One way of describing the Bohm singlet in Minkowski space-time $\mathcal{M}$ is to use the two-body Dirac equation [3]. The Bohm state becomes a $4 \times 4 = 16$ component Dirac-bispinor, which is a member of the following tensor-product Hilbert-space: $\mathcal{H}_{\text{Mink}} \equiv L^2(\mathbb{R}^3) \otimes \mathbb{C}^4 \otimes L^2(\mathbb{R}^3) \otimes \mathbb{C}^4$. The least bad choice for a position operator in Dirac theory is the one of Newton and Wigner [26], which is the Foldy-Wouthuysen transformed multiplication operator. For the spin operator we choose the Lorentz-covariant spin operator of Hilgevoord and Wouthuysen, which turns spin into a conserved quantity – in contradistinction to the usual non-Lorentz-covariant Pauli spin operator [17].

In order to describe Aspect's experiment in $\mathcal{M}$, a quantised Maxwell equation with a particle-interpretation provides an appropriate avenue to walk on [24]. The Newton-Wigner position operator does not work for photons; but the Kraus operator does [20].

(ii) In $\mathcal{M}$ the appropriate definitions of *stochastic locality* are exactly those given *supra* but without the odd frame proviso. (These definitions can also be used to apply to local deterministic and local stochastic hidden variables theories conceived as space-time theories.)

FINISH

Since we run out of space we must hit the breaks. We hope to have made a plausible case that it is not mere Technical Detail Fetishism that brings the author to such bombastic phrases as The Locality Scandal. What is at stake here is a description meeting comparatively mundane standards of informal rigour of a *Gedankenexperiment* and an actual experiment from which two of the most discussed topics in the foundations and philosophy of quantum mechanics spring: locality and separability. We also hope to have made it plausible that filling Lacuna 2 is a less trivial exercise than prima facie might be conceived.

Faculty of Physics and Astronomy
Utrecht University

ACKNOWLEDGMENTS

Helpful remarks of H.W. de Regt and D.G.B.J. Dieks are gratefully acknowledged.

REFERENCES

[1] Aspect, A., Grangier, P., Roger, G.: 1981, 'Experimental Tests of Realistic Local Theories via Bell's Theorem', *Phys Rev Lett*, **47**, 460–467; 1982, *Phys Rev Lett*, **49**, 91–94; Aspect, A., Dalibard, J., Roger, G.: 1982, 'Experimental Tests of Bell's Inequalities Using Time-Varying Analyzers', *Phys Rev Lett*, **49**, 1804–1807.

[2] Ballentine, L.: 1989, *Quantum Mechanics*. Englewood Cliffs: Prentice Hall; pp. 49–65, 74–80.

[3] Barut, A.O., Ünal, N.: 1986, 'An Exactly Soluble Relativistic Quantum Two-Fermion Problem', *J Math Phys*, **27**, 3055–3060.

[4] Bell, J.S.: 1987, *Speakable and Unspeakable in Quantum Mechanics*. Cambridge: Cambridge University Press; paper 2 from 1964, pp. 14–21.

[5] Bohm, D.: 1951, *Quantum Theory*. Englewood Cliffs: Prentice-Hall; pp. 611–623.

[6] Clifton, R. *et al.*: 1992, Relativity, Quantum Mechanics, and EPR, in Hull, D. *et al.*, eds., *PSA*, Vol. 1, East Lansing, Michigan: Philosophy of Science Association; pp. 114–128.

[7] Eberhard, P.H.: 1978, 'Bell's Theorem and the Different Concepts of Locality', *Nuov Cim*, **46B**, 392–419.

[8] Einstein, A.: 1949, Autobiographical Notes, in Schilpp, P.A., ed., *Albert Einstein: Philosopher-Scientist*. La Salle, Illinois: Open Court; p. 85.

[9] Einstein, A., Podolsky, B., Rosen, N.: 1935, 'Can Quantum-Mechanical Description of Physical Reality be Considered Complete?', *Phys Rev*, **47**, 777–780.

[10] d'Espagnat, B.: 1976, *The Conceptual Foundations of Quantum Mechanics, 2nd edn*. Reading, Massachusetts: W.A. Benjamin; pp. 76–130.

[11] van Fraassen, B.C.: 1991, *Quantum Mechanics. An Empiricist View*. Oxford: Clarendon Press; pp. 87–93, 99–105, 345–348, 363–370.

[12] Friedman, M.: 1980, *Foundations of Space-Time Theories*. Princeton: Princeton University Press; p. 32, 87, 125.

[13] Ghirardi, G.C., Rimini, A., Weber, T.: 1980, A General Argument against Superluminal Transmission through the Quantal Measurement Process, *Lett Nuov Cim*, **27**, 293–298.

[14] Ghirardi, G.C., Grassi, R.: 1994, 'Outcome Predictions and Property Attribution: the EPR Argument Reconsidered', *Stud Hist Phil Mod Sc*, **25**, 397–423.
[15] Hellman, G.: 1982, 'Stochastic Einstein Locality', *Synthese*, **53**, 461–564.
[16] Hegerfeldt, G.C.: 1974, 'Remark on Causality and Particle Localization', *Phys Rev*, **D**, 3320–3321.
[17] Hilgevoord, J., Wouthuysen, S.A.: 1963, 'On the Spin Angular Momentum of the Dirac particle', *Nucl Phys*, **40**, 1–12.
[18] Howard, D.: 1985, 'Einstein on Locality and Separability', *Stud Hist Phil Sc*, **16**, 171–201.
[19] Jones, M.R.: 1989, What Locality Isn't: A Response to Jarret, in Kafatos, M., ed., *Bell's Theorem, Quantum Theory and Conceptions of the Universe*. Dordrecht: Kluwer; pp. 77–79.
[20] Kraus, K.: 1977, Position Observables of the Photon, in Price, W.C., Chissick, S.S., eds., *The Uncertainty Principle and the Foundations of Quantum Mechanics*. London: John Wiley & Sons.
[21] Mittelstaedt, P., Stachow, E.W.: 1983, 'Analysis of the Einstein-Podolsky-Rosen Experiment by Relativistic Quantum Logic', *Int J Th Phys*, **22**, 517–540.
[22] Muller, F.A., Work in Progress (conditional on funding).
[23] Muller, F.A.: 1994, 'On Stochastic Einstein Locality in Algebraic Relativistic Quantum Field Theory', *Int J Th Phys*, **33**, 91–102.
[24] Nelson, T.J., Good, R.H.: 1969, 'Description of Massless Particles', *Phys Rev*, **179**, 1445–1449.
[25] von Neumann, J.: 1932, *Mathematical Foundations of Quantum Mechanics*, translated by R.T. Beyer, Princeton: Princeton University Press, 1955.
[26] Newton, T.D., Wigner, E.P.: 1949, 'Localised States for Elementary Systems', *Rev Mod Phys*, **21**, 400–406.
[27] Peacock, K.A.: 1991, *Peaceful Coexistence or Armed Truce? Quantum Nonlocality and the Space-Time View of the World*. PhD Thesis in Philosophy, University of Toronto.
[28] Redhead, M.L.G.: 1987, *Incompleteness, Nonlocality, and Realism. A Prolegomenon to the Philosophy of Quantum Mechanics*. Oxford: Clarendon Press; pp. 75–118.
[29] Smith, G.J., Weingard, R.: 1986, 'A Relativistic Formulation of the Einstein-Podolsky-Rosen Paradox', *Found Phys*, **17**, 149–171.

MIRKO NAVARA AND PAVEL PTÁK

UNCERTAINTY AND DEPENDENCE IN CLASSICAL AND QUANTUM LOGIC – THE ROLE OF TRIANGULAR NORMS

1. THREE TYPES OF UNCERTAINTY

In the statistical description of real systems, we often encounter the following three types of uncertainty.

1. **Probabilistic uncertainty**. This uncertainty occurs in considerations affected by unknown circumstances.

2. **Quantum uncertainty**. This type of uncertainty occurs when the observations of the system cause irreversible changes of states. A typical example is a quantum experiment.

3. **Fuzzy uncertainty**. This uncertainty occurs when we study sets of events whose truth values are not necessarily only *true* or *false*.

Various mathematical structures have been invented and pursued in the study of systems with these three types of uncertainties. In this note, we take up the Boolean logics for the probabilistic uncertainties, the quantum logics for quantum uncertainties, and the fuzzy logics for fuzzy uncertainties. Obviously, the quantum logics and the fuzzy logics are both generalizations of Boolean logics. A common generalization of the quantum and fuzzy logics seems to be desirable and, to our knowledge, does not seem to be available in a straightforward way [4, 11, 19]. We raise the question of dependence/independence of two events in the respective logics. The three above-mentioned types of uncertainty correspond to three types of dependence in these structures.

2. PRELIMINARIES – TRIANGULAR NORMS

Triangular norms are usually viewed as fuzzy generalizations of the Boolean conjunction. However, they were studied in the early sixties in the area of probabilistic metric spaces (see [22]), and some of them even earlier. Let us recall their basic properties relevant to our considerations.

A *t-norm* (triangular norm) is an operation $T : [0,1]^2 \to [0,1]$ which is commutative, associative, monotone in each component, and which satisfies

M.L. Dalla Chiara et al. (eds.), Language, Quantum, Music, 249–261.

the boundary condition $T(1,u) = u$ (see e.g. [3,22]). The function $\neg : u \mapsto 1-u$ is the standard fuzzy negation. The *dual t-conorm* to T is the operation $S : [0,1]^2 \to [0,1]$ defined by the (de Morgan) formula $S(u,v) = \neg T(\neg u, \neg v) = 1 - T(1-u, 1-v)$.

The *Frank family of t-norms* T_s, $s \in [0,\infty]$, will play an essential role in the sequel. For $s \in (0,\infty) \setminus \{1\}$, the Frank t-norms are defined by the formula

$$T_s : (u,v) \mapsto \log_s\left(1 + \frac{(s^u-1)(s^v-1)}{s-1}\right)$$

The limit cases coincide with the most frequently used t-norms:

$$T_0 = T_{\mathbf{M}} : (u,v) \mapsto \min(u,v) \text{ (minimum t-norm)},$$
$$T_\infty = T_{\mathbf{L}} : (u,v) \mapsto \max(u+v-1,0) \text{ (Łukasiewicz t-norm)},$$
$$T_1 = T_{\mathbf{P}} : (u,v) \mapsto u \cdot v \text{ (product t-norm)}.$$

The following property, characteristic for the Frank t-norms, was proved in [7].

Theorem 2.1 : *Let T be a t-norm non-decomposable to a nontrivial ordinal sum and S its dual t-conorm. The equality*

$$T(u,v) + S(u,v) = u + v$$

is satisfied for all $u, v \in [0,1]$ iff T belongs to the Frank family of t-norms.

We shall need also the following property of Frank t-norms.

Theorem 2.2 : *For each $s \in [0,\infty]$ and each $u, v \in [0,1]$, the following equality holds:*

$$T_{1/s}(1-u,v) = v - T_s(u,v).$$

Remark 2.3: We also take into account the limit case $\{s, 1/s\} = \{0,\infty\}$. The latter theorem gives also one of the implications in Th. 2.1 as an easy consequence:

$$T_s(u,v) = v - T_{1/s}(1-u,v) = v - (1-u) + T_s(1-u,1-v) = v + u - S_s(u,v),$$

where S_s is the dual t-conorm to T_s.

PROOF: The cases of $s \in \{0,1,\infty\}$ are trivial. In the remaining cases, we obtain:

$$T_{1/s}(1-u,v) = \log_{1/s}\left(1+\frac{(s^{-(1-u)}-1)(s^{-v}-1)}{s^{-1}-1}\right) =$$

$$= -\log_s\left(\frac{1-s(s^{u-1}-1)(s^{-v}-1)}{s-1}\right) = -\log_s\frac{s-1-s^{u-v}+s^{1-v}+s^u-s}{s-1} =$$

$$= -\log_s\left(s^{-v}\cdot\frac{-s^v-s^u+s+s^{u+v}}{s-1}\right) = v-\log_s\frac{-s^v-s^u+s+s^{u+v}}{s-1} =$$

$$= v-\log_s\left(1+\frac{(s^u-1)(s^v-1)}{s-1}\right) = v-T_s(u,v)$$

The following property of the Frank family of t-norms was proved by Frank and Takacs (see [3,7]).

Theorem 2.3 : *For fixed $u, v \in (0,1)$, the function*

$$F_{u,v} : s \mapsto T_s(u,v)$$

is continuous and strictly decreasing. It is a bijection of $[0,\infty]$ and $[T_{\mathbf{L}}(u,v), T_{\mathbf{M}}(u,v)]$.

3. UNCERTAINTY IN CLASSICAL LOGIC

In the classical logic, the event structure is described by a Boolean algebra L. (Countable operations will not be of interest in this paper because we will only deal with finite subalgebras.)

A *state* on L is a mapping $m : L \to [0,1]$ such that:

(s1) $m(0) = 0, \quad m(1) = 1$

(s2) $\forall x, y \in L: \; m(x \vee y) + m(x \wedge y) = m(x) + m(y)$ (the valuation property).

Let $a, b \in L$. The (Boolean) subalgebra $L_{a,b}$ of L generated by a, b is a homomorphic image of the free Boolean algebra F_{BA} with 2 free generators (F_{BA} is isomorphic to 2^4). In order to describe the state space of $L_{a,b}$, it is sufficient to characterize the state space of F_{BA}. We denote by x, y the free generators of F_{BA}.

Let m be a state on F_{BA}. An easy calculation shows that m is uniquely determined by its values on x, y, and $x \wedge y$. While $m(x)$, $m(y)$ can be arbitrary numbers from $[0,1]$, the value $m(x \wedge y)$ is constrained by some inequalities.

Proposition 3.1. Let F_{BA} be the free Boolean algebra with 2 free generators x, y.

1. If m is a state on F_{BA}, then

$$T_{\mathbf{L}}(m(x), m(y)) \leq m(x \wedge y) \leq T_{\mathbf{M}}(m(x), m(y)).$$

2. If $u, v, w \in [0, 1]$ such that

$$T_{\mathbf{L}}(u, v) \leq w \leq T_{\mathbf{M}}(u, v),$$

then there is a state m on F_{BA} such that $m(x) = u$, $m(y) = v$, $m(x \wedge y) = w$.

PROOF: 1. As

$$(x \wedge y) \vee (x \wedge y') = x,$$

$$(x \wedge y) \wedge (x \wedge y') = 0,$$

we obtain

$$m(x \wedge y) + m(x \wedge y') = m(x) + m(0) = m(x)$$

and $m(x \wedge y) \leq m(x)$. Similarly, $m(x \wedge y) \leq m(y)$. Thus

$$m(x \wedge y) \leq \min(m(x), m(y)) = T_{\mathbf{M}}(m(x), m(y)).$$

As

$$m(x) + m(y) - m(x \wedge y) = m(x \vee y) \leq 1,$$

we get

$$m(x \wedge y) \geq m(x) + m(y) - 1.$$

This, together with $m(x \wedge y) \geq 0$, gives

$$m(x \wedge y) \geq T_{\mathbf{L}}(m(x), m(y)).$$

2. We define $m(x \wedge y) = w$, $m(x \wedge y') = u - w$, $m(x' \wedge y) = v - w$, $m(x' \wedge y') = 1 - u - v + w$, and extend m uniquely to a state on F_{BA}. □

We see that the minimum t-norm $T_{\mathbf{M}}$ gives the best upper estimate of $m(x \wedge y)$ corresponding to the case of maximal positive dependence between x and y. The Łukasiewicz t-norm $T_{\mathbf{L}}$ gives the best lower estimate corresponding to the maximal negative dependence. Also the product t-norm is found to play an important role: The events x, y are stochastically independent iff $m(x \wedge y) = m(x) \cdot m(y) = T_{\mathbf{P}}(m(x), m(y))$.

Suppose that values $m(x), m(y)$ of a state m on $F_{\mathbf{BA}}$ are given. If $m(x) \in \{0,1\}$ or $m(y) \in \{0,1\}$, then m is uniquely determined. If $m(x), m(y) \in (0,1)$, there remains one degree of freedom. Obviously, the value $m(x \wedge y)$ is sufficient to determine the state m uniquely. However, it is reasonable to search for a parameter which attains its extreme values exactly at the bounds given by Prop. 1 and which changes its sign when an event is replaced with its complement. Let us suggest the following notion.

Definition 3.2: A *degree of probabilistic dependence*, p, is a mapping which assigns a value $p_m(x, y)$ to each state m and events x, y such that $m(x), m(y) \in (0,1)$, and which is subject to the following axioms:

(p1) $p_m(x, y) = p_m(y, x)$,
(p2) $p_m(x', y) = -p_m(x, y)$,
(p3) if $m(x_1) = m(x_2)$, $m(y_1) = m(y_2)$ and $m(x_1 \wedge y_1) < m(x_2 \wedge y_2)$, then $p_m(x_1, y_1) < p_m(x_2, y_2)$,
(p4) $p_m(x, y) = 0$ iff $m(x \wedge y) = T_{\mathbf{P}}(m(x), m(y))$,
(p5) $p_m(x, y) = 1$ iff $m(x \wedge y) = T_{\mathbf{M}}(m(x), m(y))$,
(p6) $p_m(x, y) = -1$ iff $m(x \wedge y) = T_{\mathbf{L}}(m(x), m(y))$.

Let us comment on the above notion. If $m(x) \in \{0,1\}$ or $m(y) \in \{0,1\}$, then $p_m(x, y)$ is undefined. The axioms express the following requirements: The value $p_m(x, y)$ is zero if and only if x, y are stochastically independent. For fixed $m(x), m(y) \in (0,1)$ and $m(x \wedge y)$ ranging in $[T_{\mathbf{L}}(m(x), m(y)),\ T_{\mathbf{M}}(m(x), m(y))]$, the values $p_m(x, y)$ cover the whole interval $[-1, 1]$. Moreover, $p_m(x, y)$ increases with the increase of $m(x \wedge y)$. If one of the arguments is replaced by its complement, the degree of probabilistic dependence changes its sign. The axiom (p2) has the following motivation: Let x_1, x_2, y_1, y_2 satisfy the assumption of (p3) and let

$$m(x_2 \wedge y_2) - m(x_1 \wedge y_1) = \varepsilon.$$

Then

$$m(x_2' \wedge y_2) - m(x_1' \wedge y_1) = -\varepsilon,$$

$$m(x_2 \wedge y_2') - m(x_1 \wedge y_1') = -\varepsilon,$$

$$m(x_2' \wedge y_2') - m(x_1' \wedge y_1') = \varepsilon.$$

Therefore an increase ε of the degree of probabilistic dependence of x, y should be considered as a decrease of the degree of probabilistic dependence of x', y.

The axiom (p2) makes the role of x, x' 'antisymmetric'. As a consequence of (p2), we obtain a weaker condition

(p2–) $p_m(x', y') = p_m(x, y)$.

It seems plausible that these axioms for a degree of probabilistic dependence are well motivated and natural. Obviously, a question arises whether such a mapping always exists.

The most natural idea is to try a correlation-like parameter. The correlation of two random variables ξ, η on L is

$$corr(\xi, \eta) = \frac{E(\xi \cdot \eta) - E\xi \cdot E\eta}{\sqrt{E(\xi - E\xi)^2 \cdot E(\eta - E\eta)^2}},$$

where E denotes the mean value. If ξ, η attain only the values 0,1 and if we identify $E\xi, E\eta, E(\xi \cdot \eta)$ with $m(x), m(y), m(x \wedge y)$, resp., we obtain the formula

$$corr_m(x, y) = \frac{m(x \wedge y) - m(x) \cdot m(y)}{\sqrt{m(x) \cdot m(x') \cdot m(y) \cdot m(y')}}.$$

The parameter *corr* (taken for p) satisfies (p1)–(p4). However, (p5) and (p6) are not satisfied unless $m(x) = m(y)$. Therefore, a modified parameter is desirable.

Another attempt – a linear interpolation of the bounds from Prop. 1 – gives us the formula

$$lin_m(x, y) = \frac{2 \cdot m(x \wedge y) - T_{\mathbf{M}}(m(x), m(y)) - T_{\mathbf{L}}(m(x), m(y))}{T_{\mathbf{M}}(m(x), m(y)) - T_{\mathbf{L}}(m(x), m(y))}.$$

The parameter *lin* satisfies (p1)–(p3) and (p5)–(p6), but it does not satisfy (p4).

We see that a more involved formula for the degree of probabilistic dependence is needed. In view of the fact that the values of m on $F_{\mathbf{BA}}$ are uniquely determined by $m(x)$, $m(y)$ and $m(x \wedge y)$, we can express the intended degree of probabilistic dependence in the form

$$p_m(x, y) = f(m(x), m(y), m(x \wedge y)),$$

where f is a suitable function (called the *generating function* of p). The function f is defined for all triples of arguments u, v, w such that $u, v \in (0, 1)$ and $w \in [T_{\mathbf{L}}(u, v), T_{\mathbf{M}}(u, v)]$. The axioms (p1)–(p6) for a degree of probabilistic dependence p are then equivalent to the following conditions for f:

(f1) $f(u,v,w) = f(v,u,w)$,
(f2) $f(1-u,v,v-w) = -f(u,v,w)$,
(f3) $w_1 < w_2 \Longrightarrow f(u,v,w_1) < f(u,v,w_2)$,
(f4) $f(u,v,T_{\mathbf{P}}(u,v)) = 0$,
(f5) $f(u,v,T_{\mathbf{M}}(u,v)) = 1$,
(f6) $f(u,v,T_{\mathbf{L}}(u,v)) = -1$.

The property (p2–) corresponds to
(f2–) $f(1-u,1-v,1-u-v+w) = f(u,v,w)$.

Let us suggest to construct the degree of probabilistic dependence as follows: Let the contours of the degree of probabilistic dependence be the graphs of Frank t-norms. In other words, let $p_m(x,y)$ attain a constant value $g(s)$ (depending only on s but not on x,y,m) for all arguments satisfying $m(x \wedge y) = T_s(m(x),m(y))$. In terms of the generating function f of p, it is required that $f(u,v,T_s(u,v)) = g(s)$ for all $u,v \in (0,1)$. Using the inverse, $F_{u,v}^{-1}$, of the function $F_{u,v}$ from Th. 2.3, we can substitute $s = F_{u,v}^{-1}(w)$, $w = T_s(u,v)$. We obtain

(F) $f(u,v,w) = g(F_{u,v}^{-1}(w))$.

The following theorem tells us when the latter formula gives rise to (the generating function of) a degree of probabilistic dependence.

Theorem 3.1: *The function f defined by* (F) *is a generating function of a degree of probabilistic dependence iff $g : [0,\infty] \to [-1,1]$ is a decreasing bijection such that*
(g2) $g(1/s) = -g(s)$

for all $s \in (0,\infty)$. The corresponding degree of probabilistic dependence, d, is defined by setting

$$d_m(x,y) = g(F_{m(x),m(y)}^{-1}(m(x \wedge y))).$$

Alternatively, $d_m(x,y) = g(s_{x,y})$, where $s_{x,y}$ is the unique element of $[0,\infty]$ such *that $T_{s_{x,y}}(m(x),m(y)) = m(x \wedge y)$.*

PROOF: 1. Let g be a function satisfying the conditions of the theorem and let f be defined by (F). We shall verify conditions (f1)–(f6).

Condition (f1) follows from the commutativity of T_s. As g is a decreasing bijection of $[0,\infty]$ onto $[-1,1]$, $g(0) = 1$ and $g(\infty) = -1$, which ensures (f5), (f6). Taking $s = 1$ in (g2), we obtain $g(1) = -g(1) = 0$, which implies (f4). The functions g, $F_{u,v}$ and $F_{u,v}^{-1}$ are strictly decreasing, so f being a composition of two of them is strictly increasing in its third argument – condition (f3) is verified.

It remains to check the validity of (f2), i.e., it remains to verify the equality

$$g(F^{-1}_{1-u,v}(v-w)) = -g(F^{-1}_{u,v}(w)).$$

In the equality from Th. 2.2, we substitute $w = T_s(u, v)$, $s = F^{-1}_{u,v}(w)$, and obtain

$$T_{1/s}(1-u, v) = v - w,$$

$$F^{-1}_{1-u,v}(v-w) = 1/s,$$

$$g(F^{-1}_{1-u,v}(v-w)) = g(1/s) = -g(s) = -g(F^{-1}_{u,v}(w)),$$

which we were to prove.

2. Let (F) define a generating function of a degree of probabilistic dependence. To prove that the properties of g are necessary, it is sufficient to compare them with the considerations from the first part of the proof. □

Remark 3.4: Let $h : [0,1] \to [0,1]$ be a decreasing bijection. We may define $g : [0,\infty] \to [-1,1]$ by the formula

$$g(s) = \begin{cases} h(s) & \text{if s} \le 1, \\ -h(1/s) & \text{if s} > 1. \end{cases}.$$

Then g satisfies the conditions of Th. 3.1 and formula (F) defines a generating function of a degree of probabilistic dependence. Thus, Th. 3.1 gives a family of degrees of probabilistic dependence which differ by a nonlinear transformation of the range $[-1,1]$.

However, Th. 3.1 probably does not give all solutions of (p1)–(p6). (No other examples are known to us at this moment.) If a degree of probabilistic dependence is defined by means of (F), the Frank t-norms cannot be replaced by any other family of t-norms. Indeed, (f2–) implies

$$F^{-1}_{1-u,1-v}(1-u-v+w) = F^{-1}_{u,v}(w).$$

Denote this value by s. This means that

$$T_s(u, v) = w,$$

$$T_s(1-u, 1-v) = 1-u-v+w.$$

Using the dual t-conorm, S_s, in the latter equality,

$$1 - S_s(u, v) = 1 - u - v + w,$$

we obtain the formula

$$T_s(u, v) + S_s(u, v) = u + v.$$

This is characteristic for the Frank t-norms (Th. 2.1).

4. UNCERTAINTY IN QUANTUM LOGICS

In quantum logic, the phenomenon of noncompatibility requires a more general structure than Boolean algebras. Recall that two events are called *noncompatible* if they are not simultaneously observable. In the corresponding mathematical model, two events are noncompatible if they do not belong to a Boolean subalgebra of the event structure (for more detailed exposition, see [1, 9]).

The event structure of a quantum system is often supposed to be an *orthomodular lattice* (OML), L. In comparison with a Boolean algebra, in OMLs we relax the absorption laws

$$a \wedge (a' \vee b) = a \wedge b, \quad a \vee (a' \wedge b) = a \vee b$$

as well as the distributivity law. We replace them by the *orthomodular law*

$$a \leq b \Longrightarrow b = a \vee (a' \wedge b).$$

Formally, an OML is a bounded lattice L with a unary operation $' : L \to L$ (*orthocomplementation*) such that it satisfies the following conditions

1. $a'' = a,$

2. $a \leq b \Longrightarrow b' \leq a',$

3. $a \vee a' = 1$

and the orthomodular law. A *state* on an OML can be defined by conditions (s1), (s2) in the same way as in Section 3. Thus, we say that $m : L \to [0, 1]$ is a state if

(s1) $m(0) = 0, \quad m(1) = 1$

(s2) $\forall x, y \in L: \; m(x \vee y) + m(x \wedge y) = m(x) + m(y)$

(It should be noted that these "modular-like" states have been investigated by several authors, see e.g. [2, 14, 18, 20, 23], etc.).

We shall make use of free algebras. Let L be an OML and let $a, b \in L$. The subalgebra (sub-OML) $L_{a,b}$ of L generated by a, b is a homomorphic image of the free OML F_{OML} with 2 free generators. In order to describe the state space of $L_{a,b}$, it is sufficient to characterize the state space of F_{OML}. Let us denote by x, y the free generators of F_{OML}. The structure of F_{OML} is described in [1, 9]. The *commutator*,

$$c(x,y) = (x \vee y) \wedge (x \vee y') \wedge (x' \vee y) \wedge (x' \vee y'),$$

plays an important role in the description. (In a Boolean algebra, the commutator is always zero.) It is known that F_{OML} is isomorphic to the direct product $B \times M$, where $B = \{z \in F_{\mathrm{OML}}: z \leq \mathrm{c(x,y)}'\}$ and $M = \{z \in F_{\mathrm{OML}}: z \leq \mathrm{c(x,y)}\}$. (The intervals B, M are given the natural OML structure inherited from L.) The factor B is a Boolean algebra isomorphic to 2^4, freely generated by $x \wedge \mathrm{c(x,y)}'$, $y \wedge \mathrm{c(x,y)}'$. The factor M is the modular ortholattice $\{0, \mathrm{c(x,y)}, x \wedge \mathrm{c(x,y)}, x' \wedge \mathrm{c(x,y)}, y \wedge \mathrm{c(x,y)}, y' \wedge \mathrm{c(x,y)}\}$, known as MO2. (Its maximal element is $1_M = \mathrm{c(x,y)}$.)

Let m be a state on F_{OML}. It is of the form

$$m(z) = (1 - q_m(x,y)) \cdot m_B(z \wedge c(x,y)') + q_m(x,y) \cdot m_M(z \wedge c(x,y))$$

for some $q_m(x,y) \in [0,1]$ and states m_B, m_M, on B, M, resp. As m_B is a classical state, m_M can be interpreted as a "purely non-Boolean state". A nonzero value of the commutator $c(x,y)$ indicates the presence of quantum effects in the observations of x and y, and $m(\mathrm{c(x,y)})$ represents their probability. As the projections $x \wedge \mathrm{c(x,y)}, y \wedge \mathrm{c(x,y)}$ (of x, y to M) have the same probability proportional to $m(\mathrm{c(x,y)})$ (see the theorem below), we may consider the parameter $q_m(x,y) = m(\mathrm{c(x,y)})$ as the *degree of quantum dependence* between x and y at state m.

Theorem 4.1 (see also [5, 10, 14, 21]) *The OML M (i.e. MO2) admits only one state, m_M. This state attains the value $1/2$ on all the elements $x \wedge c(x,y)$, $x' \wedge c(x,y)$, $y \wedge c(x,y)$, $y' \wedge c(x,y)$.*

PROOF: Let u, v be arbitrary two different elements of $\{x \wedge \mathrm{c(x,y)}, x' \wedge \mathrm{c(x,y)}, y \wedge \mathrm{c(x,y)}, y' \wedge \mathrm{c(x,y)}\} = M \setminus \{0, 1\}$. We apply the valuation property (s2) to u, v. As $u \vee v = 1_M$, $u \wedge v = 0$, we obtain

$$m(u) + m(v) = m(1_M) + m(0) = 1.$$

This equality for all combinations of $u, v \in M \setminus \{0, 1\}$ implies that m attains $1/2$ on all these elements. □

Let us comment on the previous result. If we require the condition (s2) to be satisfied only for $x \leq y'$, we obtain a weaker definition of a state. This definition is, however, often adopted in quantum mechanics (see [8,17]). In the case of Boolean algebras, both definitions coincide. With this weaker definition, the values $m(x \wedge \mathrm{c}(x,y))$, $m(y \wedge \mathrm{c}(x,y))$, may be any numbers from the interval $[0, m(\mathrm{c}(x,y))]$ and we obtain two more degrees of freedom for weaker states.

The state m_B is a classical state subject to the characterization in the preceding section. As a consequence, each state m on F_{OML} is uniquely determined by $m(x)$, $m(y)$, $m(x \wedge y)$ and $m(\mathrm{c}(x,y))$. The parameter $m(\mathrm{c}(x,y)) = q_m(x,y)$ is the degree of quantum dependence. Again, the parameter $m(x \wedge y)$ can be replaced by a more appropriate degree of probability dependence, $p_m(x,y)$, defined now as the degree of probability dependence of the classical state m_B. Explicitly, $p_m(x,y) = p_{m_B}(x \wedge \mathrm{c}(x,y)', y \wedge \mathrm{c}(x,y)')$.

5. UNCERTAINTY IN FUZZY LOGIC – A PERSPECTIVE

Despite many studies in recent years, the notion of fuzzy logic does not seem to be standardized. Various mathematical structures – MV-algebras [12], T-tribes [3], fuzzy quantum spaces [6, 16], etc. – are suggested to represent events in a fuzzy logic. Also states on fuzzy logics – fuzzy measures – allow for different generalizations (some of them being analogous to the definition of Section 3). Therefore our basic question on degrees of dependence leads us to a series of questions on various definitions of fuzzy logics and states. What is typical for most of them is the absence of the law of contradiction and the excluded middle law. Thus

$$a \wedge a' = 0, \quad a \vee a' = 1$$

need not hold. (Even if these laws are satisfied in some mathematical models, all fuzzy models admit that there is a nonzero element a such that $a \leq a'$. This is, however, excluded in both classical and quantum logic.) For most of the definitions of fuzzy measure, m, the value $m(a \wedge a')$ may be nonzero. If this is the case, the values of m on a fuzzy sublogic generated by a, b depend on new degrees of freedom (expressed, e.g. by $m(a \wedge a')$ and $m(b \wedge b')$) which are typical for a fuzzy logic.

Among the mathematical structures describing fuzzy logic, only few are equational classes (e.g. MV-algebras). Hence not all fuzzy logics allow for the existence of a "free fuzzy logic with 2 free generators". Those fuzzy logics which do – like e.g. MV-algebras – could be treated analogously to our previous analysis. In general, the question of dependence of two events in a fuzzy logic needs a further study. It is conceivable that it may help to clarify some properties of various concepts of fuzzy logic. Also, a contribution to the investigation of fuzzy quantum logics, which are common generalizations of quantum and fuzzy logics – see [4, 11, 19], could be achieved.

6. CONCLUSION

In a classical (=Boolean) logic, we have defined a degree of probabilistic dependence between two events, a, b. It corresponds to one degree of freedom in the description of a state m with fixed values $m(a), m(b)$. A reasonable degree of dependence can be obtained using the Frank family of t-norms – a family of connectives which were typical for fuzzy logical considerations but were not applied yet in classical or quantum logics.

In a quantum logic, a new degree of freedom appears, corresponding to a degree of quantum dependence. Thus, two events in a quantum logic have two independent degrees of dependence – probabilistic and quantum.

In fuzzy logic, new degrees of dependence may appear. They seem to require another generalization of the degree of probabilistic dependence. This is largely open to further research.

Faculty of Electrical Engineering
Czech Technical University, Prague

NOTE

Research supported by the project no. VS96049 of the Czech Ministry of Education and the grant no. 201/96/0117 of the Grant Agency of the Czech Republic.

REFERENCES

[1] Beran, L.: 1984, *Orthomodular Lattices. Algebraic Approach*, Academia, Praha.

[2] Birkhoff, G.: 1973, *Lattice Theory*, 3rd edition, AMS, Providence, USA.

[3] Butnariu, D., Klement, E.P.: 1993, *Triangular Norm-Based Measures and Games with Fuzzy Coalitions*. Kluwer, Dordrecht.

[4] Dalla Chiara, M.L., Cattaneo, G., Giuntini, R.: 1993, 'Fuzzy Intuitionistic Quantum Logic', *Studia Logica*, **52**, 1–24.

[5] de Lucia, P., Pták, P.: 1992, 'Quantum Probability Spaces That Are Nearly Classical', *Bulletin of the Polish Academy of Sciences – Math*, **40**, 163–173.

[6] Dvurečenskij, A., Riečan, B.: 1991, 'On Joint Distribution of Observables for F-Quantum Spaces', *Fuzzy Sets and Systems*, **39**, 65–73.

[7] Frank, M.J.: 1979, 'On the Simultaneous Associativity of $F(x,y)$ and $x+y-F(x,y)$', *Aequationes Math*, **19**, 194–226.

[8] Gudder, S.P.: 1979, *Stochastic Methods in Quantum Mechanics*. North Holland, New York.

[9] Kalmbach, G.: 1983, *Orthomodular Lattices*. Academic Press, London.

[10] Majerník, V., Pulmannová, S.: 1992, 'Bell Inequalities on Quantum Logics', *Journal of Mathematical Physics*, **33**, 2173–2178.

[11] Mesiar, R.: 1995, 'Do Fuzzy Quantum Structures Exist?' *International Journal of Theoretical Physics*, **34**, 1609–1614.

[12] Mundici, D.: 1986, 'Interpretation of AF C*-algebras in Łukasiewicz Sentential Calculus', *Journal of Functional Analysis*, **65**, 15–63.

[13] Navara, M.: 1994, 'Algebraic Approach to Fuzzy Quantum Spaces', *Demonstratio Math*, **27**, No.3, 589–600.

[14] Navara, M.: 'On Generating Finite Orthomodular Sublattices', *Tatra Mountains Mathematical Publications*, **10**, to appear.

[15] Navara, M., Pták, P.: 1993, 'P-Measures on Soft Fuzzy σ-Algebras', *Fuzzy Sets and Systems*, **56**, No. 1, 123–126.

[16] Piasecki, K.: 1985, 'Probability of Fuzzy Events Defined as Denumerable Additivity Measure', *Fuzzy Sets and Systems*, **17**, 271–284.

[17] Pták, P., Pulmannová, S.: 1991, *Orthomodular Structures as Quantum Logics*. Kluwer Academic Publishers, Dordrecht/Boston/London.

[18] Pták, P., Pulmannová, S.: 1994, 'A Measure-Theoretic Characterization of Boolean Algebras Among Orthomodular Lattices', *Comment Math Univ Carolin*, **35**, 205–208.

[19] Pykacz, J.: 1992, 'Fuzzy Set Ideas in Quantum Logics', *International Journal of Theoretical Physics*, **31**, 1765–1781.

[20] Riečanová, Z.: 1988, 'Topology in a Quantum Logic Induced by a Measure', *Proceedings of the Conference on Topology and Measure* **V**, Greifswald, 126–130.

[21] Salvati, S.: 1994, 'A Characterization of Boolean Algebras', *Ricerche di Matematica*, **43**, 357–363.

[22] Schweizer, B., Sklar, A.: 1983, *Probabilistic Metric Spaces.* North-Holland, New York.

[23] Weber, H.: 1995, 'Valuations on Complemented Lattices', *International Journal of Theoretical Physics*, **34**, 1799–1806.

[illegible] Lectures on [illegible]

[illegible]

[illegible] Handbook of [illegible] 1971 [illegible]

[illegible] (Crossroads [illegible]) [illegible]

[illegible]

[illegible] 1958 [illegible]

[illegible]

SYLVIA PULMANNOVÁ

QUANTUM LOGICS AND QUANTUM MEASUREMENTS

A classical experiment is described by a triple $(\Omega, \mathcal{F}, \mathcal{P})$, where $\mathcal{F}$ is a σ-algebra of subsets of a nonempty set Ω, and P is a probability measure on $\mathcal{F}$. $\mathcal{F}$ is interpreted as the set of all random events of the experiment.

A generalization of the classical probability theory is given in the quantum logic approach. The Kolmogorovian triple $(\Omega, \mathcal{F}, \mathcal{P})$ is replaced by a couple (L, m), where L is a quantum logic, which is interpreted as the set of all quantum random events. Most usually, L is assumed to be an orthomodular σ-lattice or poset, and m is a state on L. The quantum logic approach was initiated by the famous paper by G. Birkhoff and J. von Neumann [BN] "The logic of quantum mechanics" which appeared in 1936, where the authors suggested a more general model for the event structure of a quantum mechanical experiment than the classical Boolean algebra. A probabilistically motivated axiom system for a quantum logic as an orthomodular σ-poset has been suggested by G. Mackey in 1963 [M]. From the algebraic point of view, distributivity which holds in $\mathcal{F}$ is replaced by a weaker property – orthomodularity. For basic properties of quantum logics and their physical interpretation see [B], [K], [BC], [PP], [D1], [V].

In the traditional Hilbert space approach, a physical system is described by a Hilbert space H (complex, separable, infinite dimensional), and its quantum logic is then the lattice $L(H)$ of all closed linear subspaces (equivalently, the lattice of all orthogonal projections) of H.

Let us assume that we are given two independent physical systems with quantum logics L_1 and L_2, and we wish to regard these systems as a single one. The logic L of the joint system is usually called a 'tensor product' of L_1 and L_2, in notation $L = L_1 \otimes L_2$. In the Hilbert space approach, if two systems described by Hilbert spaces H_1 and H_2 are coupled to form a composite system, this composite system is described by the Hilbert-space tensor product $H_1 \otimes H_2$, hence the corresponding quantum logic is $L(H_1 \otimes H_2)$. Although in the category of Hilbert spaces tensor product is well-defined, there are two non-isomorphic candidates for a 'tensor product' of the Hilbert space logics $L(H_1)$ and $L(H_2)$, namely $L(H_1 \otimes H_2)$ and $L(H_1 \otimes \bar{H}_2)$, where $\bar{H}_2$ is the dual Hilbert space of H_2 (see [AD], [Ma], [P1]. In fact, in the category of orthomodular lattices no suitable 'tensor product' for the description of coupled physical systems has been found (see e.g. [RF], [A]).

A special kind of a tensor product is needed if we wish to describe one

M.L. Dalla Chiara et al. (eds.), Language, Quantum, Music, 263–268.

classical and one quantum system as a joint system. This situation arises by quantum measurements, where a quantum observable is to be measured by a measurement device, which is usually supposed to be a classical object. In the case that the classical system is described in terms of a phase space Ω and its quantum logic L_2 is modeled by the Boolean algebra B of all subsets of Ω, the tensor product $L_1 \otimes L_2$ (where L_1 is the quantum logic) has been described by several authors ([BC], [AD], [P1], [Z], [HK]) as the direct product $L = \prod_{\omega\in\Omega} L_1$ of copies of L_1 indexed by points ω in Ω. It turns out that this direct product is a special case of a more general structure, a Boolean power of a logic L with respect to a complete Boolean algebra B. In [P2], a quantum logic approach to quantum measurements is suggested, in which the joint system consisting of a quantum system and classical measuring apparatus is described in terms of a Boolean power. Some basic concepts of the theory of quantum measurements have been reformulated in frame of this approach [P3], [P4]. In this contribution, we give a brief survey of these results. We start with some necessary definitions.

An *orthomodular* (σ) *– poset* is a partially ordered set L with 0 and 1 as the smallest and the greatest element, respectively, endowed with a unary operation – orthocomplementation, such that: (i) $a'' = (a')' = a$; (ii) $a \le b \Rightarrow b' \le a'$; (iii) $a \vee a' = 1$; (iv) $(a_i)_{i\in I}$, $a_i \le a_j'$, $i \ne j \Rightarrow \bigvee_{i\in I} a_i \in L$, for any finite (countable) set I; (v) $a \le b \Rightarrow b = a \vee (a' \wedge b)$.

Condition (iv) with at most countable sets I expresses the σ-completeness of L. If (iv) holds only for finite sets N, then L is an orthomodular poset. Condition (v) is the orthomodular law. We say that $a, b \in L$ are *orthogonal* $(a \perp b)$ if $a \le b'$.

A *state* on L is a mapping $m : L \to [0, 1]$ such that $m(1) = 1$ and $a \perp b \Rightarrow m(a \vee b) = m(a) + m(b)$. A state m is σ-additive (completely additive) if $m(\bigvee_{i\in I} a_i) = \sum_{i\in I} m(a_i)$ for any countable (arbitrary) set I of mutually orthogonal elements of L such that $\bigvee_{i\in I} a_i$ exists in L.

An *observable* on L is σ-homomorphism $X : \mathcal{F} \to L$, where $(\Omega, \mathcal{F})$ is a measurable space, which is called the *value space* of X. An observable X is *discrete* if there is a countable set $\{\omega_i\}_{i\in N}$ such that $X(\{\omega_i\}_{i\in N}) = 1$. Most usually, $(\Omega, \mathcal{F}) = (R^n, \mathcal{B}(R^n))$, $\mathcal{B}(R^n)$ is Borel σ-algebra. A *spectrum* of such an observable is the smallest closet set $C \subset R^n$ such that $X(C) = 1$.

Let L be a logic, B a complete Boolean algebra. A *Boolean power* $L[B]$ is the set of all functions

$$L[B] := \{f : L \to B : f(a_1) \wedge f(a_2) = 0_B, a_1 \ne a_2; \bigvee_{a\in L} f(a) = 1_B\}.$$

A Boolean power endowed with appropriately defined partial order and orthocomplementation, is an orthomodular poset (or lattice if L is a lattice), not necessarily σ-complete. (See [Gr] for the general definition and properties of Boolean powers, and [P2] for properties of Boolean powers of orthomodular posets.)

For example, if $L = \mathcal{P}(\mathcal{A}_1)$, $B = \mathcal{P}(\mathcal{A}_2)$, where $\mathcal{P}(\mathcal{A}_i)$ $(i = 1, 2)$ are projection lattices of von Neumann algebras $\mathcal{A}_i$; and $\mathcal{A}_2$ is abelian, then

$$L[B] = \{\sum_i p_i \otimes b_i : \; p_i \in \mathcal{P}(\mathcal{A}_1); \; b_i \in \mathcal{P}(\mathcal{A}_2), \sum_i b_i = I_2\},$$

where I_2 is the identity in $\mathcal{A}_2$. (See [P3].)

If the Boolean algebra B is discrete, the Boolean power $L[B]$ is isomorphic to the direct product of copies of L indexed by atoms of B.

There are complete embeddings $\lambda : L \to L[B]$, $\beta : B \to L[B]$. Moreover, to any state m on L and every completely additive state μ on B there exists a so-called *product state* $m \otimes \mu$ on $L[B]$ such that $m \otimes \mu(\lambda(a) \wedge \beta(b)) = m(a)\mu(b)$, $a \in L, \; b \in B$.

Assume that a quantum system is described by a logic L and a measuring apparatus by a complete Boolean algebra B, and that the joint system is described by the Boolean power $L[B]$. Let $\mathcal{P}_L$ be the set of physical (σ-additive) states on L and $\mathcal{P}_B$ be the set of physical states (completely additive) on B. As physical states of $L[B]$, we will consider σ- convex combinations of product states.

Let an observable X with the value space $(\Omega, \mathcal{F})$ be fixed. (We consider only 'ordinary' observables – an analogue of PV-measures, although an analogue of POV-measures can also be defined on quantum logics, see [P5]).

Let X_A be an observable on the apparatus described by B with value space $(\Omega_A, \mathcal{F}_A)$. Let $f : \Omega \to \Omega_A$ be a measurable function (so-called *pointer function*). In analogy with [BLM], a *measurement* of the observable X is a five-tuple

$$\mathcal{M} = (B, X_A, m_A, f, V),$$

where m_A is an initial state of the apparatus, and $V : \mathcal{P} \to \mathcal{P}$ is a convexity preserving transformation of the state space of the coupled system, which describes the measuring interaction. The five-tuple $\mathcal{M}$ is classified as a measurement of X if the following equality holds:

$$m(X(F)) = V(m \otimes m_A)(\beta(X_A \circ f^{-1}(F)))$$

for all $m \in \mathcal{P}_L$, and all $F \in \mathcal{F}$. The convexity preserving transformation V can be extended to the cone $K(\mathcal{P}) := \{\alpha s : \alpha \geq 0, \; s \in \mathcal{P}\}$ by homogeneity, and by linearity to $K(\mathcal{P}) - K(\mathcal{P})$.

Let L be an orthomodular poset. Recall that $a, b \in L$ are *compatible* (aCb) if they are contained in the same Boolean subalgebra of L. The set of all elements of L which are compatible with any other element of L is the *centre* of L. If m is a state and c is a central element of L, then the mapping $s_c : L \to [0, 1]$, $s_c(a) = s(a \wedge c)$, is an additive measure on L. If $L[B]$ is a Boolean power, then the set $\beta[B]$ is contained in the centre of $L[B]$ ([P2]). This means that the coupled system is endowed with superselection rules. In comparison with the strictly Hilbert space approach, this approach is more general and has the advantage that problem of objectification of measurements (see [BLM]) does not appear here.

For every measurement $\mathcal{M}$, an *instrument* $I_{\mathcal{M}}$ is defined by

$$I_{\mathcal{M}}(F)(m) = V(m \otimes m_{\mathcal{A}})_{\beta(X_{\mathcal{A}} \circ f^{-1}(F))} \circ \lambda.$$

$I_{\mathcal{M}}(F)(m)$ is interpreted as the final nonnormalized state of the system, provided that the measurement lead to an outcome in F. From the definition of $I_{\mathcal{M}}$ and properties of $\mathcal{P}$, it follows

$$\{F_i\}_{i \in N} \subset \mathcal{F},\ F_i \cap F_j = \emptyset,\ i \neq j \ \Rightarrow\ I_{\mathcal{M}}(\bigcup_{i \in N} F_i)(m) = \sum_{i \in N} I_{\mathcal{M}}(F_i)(m)$$

for every at most countable set N. We note that in some approaches, the notion of an instrument was introduced independently of a measurement, see e.g. [Da].

Before introducing some important classes of measurements (or instruments, equivalently), we will define some special properties of quantum logics. In what follows, L denotes a quantum logic.

Let m be a state on L, $b \in L$ such that $m(b) \neq 0$. A function $p_m(a|b) : L \to [0, 1]$ satisfying

(i) $p_m(a|b)$ is a state on L and $m(b|b) = 1$;

(ii) $a, b \in L,\ aCb \ \Rightarrow\ p_m(a|b) = \frac{m(a \wedge b)}{m(b)}$

is called a *conditional state* (with respect to m and b).

We say that L *admits conditional states* with respect to a set of states S if for every $m \in S$ and $b \in L$ with $m(b) \neq 0$ there is a unique conditional state $p_m(.|b)$.

Von Neumann algebras with no I_2-factor as direct summand are examples of logics with conditional states with respect to the set of all states [CZ].

A set M of states on L is said to be *rich* if

$$a, b \in L,\ a \not\leq b \ \Rightarrow\ \exists m \in M : m(a) = 1, m(b) < 1.$$

A logic L with a rich convex set of states M is called *u-spectral* if for every convex functional $f : M \to [0, 1]$ there is a unique observable $x : \mathcal{B}(R) \to L$ such that $f(m) = m(x)$, where $m(x) = \int \lambda m(x(d\lambda))$ is the expectation of x in the state m, for all $m \in M$. Von Neumann algebras yield an example of u-spectral logics [R].

A measurement $\mathcal{M}$ or the corresponding instrument I is (cf. [BLM])

- *repeatable*, if $I(E)(I(F)(m))(1) = I(E \cap F)(m)(1)$ $(E, F \in \mathcal{F}, m \in \mathcal{P}_{\mathcal{L}})$;
- *first kind*, if $m(x(E)) = I(\Omega)(m)(x(E))$;
- *ideal*, if $a \in C(X)$, $m(a) = 1 \Rightarrow I(\Omega)(m)(a) = 1$; where $C(X)$ denotes the set of all $a \in L$ such that $aCX(E)$ for all $E \in \mathcal{F}$;
- *Lüders* for a discrete observable with the spectrum $\{\omega_i\}$, $X(\{\omega_i\}) = b_i$, if $I(\{\omega_i\})(m)(a) = m(b_i)p_m(a|b_i)$ for all $m \in \mathcal{P}_L$ and all i on a logic with conditional states with respect to $\mathcal{P}_L$ ([P2], [P3]).

In [P3], the following relations between different types of measurements have been proved (we note that if we consider measurements on u-spectral logics with rich convex set M of states, we assume $\mathcal{P}_L = M$, and if we consider measurements on logics with conditional states with respect to a set S of states, we assume $\mathcal{P}_L = S$): (i) On any logic, a repeatable measurement is of the first kind. (ii) On u-spectral logics, a first kind measurement or an ideal measurement is repeatable. (iii) On a logic with conditional states a discrete measurement is ideal iff it is Lüders, and then it is repeatable.

In [P4], some more refined ideality conditions are introduced and their relations are studied.

We note that in some more general approaches using orthoalgebras or effect algebras as 'quantum logics', tensor products are defined ([FB], [D2], [DP1]). Also in these approaches, Boolean powers correspond to tensor products of the above structures with Boolean algebras [FP], [P5], [DP2]. An example of an effect algebra is the set of all quantum mechanical effects, i.e. self-adjoint operators between 0 an I on a Hilbert space. A POV observable has its range in the set of these effects. Therefore it would be interesting to formulate a measurement theory in the frame of effect algebras. Some basic ideas have been formulated in [DP2].

Mathematical Institute
Slovak Academy of Sciences

REFERENCES

[A] Aerts, D.: 1983, The Description of One and Many Physical Systems, in C Gruber, C Piron, T Minhton, R Weil, eds., *Les Fondements de la Méchanique Quantique*. Lausanne, Switzerland, 63–148.

[AD] Aerts, D., Daubechies, I.: 1973, 'Physical Justification for Using the Tensor Product to Describe Two Quantum Systems as One Joint System', *Helv Phys Acta*, **51**, 661–675.

[B] Beran, L.: 1985, *Orthomodular Lattices. Algebraic Approach*. Reidel, Dordrecht.

[BC] Beltrametti, E., Cassinelli, G.: 1981, *The Logic of Quantum Mechanics*. Addison-Wesley, Reading, Mass.

[BN] Birkhoff, G., von Neumann, J.: 1936, 'The Logic of Quantum Mechanics', *Ann Math*, **37**, 823–843.

[BLM] Busch, P., Lahti, P., Mittelstaedt, P.: 1991, *Quantum Theory of Measurement*. Springer-Verlag, Berlin.

[CZ] Cassinelli, G. Zanghi, N.: 1983, 'Conditional Probabilities in Quantum Mechanics, *Nuovo Cimento*, **B 73**, 237–245.

[Da] Davies, B.: 1976, *Quantum Theory of Open Systems*. Academic Press, New York.

[D1] Dvurečenskij, A.: 1993, *Gleason's Theorem and its Applications*. Kluwer, Dordrecht.

[D2] Dvurečenskij, A.: 1995, 'Tensor Product of Difference Posets', *Trans Am Math Soc*, **347**, 1043–1057.

[DP1] Dvurečenskij, A., Pulmannová, S.: 1994, 'Tensor Product of D-Posets and D-test Spaces', *Rep Math Phys*, **34**, 251–275.

[DP2] Dvurečenskij, A., Pulmannová, S.: 1994, 'Difference Posets, Effects and Quantum Measurements', *Int J Theor Phys*, **33**, 819–850.

[FB] Foulis, D., Bennett, M.K.: 19??, 'Tensor Product of Orthoalgebras', *Order*, **10**, 271–282.

[FP] Foulis, D., Pták, P.: 1995, 'On the Tensor Product of a Boolean Algebra and and an Orthoalgebra', *Czech Math J*, **45**, 117–126.

[Gr] Grätzer, G.: 1968, *Universal Algebras*. Van Nostrand, Princeton, NJ.

[HK] Hellwig, K-E., Krauser, D.: 1977, 'Propositional Systems and Measurement-III. Quasitensorproducts of Certain Orthomodular Lattices', *Int J Theor Phys*, **16**, 775–793.

[K] Kalmbach, G.: 1983, *Orthomodular Lattices*. Academic press, London.

[M] Mackey, G.: 1963, *The Mathematical Foundations of Quantum Mechanics*. Benjamin, New York.

[Ma] Matolcsi, T.: 1975, 'Tensor Product of Hilbert Lattices and Free Orthodistributive Product of Orthomodular Lattices', *Acta Sci Math Szeged*, **37**, 263–272.

[P1] Pulmannová, S.: 1985, 'Tensor Product of Quantum Logics', *J Math Phys*, **26**, 1–5.

[P2] Pulmannová, S.: 1993, 'Boolean Powers and Quantum Measurements', *Rep Math Phys*, **32**, 235–250.

[P3] Pulmannová, S.: 1994, 'Quantum Measurements and Quantum Logics, *J Math Phys*, **35**, 1555–1572.

[P4] Pulmannová, S.: 1995, A Quantum Logic Description of Some Ideal Measurements, in V.P. Belavkin, O. Hirota and R.L. Hudson, eds., *Quantum Communications and Measurement*. Plenum Press, New York, 187–192.

[P5] Pulmannová, S.: 1980, 'Semiobservables on Quantum Logics', *Math Slovaca*, **30**, 351–360.

[PP] Pták, P., Pulmannová, S.: 1991, *Orthomodular Structures as Quantum Logics*. Kluwer, Dordrecht.

[RF] Randall, C.H., Foulis, D.J.: 1979, 'Tensor Products of Quantum Logics do not Exist, *Not Am Math Soc*, **26**, A-557.

[R] Rüttimann, G.: 1981, Detectable Properties and Spectral Quantum Logics, in *Interpretations and Foundations of Quantum Theory*, Proceedings of a Conference held in Marburg, 28–30 May 1979 (Bibliographisches Institut, Mannheim).

[V] Varadarajan, V.: 1985, *Geometry of Quantum Theory*. Springer-Verlag, New York.

[Z] Zecca, A.: 1978, 'On the Coupling of Logics', *J Math Phys*, **19**, 1482–1485.

JAROSŁAW PYKACZ

ATTEMPT AT THE LOGICAL EXPLANATION OF THE WAVE-PARTICLE DUALITY

1. INTRODUCTION

In the orthodox quantum logic approach to foundations of quantum mechanics originated by Birkhoff and von Neumann [1] the object called *quantum logic* is not a logic in the very sense of this word but an algebraic structure: orthomodular lattice or poset. These orthomodular structures, abstracted from structures of Hilbert spaces used to describe quantum phenomena are weakenings of Boolean algebras. They can be thought of as representing Lindenbaum algebras of systems of propositions about quantum objects, the propositions belonging to the *quantum logic proper*. Since Lindenbaum algebras of propositional systems which obey the rules of classical logic are Boolean algebras it is inferred that the *quantum logic proper*, whatever it means, is non-classical.

The Lindenbaum algebra of a propositional system is obtained by defining the partial order relation between propositions with the aid of implication and identifying equivalent propositions. Meets, joins, and complements in the obtained partially ordered set are generated, respectively, by logical operations of conjunction, disjunction and negation. This is the usual procedure. In the quantum logic approach this procedure is, in a sense, reversed since algebraic relations and operations (partial order, meets, joins, and complements) are either 'inherited' from the Hilbert space structure or introduced axiomatically on the basis of more or less physically plausible speculations and they are only later on interpreted as representing logical connectives.

The least doubtful is the interpretation of orthocomplementation as representing logical negation, although even here some objections were raised [12]. Much more discussed is the interpretation of meets and joins as representing conjunctions and disjunctions, specially in the case of noncompatible propositions (see Sec. 2 for precise definitions). Even the founding fathers of the quantum logic approach [1] were well aware of the problems that arise here.[1]

In the paper, following the suggestion published in [18], I represent propositions about quantum objects not as elements of an abstract partially ordered set but as fuzzy subsets of a set of physical states which give rise to propositional functions belonging to the domain of infinite-valued Łukasiewicz logic. Therefore, the *quantum logic proper* is recognized as a kind of infinite-

M.L. Dalla Chiara et al. (eds.), Language, Quantum, Music, 269–282.

valued (partial) Łukasiewicz logic endowed with connectives studied by Łukasiewicz [8] and Frink [5] already before the Second World War. This fact applied to the typical two-slit experiment and propositions a (resp. b) = 'a quantum object passes through the slit A (resp. B)' allows to reveal the logical background of the mysterious wave-particle duality of quantum objects.

2. ORTHOMODULAR ALGEBRAS

By an *orthomodular algebra* I mean in a present paper an orthocomplemented σ-orthocomplete orthomodular poset, i.e. a partially ordered set L which contains the smallest element O and the greatest element I, in which the orthocomplementation map $\perp : L \to L$ satisfying conditions (a)–(c) exists:

(a) $(a^{\perp})^{\perp} = a$.

(b) If $a \leq b$, then $b^{\perp} \leq a^{\perp}$.

(c) The greatest lower bound (*meet*) $a \wedge a^{\perp}$ and the least upper bound (*join*) $a \vee a^{\perp}$ with respect to the given partial order exist in L for any $a \in L$ and $a \wedge a^{\perp} = O$, $a \vee a^{\perp} = I$.

Moreover, the σ-orthocompleteness condition holds:

(d) If $a_i \leq a_j^{\perp}$ for $i \neq j$ (such elements are called *orthogonal*), then the join $\bigvee_i a_i$ exists in L,

and so does the orthomodular identity:

(e) If $a \leq b$, then $b = a \vee (a^{\perp} \wedge b) = a \vee (a \vee b^{\perp})^{\perp}$.

Elements $a, b \in L$ are called *compatible* iff there exist in L pairwise orthogonal elements a_1, b_1, and c such that $a = a_1 \vee c$ and $b = b_1 \vee c$.

The structure defined above is in the existing literature usually called *quantum logic*. I prefer to follow the terminology introduced by Burmeister and Mączyński[2] [2] for the following reasons:

1. Various forms of this structure are encountered both in the quantum mechanics and in the classical statistical mechanics, therefore labelling it with the name 'quantum' is misleading.

2. The name 'quantum logic' is also used by some authors in a purely logical sense, i.e. it denotes what I called in the Introduction the *quantum logic proper*. The proposed change of the name of the 'algebraic' quantum logic helps to distinguish these two closely connected but, nevertheless, different objects.

3. The proposed change of the name is in accordance with with the modern trend in the theory of quantum structures where more general structures are called orthoalgebras, effect algebras, quantum MV algebras, etc.

According to the standard interpretation with any physical system we can associate an orthomodular algebra L consisting of elementary ('yes-no') propositions about this system. *Probability measures* on the orthomodular algebra L, i.e. mappings $s : L \to [0,1]$ such that

(i) $s(I) = 1$,

(ii) $(\bigvee_i a_i) = \sum_i s(a_i)$ for any sequence consisting of pairwise orthogonal elements of L,

represent states of a physical system and, therefore, are usually themselves called *states on L*. A number $s(a) \in [0,1]$ is usually interpreted as a probability of obtaining the result 'yes' in an experiment designed to check the proposition a when a physical system is in a state represented by s. A set of states S is called *ordering* (another names: *full*, *order determining*) iff $s(a) \leq s(b)$ for all $s \in S$ implies $a \leq b$.

Two most standard examples of orthomodular algebras are the following:

C. Orthomodular algebra of a classical statistical system is a Boolean algebra $\mathcal{B}(\Gamma)$ consisting of Borel subsets of a phase space Γ of a system. All elements of $\mathcal{B}(\Gamma)$ are compatible, as it happens in any Boolean algebra. States of a physical system are represented by usual Kolmogorovian probability measures on $\mathcal{B}(\Gamma)$, pure states being represented by Dirac measures concentrated on one-point subsets of Γ or, equivalently, by points of Γ on which these measures are concentrated.

Q. Orthomodular algebra of a quantum system described with the aid of a Hilbert space $\mathcal{H}$ is an orthomodular lattice $L(\mathcal{H})$ of closed subspaces of $\mathcal{H}$ or, equivalently, of orthogonal projections onto these closed subspaces. Only commuting projections are compatible. States are represented by density operators, pure states being represented by projections onto one-dimensional subspaces or, equivalently, by unit vectors which determine these subspaces.

In both cases the set of all states on an orthomodular algebra is ordering.

The example '**C**' strongly supports the hypothesis that the orthomodular algebra of a physical system is a Lindenbaum algebra of the logic of propositions about this system. Indeed, any two yes-no propositions *a*,*b* about a classical statistical system define Borel subsets *A*,*B* of the phase space Γ of this system consisting of pure states for which these propositions are, respectively, true. This means that the conjunction (resp. disjunction) of a and b, which for classical systems can be always formed, define a set-theoretic intersection (resp. union) of subsets A and B, i.e. a meet (resp. join) of A and B in the Boolean algebra $\mathcal{B}(\Gamma)$.

The situation described in the example '**Q**' is not so clear since meets and joins of any elements of a lattice $L(\mathcal{H})$ always exist, as they do in any lattice. It is hard to reconcile this fact with the well-known difficulties of forming experimentally meaningful conjunctions and disjunctions of experimentally testable propositions referring to quantum observables which are not simultaneously measurable. These difficulties were most probably among the reasons for passing from orthomodular lattices to orthomodular posets. However, even in such more general framework the problem of the logical interpretation of joins and meets persists.

A new solution to this problem, suggested in [18] and developed in Sections 4 and 5 is one of the objectives of this paper since it allows to reveal the (many-valued) logical background of the wave-particle duality exhibited in the two-slit experiment.

3. FUZZY SETS

Let $\mathcal{U} \neq \emptyset$ be a fixed set called a *universe*. According to Zadeh [21] a *fuzzy set A* in $\mathcal{U}$ is defined by its membership function $\mu_A : \mathcal{U} \to [0, 1]$ in such a way that for any $x \in \mathcal{U}$ the number $\mu_A(x) \in [0, 1]$ represents the *degree of membership* of the point x to the fuzzy set A. Characteristic functions of traditional sets (called *crisp sets* in the 'fuzzy' literature) are of course special cases of membership functions. Therefore, fuzzy sets are generalizations of traditional crisp sets. Many authors identify fuzzy sets with their membership functions and write $A(x)$ instead of $\mu_A(x)$. This convention is adopted throughout the rest of the paper.

Already in his first paper on fuzzy sets [21] Zadeh proposed the following expressions for various relations between fuzzy sets and operations on them:

$$A = B \ \textit{iff}\ A(x) = B(x) \tag{1}$$

$$A \subseteq B \ \textit{iff}\ A(x) \leq B(x) \tag{2}$$

$$B = A' \ \textit{iff}\ B(x) = 1 - A(x) \tag{3}$$

$$C = A \cap B \ \textit{iff}\ C(x) = min[A(x), B(x)] \tag{4}$$

$$C = A \cup B \ \textit{iff}\ C(x) = max[A(x), B(x)] \tag{5}$$

The right-hand sides of the above-listed equivalences hold for all $x \in \mathcal{U}$.

Ten years passed since the birth of the fuzzy set theory before Giles [6] noticed the nowadays obvious fact that the relation between fuzzy sets and infinite-valued Łukasiewicz logic is exactly the same as the relation between traditional crisp sets and classical bivalent logic. In particular a value $A(x)$ which a membership function of a fuzzy set A takes in a point x can be interpreted as a truth-value of a proposition '$x \in A$' evaluated according to the rules of infinite-valued Łukasiewicz logic. In symbols

$$A(x) = \tau(\text{`}x \in A\text{'}). \tag{6}$$

Therefore, complementation (3), intersection (4), and union (5) of fuzzy sets occur to be generated by Łukasiewicz negation, conjunction, and disjunction:

$$A'(x) = \tau(\neg\text{`}x \in A\text{'}) = 1 - \tau(\text{`}x \in A\text{'}) = 1 - A(x) \tag{7}$$

$$\begin{aligned}(A \cap B)(x) &= \tau(\text{`}x \in A\text{' } \mathit{and} \text{ `}x \in B\text{'}) = \mathit{min}[\tau(\text{`}x \in A\text{'}), \tau(\text{`}x \in B\text{'})] \\ &= \mathit{min}[A(x), B(x)]\end{aligned} \tag{8}$$

$$\begin{aligned}(A \cup B)(x) &= \tau(\text{`}x \in A\text{' } \mathit{or} \text{ `}x \in B\text{'}) = \mathit{max}[\tau(\text{`}x \in A\text{'}), \tau(\text{`}x \in B\text{'})] \\ &= \mathit{max}[A(x), B(x)]\end{aligned} \tag{9}$$

Let us note that these operations form de Morgan triplet, i.e. intersection, union, and complementation (resp. conjunction, disjunction, and negation) are connected via de Morgan laws:

$$(A \cap B)' = A' \cup B' \tag{10}$$

$$(A \cup B)' = A' \cap B' \tag{11}$$

However, '*min*' (8) and '*max*' (9) expressions are not the only possible expressions for truth-values of conjunction and disjunction in the many-valued Łukasiewicz logic. Actually, there is an infinite variety of them, some of them being more and some less natural or plausible. Another pair of conjunction and disjunction (also forming de Morgan triplet together with original Łukasiewicz negation (7)) was studied by Frink [5] in 1938. Expressions for the truth-values of these connectives, called by Frink[3] *arithmetic*, which I denote by $\sqcap$ and $\sqcup$, are the following:

$$\tau(a \sqcap b) = \mathit{max}[\tau(a) + \tau(b) - 1, 0] \tag{12}$$

$$\tau(a \sqcup b) = \mathit{min}[\tau(a) + \tau(b), 1]. \tag{13}$$

Giles [6] called operations on fuzzy sets generated by (12) and (13):

$$\begin{aligned}(A \sqcap B)(x) &= \tau(\text{`}x \in A\text{'} \sqcap \text{`}x \in B\text{'}) = \mathit{max}[\tau(\text{`}x \in A\text{'}) + \tau(\text{`}x \in B\text{'}) - 1, 0] \\ &= \mathit{max}[A(x) + B(x) - 1, 0]\end{aligned} \tag{14}$$

$$(A \sqcup B)(x) = \tau(\text{`}x \in A\text{'} \sqcup \text{`}x \in B\text{'}) = min[\tau(\text{`}x \in A\text{'}) + \tau(\text{`}x \in B\text{'}), 1]$$
$$= min[A(x) + B(x), 1] \qquad (15)$$

bold operations and they appear in the literature also under the names of *Giles, bounded, truncated*, or *Łukasiewicz* operations.[4]

In the vast literature on fuzzy sets there are also other pairs of pointwisely generated sum-like and product-like operations forming de Morgan triplets with complementation (3). However, the results obtained by Mesiar [11] strongly indicate that they could not be used to produce models of orthomodular algebras of physical systems.

4. ORTHOMODULAR ALGEBRAS OF FUZZY SETS

It follows from theorems proved by Mączyński [9], [10], my previous observations [15], [16], [17], and results of [18] that any orthomodular algebra L with an ordering set of states S can be represented in the form of a family $\mathcal{L}(S)$ of fuzzy subsets of S satisfying the following conditions:

(a) $\mathcal{L}(S)$ contains the empty set $\emptyset$, i.e. such set that $\emptyset(s) = 0$ for all $s \in S$.

(b) $\mathcal{L}(S)$ is closed under the complementation (3).

(c) $\mathcal{L}(S)$ is closed under countable Giles unions of pairwise *weakly disjoined* sets,[5] i.e. if $A_i \sqcap A_j = \emptyset$ for $i \neq j$, then $\bigsqcup_i A_i \in \mathcal{L}(S)$.

(d) If $A \sqcap A = \emptyset$, then $A = \emptyset$ (equivalently: if $A \subseteq A'$, then $A = \emptyset$; if $A' \subseteq A$, then $A = S$).

Conversely, it can be inferred from the results of Mączyński [9] and myself [18] that any family of fuzzy subsets of an arbitrary universe $\mathcal{U}$ satisfying conditions (a)–(d) is an orthomodular algebra partially ordered by the inclusion of fuzzy sets (2), with the fuzzy set complementation (3) as orthocomplementation, orthogonality of elements coinciding with their weak disjointness, and an ordering set of states generated by points of the universe $\mathcal{U}$ according to the formula

$$s_x(A) = A(x). \qquad (16)$$

Therefore, any family $\mathcal{L}(\mathcal{U})$ of fuzzy subsets of an arbitrary universe $\mathcal{U}$ satisfying conditions (a)–(d) will be called *orthomodular algebra of fuzzy sets* or simply *fuzzy orthomodular algebra*. Let us note that since inclusion of fuzzy sets is defined pointwisely by the formula (2) any orthomodular algebra of fuzzy sets $\mathcal{L}(\mathcal{U})$ has an ordering set of states generated by points of the universe $\mathcal{U}$ according to the formula (16).

The above-quoted results from [9] and [18] can be summarized as follows:

Proposition 1 *Any orthomodular algebra L with an ordering set of states S is isomorphic to an orthomodular algebra of fuzzy sets $\mathcal{L}(S)$, i.e. to a family of fuzzy subsets of S satisfying conditions (a) - (d). The isomorphism is given by the formula:*

$$L \ni a \mapsto A \in \mathcal{L}(S), A(s) = s(a) \text{ for any } a \in L, s \in S. \tag{17}$$

Usually in computations performed with probability measures on orthomodular algebras not joins of arbitrary elements but only joins of pairwise orthogonal elements are utilized. However, it follows from the results of Mesiar [11] that in any orthomodular algebra of fuzzy sets joins of pairwise orthogonal, i.e. weakly disjoint sets coincide with their Giles unions. Therefore, it seems worthy to study further relations between Giles and order-theoretic operations on fuzzy orthomodular algebras.

The following proposition shows that the existence of Giles operations for any pair A, B of elements of a fuzzy orthomodular algebra implies the existence of order-theoretic operations. Moreover, it implies that the elements A and B are in this case compatible and that Giles operations on them coincide with order-teoretic ones.

Proposition 2 *Let $\mathcal{L}(\mathcal{U})$ be an orthomodular algebra of fuzzy subsets of a universe $\mathcal{U}$ and let $A, B \in \mathcal{L}(\mathcal{U})$. Then $A \sqcap B \in \mathcal{L}(\mathcal{U})$ iff $A \sqcup B \in \mathcal{L}(\mathcal{U})$ and in this case A and B are compatible, $A \sqcap B = A \wedge B$, and $A \sqcup B = A \vee B$.*

In order to keep the size of this paper within reasonable limits the lengthy proof of this proposition, as well as proofs of other propositions of this section will be published in the forthcoming paper [19].

Since the existence of Giles union or intersection of two elements of a fuzzy orthomodular algebra implies their compatibility, we obtain a straightforward corollary:

Corollary 1 *If $A \sqcap B$ or $A \sqcup B$ exists in a fuzzy orthomodular algebra $\mathcal{L}(\mathcal{U})$ for all $A, B \in \mathcal{L}(\mathcal{U}), A \neq B$, then $\mathcal{L}(\mathcal{U})$ is a Boolean algebra.*

Although the existence of $A \sqcap B$ and $A \sqcup B$ in $\mathcal{L}(\mathcal{U})$ implies that A and B are compatible, the converse implication is, in general, not true. In order to see this let us study the following example:

Example 1 Let X be a triangle with vertices a, b, c and let $\mathcal{L}(X)$ consist of all fuzzy subsets of X with membership functions being affine functions that assume on all vertices of X values 0 or 1 only. It was proved in [14] that such $\mathcal{L}(X)$ is a Boolean algebra, therefore, all its elements are compatible. Let us denote A, B elements of $\mathcal{L}(X)$ such that $A(a) = 1$, $A(b) = A(c) = 0$, $B(b) = 0$, $B(a) = B(c) = 1$, and let us construct their Giles union $A \sqcup B$ and intersection $A \sqcap B$. Since $(A \sqcup B)(a) = min(1 + 1, 1) = 1$, $(A \sqcup B)(b) = min(0 + 0, 1) = 0$, $(A \sqcup B)(c) = min(0 + 1, 1) = 1$, and $(A \sqcap B)(a) = max(1 + 1 - 1, 0) = 1$, $(A \sqcap B)(b) = max(0 + 0 - 1, 0) = 0$, $(A \sqcap B)(c) = max(0 + 1 - 1, 0) = 0$, $A \sqcup B$ and $A \sqcap B$ would have belonged to $\mathcal{L}(X)$ iff their membership functions were affine functions. But this is impossible: Let d be a midpoint of the edge ab. Obviously $A(d) = B(d) = 0.5$, so $(A \sqcup B)(d) = min(0,5 + 0,5, 1) = 1$ and $(A \sqcap B)(d) = max(0,5 + 0,5 - 1, 0) = 0$, but affine functions $(A \sqcup B)(x)$ and $(A \sqcap B)(x)$ coinciding with affine functions $A \sqcup B$ and $A \sqcap B$ on the endpoints of the edge ab must coincide with them on the whole edge ab. Therefore, $(A \sqcup B)(x)$ and $(A \sqcap B)(x)$ are not affine functions and cannot belong to $\mathcal{L}(X)$.

Although order-theoretic operations on elements of a fuzzy orthomodular algebra $\mathcal{L}(\mathcal{U})$ coincide with Giles operations if only the latter are defined, there are cases when order-theoretic operations are defined but Giles operations are not. One particular situation of this type was described in Example 1. The other concerns pairs consisting of identical elements: Order-theoretic operations are idempotent, i.e. $A \wedge A = A \vee A = A \in \mathcal{L}(\mathcal{U})$ for any $A \in \mathcal{L}(\mathcal{U})$. On the contrary, neither $A \sqcap A$, nor $A \sqcup A$ belongs to $\mathcal{L}(\mathcal{U})$ for any 'genuine fuzzy' element of $\mathcal{L}(\mathcal{U})$:

Proposition 3 *If $\mathcal{L}(\mathcal{U})$ is an orthomodular algebra of fuzzy subsets of a universe $\mathcal{U}$, then $A \sqcap A$ and $A \sqcup A$ exist in $\mathcal{L}(\mathcal{U})$ iff A is a crisp subset of $\mathcal{U}$, i.e. iff $A(x) \in \{0, 1\}$ for all $x \in \mathcal{U}$. In this (and only in this) case Giles operations are idempotent, i.e. $A \sqcap A = A \sqcup A = A$.*

The proof will be published in [19].

Let us note that a fact that Giles intersections and unions of some elements of fuzzy orthomodular algebras do not have to exist in these algebras (even if they are lattices) is rather a virtue than a drawback of these operations. Actually, this fact may help to solve the problem of interpretation of meets and joins of noncompatible elements in orthomodular algebras with ordering sets of states: If we assume, as it was suggested in [18], that conjunctions and disjunctions of propositions about quantum system are represented in a fuzzy orthomodular algebra $\mathcal{L}(S)$ not by meets and joins but by Giles operations, then there is nothing wrong if $\mathcal{L}(S)$ is a lattice in which meets and joins of all elements have to exist provided that Giles intersections and unions of noncompatible elements do not exist in $\mathcal{L}(S)$. The close links between Giles and order-theoretic operations established in Proposition 2 also explain why for so many years meets and joins were (I believe wrongly) interpreted as proper algebraic representations of quantum-logical conjunctions and disjunctions. The detailed discussion of all these problems will be published in [19].

5. INFINITE-VALUED LOGICS ASSOCIATED WITH ORTHOMODULAR ALGEBRAS

According to what was said about relations between fuzzy sets and infinite-valued Łukasiewicz logic, the isomorphism of an abstract orthomodular algebra L with an ordering set of states S to an orthomodular algebra $\mathcal{L}(S)$ of fuzzy subsets of S established by Proposition 1 allows to associate with L a family of infinite-valued propositions. More precisely, to any element a of an orthomodular algebra L there corresponds a propositional function $a(\cdot)$ defined on the set of states S of L such that

$$\tau(a(s)) = A(s) = s(A) \in [0, 1] \tag{18}$$

where $A \in \mathcal{L}(S)$ is a fuzzy subset of S isomorphic to $a \in L$. The most natural choice of logical connectives to be used in the emerging propositional calculus is de Morgan triplet consisting of Łukasiewicz negation

$$\tau(\neg a) = 1 - \tau(a) \tag{19}$$

and Frink's arithmetic connectives (12) and (13) since they generate via formulas (7), (14), and (15) operations on fuzzy sets used in the definition of a fuzzy orthomodular algebra. Indeed, let us call two propositions *exclusive* or say that they *exclude each other* iff their arithmetic conjunction is a false proposition, i.e. iff

$$\tau(a \sqcap b) = \tau(F) = 0. \tag{20}$$

Let us also call two propositional functions defined on the same domain *exclusive* or say that they *exclude each other* iff the formula (20) holds for all arguments in their common domain in which case we say that their arithmetic conjunction is the *always-false propositional function*. Then the following proposition is a straightforward consequence of the definition of an orthomodular algebra of fuzzy sets.

Proposition 4 *With any orthomodular algebra $\mathcal{L}(\mathcal{U})$ consisting of fuzzy subsets of a universe $\mathcal{U}$ (therefore, by Proposition 1 also with any abstract orthomodular algebra with an ordering set of states) we can associate a family $\tilde{\mathcal{L}}$ of infinite-valued propositional functions with arguments ranging over the universe $\mathcal{U}$ such that:*

(a') The always-false propositional function belongs to $\tilde{\mathcal{L}}$.

(b') If a propositional function belongs to $\tilde{\mathcal{L}}$, its negation also belongs to $\tilde{\mathcal{L}}$.

(c') *Arithmetic disjunction of any countable family of propositional functions which exclude each other belongs to* $\tilde{\mathcal{L}}$.

(d') *The always-false propositional function is the only propositional function in* $\tilde{\mathcal{L}}$ *which excludes itself.*

Let us note (cf. [5], [17], [18]) that algebraic conjunction (12) and disjunction (13) can be obtained from Łukasiewicz negation (19) and implication[6]

$$\tau(a \rightarrow b) = min[1 - \tau(a) + \tau(b), 1] \tag{21}$$

by the following formulas:

$$\tau(a \sqcup b) = \tau(\neg a \rightarrow b) \tag{22}$$

$$\tau(a \sqcap b) = \tau(\neg(\neg a \sqcup \neg b)) = \tau(\neg(a \rightarrow \neg b)) \tag{23}$$

which suggests that the set of logical connectives suitable to deal with propositional functions representing elements of a fuzzy orthomodular algebra could be enlarged to contain also Łukasiewicz implication (21). If we do it and say that that a proposition *a surely implies b* iff Łukasiewicz implication $a \rightarrow b$ is a true proposition, i.e. iff

$$\tau(a \rightarrow b) = \tau(T) = 1 \tag{24}$$

with an obvious generalization of this notion to propositional functions defined on the common domain, then we can express two remaining equivalent forms of the condition (d) of the definition of an orthomodular algebra of fuzzy sets in the following way:

(d') *The always-false propositional function is the only propositional function in* $\tilde{\mathcal{L}}$ *which surely implies its own negation.*

(d") *The always-true propositional function is the only propositional function in* $\tilde{\mathcal{L}}$ *which is surely implied by its own negation.*

6. ATTEMPT AT THE (MANY-VALUED) LOGICAL EXPLANATION OF THE WAVE-PARTICLE DUALITY

Let us note that since expressions for truth-values of both arithmetic conjunction (12) and arithmetic disjunction (13) contain the sum $\tau(a) + \tau(b)$, we can distinguish the following four cases:

(1a) If $0 \leq \tau(a) + \tau(b) < 1$, then $\tau(a \sqcap b) = 0$ and $0 \leq \tau(a \sqcup b) < 1$.

(1b) If $\tau(a) + \tau(b) = 1$, then $\tau(a \sqcap b) = 0$ and $\tau(a \sqcup b) = 1$.

(2a) If $1 < \tau(a) + \tau(b) < 2$, then $0 < \tau(a \sqcap b) < 1$ and $\tau(a \sqcup b) = 1$.

(2b) If $\tau(a) + \tau(b) = 2$ (i.e. $\tau(a) = \tau(b) = 1$), then $\tau(a \sqcap b) = 1$ and $\tau(a \sqcup b) = 1$.

Therefore, we obtain the following proposition:

Proposition 5 *For any two many-valued propositions a, b either*

(1a) their arithmetic conjunction is false, i.e. $\tau(a \sqcap b) = 0$, and their arithmetic disjunction is not true, i.e. $\tau(a \sqcup b) < 1$, or

(1b) their arithmetic conjunction is false, i.e. $\tau(a \sqcap b) = 0$, and their arithmetic disjunction is true, i.e. $\tau(a \sqcup b) = 1$, or

(2a) their arithmetic conjunction is neither false nor true, i.e. $0 < \tau(a \sqcap b) < 1$, and their arithmetic disjunction is true, i.e. $(a \sqcup b) = 1$, or

(2b) both their arithmetic conjunction and disjunction are true.

Let us note that besides the trivial possibility $\tau(a) = \tau(b) = 0$ which obviously yields $\tau(a \sqcap b) = \tau(a \sqcup b) = 0$ only possibilities (1b) and (2b) are allowed by classical bivalent logic. Therefore, many-valued logic allows to describe behaviour of quantum systems in a better way than it can be done with the use of classical bivalent logic. We shall use this possibility to cast some light on the famous wave-particle duality exhibited in the two-slit experiment.

Let $a(s)$ (resp. $b(s)$) denote a proposition ‘a quantum object in a state s passes through the slit A (resp. B)’. If we interpret arithmetic conjunction $a(s) \sqcap b(s)$ as representing compound proposition ‘a quantum object in a state s passes through the slit A *and* through the slit B’ and arithmetic disjunction $a(s) \sqcup b(s)$ as representing compound proposition ‘a quantum object in a state s passes through the slit A *or* through the slit B’, then truth-values of these compound propositions can be used to decide whether a quantum object in the state s reveals its wave-like or particle like properties:

- $\tau(a(s) \sqcap b(s)) = 0$ and $\tau(a(s) \sqcup b(s)) = 1$ means that it is impossible to state experimentally that a quantum object in a state s passes simultaneously through the slit A *and* through the slit B, although we are sure that in any experiment it will pass *either* through the slit A *or* through the slit B (i.e. this *or* is exclusive). Therefore, an object in a state s reveals purely particle-like behaviour.

– if $\tau(a \sqcap b) = 1$ (in which case also $\tau(a \sqcup b) = 1$), we are sure that an object passes simultaneously through the slit A *and* through the slit B, so it behaves like a pure wave.

These two cases, corresponding to possibilities (1b) and (2b) of Proposition 5, can be equally well described with the aid of classical bivalent logic. However, Proposition 5 allows also possibilities (1a) and (2a) which in the studied case of the two-slit experiment can be interpreted as follows:

(1a) We are not sure whether an object passes through the slit A *or* B, although we are sure that it does not pass simultaneously through A *and* through B. This possibility seems to correspond to a non-perfect experiment in which an object behaving like a pure particle can pass either through the slit A *or* through the slit B (but, because $\tau a(s) \sqcap b(s) = 0$, not simultaneously through both slits, so this *or* is again exclusive), although it is not sure whether it passes at all.

(2a) We are sure that an object passes through the slit A *or* through the slit B (or through both slits since this *or* is non-exclusive) which allows both wave-like and particle-like behaviour. However, contrary to the case (2b) in which an object behaves like a pure wave we cannot be sure that it passes through both slits simultaneously (since $\tau(a \sqcap b) < 1$), therefore, its behaviour is not 100% wave-like. On the other hand its behaviour is also not 100% particle-like since, contrary to the case (1b) we are also not sure that it does not pass through both slits simultaneously (since $\tau(a \sqcap b) > 0$). Therefore, an object reveals partial wave-like and partial particle-like behaviour, i.e. it behaves simultaneously but not 'to the full extent' like a wave and like a particle. This kind of behaviour is not allowed by the orthodox Copenhagen paradigm based on classical bivalent logic. Therefore, it was a kind of surprise when Wooters and Zurek [20] demonstrated that quantum mechanical formalism allows to obtain in a two-slit experiment non-perfect knowledge about the path of a quantum object which shows its non-100% particle-like behaviour simultaneously with non-perfect interference pattern which demonstrates that this behaviour is also non-100% wave-like. Predictions of Wooters and Zurek were confirmed by Mittelstaedt, Prieur and Schieder [13] in a carefully performed experiment in which they found that: "A photon possesses simultaneously particle properties and wave properties. ... Even for a very high value of the particle property ... there is still a nonvanishing amount of the wave property ... which leads to an observable and significant interference pattern." This experimental result strongly supports the considerations of Section 5 according to which the proper logic of a quantum system is a version of infinite-valued Łukasiewicz logic endowed with Łukasiewicz negation (19), implication (21) and partially defined arithmetic conjunction (12) and disjunction (13).

ACKNOWLEDGMENTS

The author is very grateful to the Organizers of the 10th International Congress of Logic, Methodology, and Philosophy of Science (Firenze, August 1995) where the first version of this paper was presented for their invitation and financial support. This work and the participation in the Congress was also supported by the University of Gdańsk grants BW/5100-5-0044-5 and BW/5100-5-0065-6. The final version of the paper was prepared thanks to the University of Gdańsk grant BW/5100-5-0276-7.

Institute of Mathematics
University of Gdańsk

NOTES

1 The detailed discussion of various solutions proposed to solve these problems up to the early seventies can be found in the Chapter 8 of M. Jammer's book *The Philosophy of Quantum Mechanics* (Wiley Interscience, New York, 1974)
2 Actually, Burmeister and Mączyński defined an orthomodular algebra by adopting another set of purely 'algebraic' axioms. However, they proved the full equivalence of their definition to the definition of an orthomodular poset. Another way of defining an orthomodular poset in a purely 'algebraic' way is to add some constraints to the definition of an orthoalgebra (see, e.g. [3], [4], [7]).
3 Some authors ascribe the very definition of these operations to Łukasiewicz but I could not find their definitions in [8]. Actually, Frink in [5] also did not give an explicit expression for their truth-values but they can be easily inferred from the other formulas of his paper.
4 It seems that Giles was not aware of Frink's 1938 paper [5]. Most probably these operations were rediscovered many times by various authors.
5 The definition of weakly disjoint fuzzy sets was introduced by Giles in [6].
6 In the Łukasiewicz propositional calculus negation and implication are usually treated as basic connectives, conjunction, disjunction, and equivalence being defined with their aid.

REFERENCES

[1] Birkhoff, G., von Neumann, J.: 1936, 'The Logic of Quantum Mechanics', *Ann Math*, **37,** 823–843.
[2] Burmeister, P., Mączyński, M.: 1994, 'Orthomodular (Partial) Algebras and their Representations', *Demonstratio Mathematica*, **27**, 701–722.
[3] Foulis, D.J., Bennett, M.K.: 1994, 'Effect Algebras and Unsharp Quantum Logics', *Found Phys*, **24**, 1331–1352.
[4] Foulis, D.J., Greechie, R.J., Rüttimann, G.T.: 1992, 'Filters and Supports in Orthoalgebras', *Int J Theor Phys*, **37,** 789–807.
[5] Frink, O.: 1938, 'New Algebras of Logic', *Am Math Monthly*, **45**, 210–219.
[6] Giles, R.: 1976, 'Łukasiewicz Logic and Fuzzy Set Theory', *Int J Man-Machine Studies*, **8**, 313–327.
[7] Giuntini, R., Greuling, H.: 1989, 'Toward a Formal Language for Unsharp Properties', *Found Phys*, **19**, 931–945.
[8] Łukasiewicz, J.: 1970, *Selected Works*. North-Holland, Amsterdam.
[9] Mączyński, M.J.: 1973, 'The Orthogonality Postulate in Axiomatic Quantum Mechanics', *Int J Theor Phys*, **8**, 353–360.

[10] Mączyński, M.J.: 1974, 'Functional Properties of Quantum Logics', *Int J Theor Phys*, **11**, 149–156.

[11] Mesiar, R.: 1994, 'h-Fuzzy Quantum Logics', *Int J Theor Phys*, **33**, 1417–1425.

[12] Mielnik, B.: 1976, Quantum Logic: Is it Necessarily Orthocomplemented?, in M. Flato et al., eds., *Quantum Mechanics, Determinism, Causality, and Particles*. Reidel, Dordrecht, 117–135.

[13] Mittelstaedt, P., Prieur, A., Schieder, R.: 1987, 'Unsharp Particle-Wave Duality in a Photon Split-Beam Experiment', *Found Phys*, **17**, 891–903.

[14] Pykacz, J.: 1983, 'Affine Mączyński Logics on Compact Convex Serts of States', *Int J Theor Phys*, **22**, 97–106.

[15] Pykacz, J.: 1987, Quantum Logics as Families of Fuzzy Subsets of the Set of Physical States, *Preprints of the Second International Fuzzy Systems Association Congress, Tokyo, July 20–25, 1987*, Vol. 2, 437–440.

[16] Pykacz, J.: 1987, 'Quantum Logics and Soft Fuzzy Probability Spaces', *Bulletin pour les Sous-Ensembles Flous et leurs Applications*, **32**, 150–157.

[17] Pykacz, J.: 1992, 'Fuzzy Set Ideas in Quantum Logics', *Int J Theor Phys*, **31**, 1767–1783.

[18] Pykacz, J.: 1994, 'Fuzzy Quantum Logics and Infinite-Valued Łukasiewicz Logic', *Int J Theor Phys*, **33**, 1403–1416.

[19] Pykacz, J., manuscript in preparation.

[20] Wooters, W.K., Zurek, W.H.: 1979, 'Complementation in the Double-Slit Experiment: Quantum Nonseparability and a Quantitative Statement of Bohr's Principle', *Phys Rev D*, **19**, 473–484.

[21] Zadeh, L.A.: 1965, 'Fuzzy Sets', *Information and Control*, **8**, 338–353.

DENNIS DIEKS

THE BOHR-EINSTEIN PHOTON BOX DEBATE

1. INTRODUCTION

According to Bohr's later report [1], at the 1930 Solvay conference Einstein made a final attempt at demonstrating that the actual possibilities of measurement and prediction, as allowed by quantum mechanics itself, lead to violations of the uncertainty relations. If this were correct, the conclusion would be inevitable that quantum mechanics is incomplete: a more precise description would be possible for an individual system than given by the quantum mechanical state. According to Bohr Einstein devised a thought-experiment with which it would be possible to predict both the precise energy of a photon and its exact time of arrival at a detector, in conflict with a version of the energy-time uncertainty relation. Bohr countered the proposal by arguing that if the details of the measuring procedure are properly considered, taking into account the quantum mechanical character of the measuring device itself, it turns out that the photon energy and the instant of the photon's emission from the source (a 'photon box') cannot both be perfectly fixed.

Bohr's report appeared in the famous Einstein-volume of the Library of Living Philosophers, 19 years after the Solvay Conference. According to this account, Einstein was forced to admit that there is no internal inconsistency in the Copenhagen interpretation and had consequently to develop a new strategy of attack. But in spite of Bohr's claim that the issue was definitively settled, also for Einstein, and in spite of the reputation Bohr's reply has as the decisive vindication of the Copenhagen position, the details of Bohr's reply to Einstein have been giving rise to confusion and criticism. There are mutually conflicting interpretations of exactly how Bohr intended the photon-box Gedankenexperiment to take place and of how the symbols he uses should be interpreted. Moreover, several authors have voiced doubts about the validity and consistency of Bohr's arguments (see for a small sample of comments [2, 4, 6, 7, 8, 9] and [5], Sec. 5.4, for an overview of the discussion up to 1974).

Staying close to Bohr's text, but filling in details and discussing points not mentioned by Bohr himself, we shall give a more precise quantum mechanical version of Bohr's reasoning. It will become clear that although Bohr's conclusion is correct, the details of his argument pertain to a quite specific case. There is therefore room for doubt concerning the general validity of the conclusion.

M.L. Dalla Chiara et al. (eds.), Language, Quantum, Music, 283–292.

This raises the historical question of why Einstein never responded. Can it really be the case that Einstein was converted and as a result directed his energy to completely new possible obstacles for quantum mechanics, as suggested by Bohr? We will argue that it is more plausible that Einstein was partly misunderstood by Bohr. Probably Einstein already in 1930 no longer seriously doubted the consistency of the Copenhagen interpretation. He seems to have been one step ahead: Einstein had already begun to focus on the difficulties complementarity entails for the notion that spatially separated systems possess their own autonomous physical states.

2. THE PHOTON BOX EXPERIMENT

The general idea of Einstein's Gedankenexperiment, as Bohr understood it, is made clear enough in Bohr's report.

> Einstein proposed [a] device . . ., consisting of a box with a hole in its side, which could be opened or closed by a shutter moved by means of a clock-work within the box. If, in the beginning, the box contained a certain amount of radiation and the clock was set to open the shutter for a very short interval at a chosen time, it could be achieved that a single photon was released through the hole at a moment known with as great accuracy as desired. Moreover, it would apparently also be possible, by weighing the whole box before and after this event, to measure the energy of the photon with any accuracy wanted, in definite contradiction to the reciprocal indeterminacy of time and energy quantities in quantum mechanics ([1], pp. 225–226).

It is important to remark that if the clock opens the shutter when its hands reach a certain position on the dial, this does not immediately yield the time of opening as indicated by the laboratory clocks. In order to know this latter time, the synchronisation between the clock in the box and the laboratory clocks has to be determined – our later discussion will show the relevance of this point.

Bohr's discussion ([1], pp. 226–227) uses a concrete experimental set-up. In it, the box is suspended in a spring-balance; it has a pointer 'to read its position on a scale fixed to the balance support'. The argument then focuses on the 'accuracy' with which this position is 'determined', and the related (in)accuracies in other quantities. As Bohr says, ". . . any determination of this position with a given accuracy Δq will involve a minimum latitude Δp in the control of the momentum of the box connected with Δq by the relation (3)" (i.e. the uncertainty relation $\Delta q.\Delta p \approx h$).

It is clear from this that what is actually indicating the result of the weighing procedure is the position of the pointer, which is rigidly attached to the box[1]. The pointer does not possess a precise position and momentum; in both quantities there is a certain indeterminacy (or latitude). About the momentum latitude Bohr says:

> This latitude must obviously again be smaller than the total impulse which, during the whole interval T of the balancing procedure, can be given by the gravitational field to a body with a mass Δm, or

$$\Delta p \approx \frac{h}{\Delta q} < T.g.\Delta m,$$

where g is the gravity constant. The greater the accuracy of the reading q of the pointer, the longer must, consequently, be the balancing interval T, if a given accuracy Δm of the weighing of the box with its content shall be obtained.

The intuition clearly is that all differences in the final situation come about by virtue of differences in the gravitational force on the box, which depends on the box's mass. If a mass difference Δm is to be measurable, a difference in gravitational force $\Delta m.g$ should lead to a clear-cut difference in the final situation. For this, the difference in momentum transferred during the weighing should be greater than the initial latitude Δp: otherwise the difference in momentum caused by the difference Δm would be drowned in the already existing momentum latitude of the pointer position[2].

Now we come to the essential point of Bohr's reply to Einstein.

According to general relativity theory, a clock, when displaced in the direction of the gravitational force by an amount of Δq, will change its rate in such a way that its reading in the course of a time interval T will differ by an amount ΔT given by the relation

$$\frac{\Delta T}{T} = \frac{1}{c^2} g \Delta q.$$

By comparing (6) and (7) (i.e. the equations for Δq and for ΔT) we see, therefore, that after the weighing procedure there will in our knowledge of the adjustment of the clock be a latitude

$$\Delta T > \frac{h}{c^2 \Delta m}.$$

Together with the formula (5) (i.e. $E = mc^2$), this relation again leads to $\Delta T.\Delta E > h$, in accordance with the indeterminacy principle. Consequently, a use of the apparatus as a means of accurately measuring the energy of the photon will prevent us from controlling the moment of its escape.

The indeterminacy Δq during the balancing procedure leads to an indeterminacy ΔT that is proportional to T.[3] To see the importance of ΔT, suppose that the clock in the box has been constructed in such a way that it opens the shutter at 12 o'clock (when its hands reach the 12 o'clock mark on its own dial). To determine at what time the photon has escaped, we look where the hands of the clock are in the final situation, say at 12.30 h laboratory time. Suppose that the time indicated on the clock itself is 12.25 h. That would apparently mean that the photon has escaped 25 minutes earlier, 12.05 h as indicated by the laboratory clocks. But we have to take into account the indeterminacy in the rate of the clock between the escape of the photon and the final reading of the clock (which is an interval at least of duration T). This introduces an uncertainty $Tg\Delta q/c^2$ in the moment of the escape of the photon.

The quantities Δp and Δq stand for the quantum indeterminacies in the position and momentum of the box, when it is indicating a position on the scale. ΔT represents the indeterminacy in the position of the hands of the laboratory clocks, associated with the situation in which the clock in the box, exactly T seconds ago in its own proper time, opened the shutter. These quantum indeterminacies pertain to the individual case. However, they are related to statistical spreads ΔE and ΔT in the predictions of measurement results, at a distant detector, of the photon's energy and arrival time. These spreads are empirically realized in repeated runs of the experiment with the same final pointer position q and same value of T.[4]

Obviously, what Bohr presents is not a *derivation* of the time-energy uncertainty relation. Rather, it is an argument to show that that relation cannot be circumvented even by Einstein's sophisticated measurement procedure. To this end, Bohr treats the measuring device quantum mechanically, taking into account uncertainty relations, and analyzes the dynamical process by which a correlation is established between the photon's mass and instant of escape on the one hand and the pointer position and the hands of the clock on the other. However, Bohr's discussion is qualitative, and it would be helpful to see a more precise version of it.

3. QUANTUM MECHANICAL DESCRIPTION OF THE EXPERIMENT

We start our considerations in the situation in which the first weighing has taken place. The box and its contents are in a stationary position; the gravitational force and the pull of the spring compensate each other. The position and momentum of the centre of mass of the box will only be well-defined up to certain indeterminacies, and we assume $\Delta q.\Delta p = h$, in order to be in accordance with Bohr's calculation. This does not imply that the energy of the box is undetermined: because $E \approx Mc^2 + p^2/2M$, with M the very large rest mass of the box, a considerable indeterminacy in momentum is compatible with a very small ΔE. We can therefore consider the box, together with its contents, to have a well-defined energy before the photon's escape.

When the photon escapes, the box begins to move upward. Bohr mentions that after some time the box can be brought down again by attaching a suitable load to it; he does not go into the details of this procedure and does not refer to it in his formulas. In our analysis we will follow Bohr on this point. Indeed, the essential thing is to see how a correlation is established between the pointer position on the one hand and the photon mass on the other. For this it is sufficient to consider the upward motion of the box.

The spring compensates the original weight of the box, so the force which is responsible for the motion is mg, with m the mass of the escaped photon and g the acceleration of gravity. We make the unessential simplification that this force is constant during the motion. Let us now consider how the wavefunction of the box changes with time when the gravitational force has become effective.

In the Heisenberg-picture we have the following equations of motion for vertical position and momentum:

$$\dot{q}(t) = p/M, \;\; \dot{p}(t) = mg. \tag{1}$$

It follows that

$$p(t) = p(0) + mgt \qquad q(t) = q(0) + p(0)t/M + mgt^2/2M. \tag{2}$$

Taking $\langle q(0)\rangle = \langle p(0)\rangle = 0$, we find:

$$\langle q(t)\rangle = mgt^2/2M, \text{and } \langle p(t)\rangle = mgt. \tag{3}$$

In these equations M stands for the (very large) mass of the box. It is simple to show from these formulas that $\Delta p(t) = \Delta p(0)$, so the indeterminacy in momentum stays constant during the motion. The spread in position, $\Delta q(t)$, does not remain constant; it becomes larger. If we assume that $d\Delta q/dt = 0$ in the initial situation (otherwise different opening instants of the shutter would correspond to box wavefunctions with different Δq), we find that:

$$\{\Delta q(t)\}^2 = \{\Delta q(0)\}^2 + \frac{t^2}{M^2}\{\Delta p(0)\}^2. \tag{4}$$

After the force of gravity has been acting for a period T, the initial pointer position wavefunction has evolved into a wave function with

$$\langle q(T)\rangle = \frac{1}{2}\frac{m}{M}gT^2 \;\text{ and }\; \langle p(T)\rangle = mgT, \tag{5}$$

dependent on the mass m of the escaped photon. This makes it clear that the pointer has indeed become correlated to the photon's mass.

In order that the final situation be different, depending on whether the mass of the escaped photon was m_1 or m_2, the corresponding wavefunctions ψ_1 and ψ_2 must be (nearly) orthogonal. Since it is clear from Bohr's text that the *position* of the pointer is the 'pointer variable', the indeterminacy of position in the final pointer state should be smaller than the distance between two states corresponding to mass m_1 and m_2, respectively. In other words,

$$\Delta q(T) < \frac{1}{2}\frac{\Delta m}{M}gT^2, \tag{6}$$

where $\Delta m = |m_1 - m_2|$ is the attainable accuracy of the weighing procedure. In view of Eq. (4) we have:

$$\{\Delta q(0)\}^2 + \frac{T^2}{M^2}\{\Delta p\}^2 < \left(\frac{1}{2}\frac{\Delta m}{M}gT^2\right)^2. \tag{7}$$

Here we see why $\Delta q(0)$ cannot be made arbitrarily small; smaller values of $\Delta q(0)$ lead to larger values of Δp and therefore to less accuracy of the experiment. Since the two terms on the left-hand side are positive, they each have to be smaller than the term appearing at the right-hand side. It follows that

$$\Delta p < \Delta mgT/2. \tag{8}$$

Eq. (8) implies the inequality $\Delta p < \Delta mgT$, the condition used by Bohr. Our justification of the inequality is more satisfying, because it directly relates to the way the balance records the measurement result, namely by means of the pointer. By contrast, Bohr's remarks about the distinguishability with respect to momentum may create the impression (see [7]) that momentum instead of position is the pointer variable—in contradiction to the remainder of Bohr's text. However, the conflict is more apparent than real, because the position and momentum of the box are correlated (see formula 5).

We now turn to the calculation of ΔT. According to the general theory of relativity, the relation between the proper time indicated by a clock and co-ordinate time is, in the case of low velocities and weak field, given by:

$$d\tau = \sqrt{g_{00}}dt \tag{9}$$

with

$$g_{00} \approx 1 - \frac{2g}{c^2}q \tag{10}$$

Here q stands for the clock's vertical position in the gravitational field and g is the acceleration of gravity. We find therefore:

$$\int dt = \int d\tau(1 + \frac{g}{c^2}q). \tag{11}$$

This formula gives us the connection between lapses of time indicated by laboratory clocks and by the moving clock we are considering, respectively. If the right-hand side if the equation is a quantum mechanical quantity, with an indeterminacy attached to it, it follows that the same is true for the left-hand side:

$$\Delta \int dt = \frac{g}{c^2}\Delta \int q d\tau. \tag{12}$$

The interpretation of this expression is that there is a quantum indeterminacy in the positions of the hands of the laboratory clocks, corresponding to an uncertainty in laboratory time, given a value of $\int d\tau$.

Substituting the solution we found for $q(t)$, Eq. (2), we find

$$\Delta T = \frac{g}{c^2}\Delta \int_0^T \left\{ q(0) + \frac{p(0)t}{M} + \frac{mgt^2}{2M} \right\} dt \tag{13}$$

$$= \frac{g}{c^2}\left(\Delta^2 q(0) T^2 + \frac{1}{4}\frac{\Delta^2 p(0)}{M^2} T^4 \right)^{\frac{1}{2}}. \tag{14}$$

It is now possible to determine a lower bound for the uncertainty product $\Delta E.\Delta T$, with $\Delta E = c^2 \Delta m$:

$$\Delta E.\Delta T > c^2 \frac{2M\Delta q(T)}{gT^2}\frac{g}{c^2}\left(\Delta^2 q(0) T^2 + \frac{1}{4}\frac{\Delta^2 p(0)}{m^2} T^4 \right)^{\frac{1}{2}}. \tag{15}$$

Using the expression (4) for $\Delta q(T)$, and making use of the inequality $a^2 + b^2 \geq 2ab$, we find:

$$\Delta E.\Delta T > 3\Delta q(0)\Delta p(0) = 3h. \tag{16}$$

This result is in essential agreement with Bohr's conclusion. The small numerical discrepancy is due to the fact that in our calculation ΔT is larger than according to Bohr's qualitative reasoning. We may therefore consider the above as a more precise version of Bohr's argument. However, our expressions are more complicated than the ones occurring in Bohr's exposition, and show that the calculation depends on the details of the experimental set-up. If the box does not accelerate uniformly, but feels a force from the spring to which it is attached, the formulas for Δx, Δp and ΔT change drastically. To see exactly how much the analysis is altered when other experimental conditions are considered, we now investigate the case in which the box executes a harmonic motion.

As before, we assume that the box is stationary before the photon's emission; after the emission of the photon (mass m) the box (mass M) moves periodically up and down. Let us suppose that before the emission the pointer connected to the box was described by a wave packet centred at $q = 0$, and that he spring is attached to the yoke at height a. The Heisenberg equations of motion for the box can again be solved explicitly:

$$q(t) = q_0 \cos \omega t + \frac{1}{M\omega} p_0 \sin \omega t + \frac{ma}{M+m}(1 - \cos \omega t), \tag{17}$$

$$p(t) = p_0 \cos \omega t - M\omega(q_0 - \frac{ma}{M+m}) \sin \omega t, \tag{18}$$

$$\text{with } \omega^2 = \frac{(M+m)g}{Ma.} \tag{19}$$

For the expectation values of position and momentum we find:

$$< q(t) > = \frac{ma}{M+m}(1 - \cos \omega t), \tag{20}$$

$$< p(t) > = \frac{Mma\omega}{M+m} \sin \omega t. \tag{21}$$

The way the pointer moves therefore depends on the mass of the escaped photon. In order that the pointer position at time T makes it possible to discriminate between photon masses with a difference Δm, we should have:

$$\Delta q(T) < \frac{\Delta m.a(1 - \cos \omega T)}{M+m}. \tag{22}$$

This leads to a bound on the precision with which the photon energy can be determined:

$$\Delta E = c^2 \Delta m > \frac{(M+m)c^2 \Delta q(T)}{a(1 - \cos \omega T)}. \tag{23}$$

For the time dependence of the position indeterminacy we find:

$$\Delta^2 q(T) = < q_0^2 > \cos^2 \omega T + < p_0^2 > \frac{\sin^2 \omega T}{M^2 \omega^2}. \tag{24}$$

At this point a significant difference compared to Bohr's formulas should be noted. In our new example ΔE changes periodically and does not approach zero when $T \longrightarrow \infty$ (by contrast, Bohr in his text stresses that the accuracy becomes better with growing T).

The 'redshift' ΔT can also be calculated explicitly:

$$\Delta T = \frac{g}{c^2} \Delta \int q(\tau) d\tau = \frac{g}{c^2} \{ < q_0^2 > \frac{\sin^2 \omega T}{\omega^2} + \frac{< p_0^2 >}{M^2 \omega^4} (1 - \cos \omega T)^2 \}^{\frac{1}{2}}. \tag{25}$$

Again, this differs significantly from Bohr's expression. The time uncertainty varies periodically, and is zero when $\omega T = k.2\pi$.

Combining these results and applying rather long but elementary algebraic manipulations, we finally arrive at a lower bound for the uncertainty product:

$$\Delta T.\Delta E > \frac{g(M+m)}{aM\omega^2}\Delta q_0 \Delta p_0 = h. \tag{26}$$

The uncertainty relation is therefore obeyed. But we see that the uncertainty product in this case takes on lower values than in the previous example. Moreover, the various quantities now behave quite differently from their counterparts in the first example, and several of the comments Bohr made fail to apply.

4. CONCLUSION

For someone who seriously doubted the consistency of the Copenhagen interpretation, Bohr's argumentation could hardly have been convincing. Indeed, we have already mentioned that several commentators have expressed complaints about the lack of precision and clarity in Bohr's text. It is true that our 'quantum mechanical reconstruction' has vindicated Bohr's conclusion for two specific cases. But for a stubborn opponent of Copenhagen even these more precise calculations would not have sufficed. The details of the weighing procedure and the behaviour of the various expressions vary greatly from case to case, and it is not easy to see how a general consistency proof could be found this way. Why then did Einstein never respond?

In all probability, the answer to this historical riddle is that Bohr in his report misrepresented Einstein. In 1930 Einstein already accepted the consistency of the Copenhagen interpretation as very likely. Instead of wanting to prove a violation of the uncertainty relations, Einstein with his photon box experiment hinted at the possibility that a *choice* can be made, even after the photon's escape, regarding whether the photon's energy or its time of arrival at a distant detector can be predicted. This would mean that the two great men at the 1930 Solvay Conference were not on the same wave length and to some extent talked at cross-purposes.

This interpretation is confirmed by a letter from Ehrenfest to Bohr. Ehrenfest visited Einstein in 1931 and, as Ehrenfest immediately wrote to Bohr, Einstein on that occasion was very emphatic that he had never intended the photon box as an example of a violation of the uncertainty relations. Ehrenfest to Bohr, 9 July 1931: "He said to me that, for a long time already, he absolutely no longer doubted the uncertainty relations, and that he thus, e.g. had BY NO MEANS invented the 'weighable light-flash box' 'contra uncertainty relation', but for a totally different purpose." [3]

It is a pity that we cannot relive this piece of history, and that some of its aspects will forever remain uncertain.

Foundations of the Natural Sciences
Utrecht University

NOTES

[1] In ref. [7] it is assumed that the *momentum* of the box is measured. This is consistent with Bohr's text if the pointer's position can be considered as an indicator of the momentum of the box. It is shown in Section 3 that this can be justified.

[2] In [4] Δp is incorrectly taken to be the momentum imparted to the box during the weighing procedure.

[3] Treder [9] argues that the use of an electric field, instead of gravity, would have avoided the indeterminacy ΔT. However, by means of an electric field we can only determine the electric charge in the box (or the number of charged particles in the box) and not the mass. Treder's proposal to multiply the total electric charge by the mass/charge ratio of the particles in the box does not help, because we would need a determination of the mass to find that ratio. We cannot use the rest mass, as that would give us the total rest mass, a quantity without quantum indeterminacy and not subject to uncertainty relations. The objection put forward by Landé [6] and Popper [8], namely that the introduction of the red-shift formula implies an illegitimate appeal to the general theory of relativity, was convincingly answered in [10]. See also the discussion in [5], pp. 137–138.

[4] The relevant spreads should be distinguished from the statistical spreads found in repetitions of the initial situation, regardless of the later measurement outcomes. The latter obey an energy-time uncertainty relation; see [2] for a modern quantum mechanical treatment. The idea of the photon box experiment, as discussed by Bohr, is that one may *nevertheless* make precise predictions of both the photon's energy and its time of arrival by making cunning use of the results of later measurements on the box.

REFERENCES

[1] Bohr, N.: 1949 Discussion with Einstein on Epistemological Problems in Atomic Physics, in P.A. Schilpp, ed., *Albert Einstein: Philosopher-Scientist.* Open Court, La Salle, 199–241.

[2] Busch, P.: 1990 'On the Energy-Time Uncertainty Relation', *Foundations of Physics*, **20**, 1–32.

[3] Howard, D.: 1990 Nicht Sein Kann Was Nicht Sein Darf, in A.I. Miller, ed., *Sixty-Two Years of Uncertainty.* New York, Plenum, 61–111.

[4] Hughes, R.J.: 1990 'The Bohr-Einstein 'Weighing-of-Energy' Debate and the Principle of Equivalence', *American Journal of Physics*, **58**, 826–828.

[5] Jammer, M.: 1974, *The Philosophy of Quantum Mechanics.* New York, Wiley & Sons.

[6] Landé, A.: 1965, *New Foundations of Quantum Mechanics.* Cambridge, Cambridge University Press, 123.

[7] Peierls, R.: 1979, *Surprises in Theoretical Physics.* Princeton, Princeton University Press, 36–39.

[8] Popper, K.R.: 1959, *The Logic of Scientific Discovery.* London, Hutchinson & Co, Appendix xi.

[9] Treder, H.-J.: 1971, The Einstein-Bohr Box Experiment, in W. Yourgrau, A. van der Merwe, eds., *Perspectives in Quantum Theory.* Cambridge, Mass., MIT Press, 17–24.

[10] Unruh, W.G., Opat, G.I.: 1979, 'The Bohr-Einstein 'Weighing of Energy' Debate', *American Journal of Physics*, **47**, 743–744.

ARCANGELO ROSSI

KANTIANISM AND PHYSICS FROM THE 19TH TO THE 20TH CENTURY

1. INTRODUCTION

The great influence exerted by Kantianism on 19th century physics can be firstly singled out in the process of mathematization which Kant fostered, in particular in *Metaphysical Foundations of Natural Science* (from now on *MF*) (1786),[1] requiring, since its *Preface*, that chemistry assumed, as a special kind of physics, mathematical clothes in order to become a true science, a science of precision as general physics was already, thus evoking A.L. Lavoisier's contemporary 'Newtonian' reform of chemistry. Secondly, Kant's influence on physics can also be seen in the unification of physics Kant himself proposed in *MF* and physical science developed afterwards, that is, since the early forties of the 19th century, and which was dynamical in character. By stressing the role of movements and tensions in matter, Kant in fact paved the way to the affirmation of the energy concept in subsequent 19th century physics. In fact, in *MF*, he had even suggested to consider all physical matter as only consisting of elementary dynamical 'Newtonian' interactions, attractive and repulsive in character (essentially cohesion and impenetrability), among parts of a unique absolute physical space. Thus, he reduced the essence of material substances, deprived of the absolute impenetrability and bulk they were credited with by previous metaphysics, to their dynamical manifestations and their mutual equilibrium, according to mathematical laws expressing global processes rather than local sequences of physical actions as in previous physics of forces.[2] In fact, in order to transform this Kantian philosophical intuition into a unitary physical science, just the concept of energy had to be developed, beyond the tendency, dominating in the four early decades of the 19th century, to accept only partial unifications of domains of experience by using mathematics and distinguishing among different forces referring to different forms of experience and experimental data (such as mechanical movements, chemical, optical, electrical, magnetic and heat phenomena). Indeed, they unified them only mathematically and methodologically, rather than on the base of a structural reality underlying the various phenomenical appearances. Truly, to be faithful to Kant's unification program, this structural reality had not to be conceived as a *Ding an sich*, to use Kant's terminology,[3] that is as an ontological substance irreducible to appearances, as it had instead been

M.L. Dalla Chiara et al. (eds.), Language, Quantum, Music, 293–301.

conceived in previous philosophies of nature by Cartesians, Leibnizians and Newtonians, in terms of *matìere subtile* or ether, monads and atoms, even in the secularised, that is empiricized forms assumed in Laplac's system of nature or in Young's and Fresnel's ondulatory theory of light. Rather, M. Faraday's example had to be followed when he put at the base of the various dynamical appearances, instead of material atoms and ether, a unique field of forces in the form of a web of lines filling up all physical space. But Faraday's lines of forces had the defect to be not fully defined in mathematical terms in order to be accepted from a Kantian viewpoint, even if they seemed both empirically intuitive and universal enough to be conceived a priori.[4] On the other side, R. Mayer's attempt of constructing a unitary view of nature in terms of purely mathematical equivalences among dynamical empirical manifestations (forces), without any intuitive unitary representation of the nature of the forces themselves,[5] contradicted the 'constructive' Kantian approach of *MF*, which started from elementary and intuitively representable components of nature to reproduce phenomena.

2. NEO-KANTIAN RENAISSANCE IN PHYSICS: HELMHOLTZ TO MACH

It was H. von Helmholtz who instead explicitely tried, around the half of the 19th century, to translate the new phenomenological and energetic view of nature into Kantian terms through a recovery of the 'constructive' approach typical of Kant. At any rate, the minima from which the construction had to be accomplished for him, as both empirically intuitive and a priori, that is universal, were, instead of the web of lines of forces of Faradayan physics, the more mathematically definite centers of force or poles acting at a distance as in Kant's *MF*, according to the atomistic continental tradition of Laplacian physics which tried to empiricize classical atomism. But, much more than in Laplacian physics, which still maintained strong ontological atomistic components, Helmholtz's atomic centers of force were just minimum presuppositions for the 'critical' reconstruction of experience, intuitive enough to be a firm ground for the 'construction', but almost devoid of substantial properties other than strictly spatial and dynamical ones.[6] So, Helmholtz went back to the Lockean sensistic conception of the primary properties of matter as a 'physiological' translation of Kant's 'transzendentale', by asserting that geometrical and dynamical experiences were the primary pre-linguistic symbols, actively elaborated by our senses as the essential, elementary base of our knowledge, not just passive images of the outside world. Instead, the mathematical laws of force, as the conservation principles stably linking together our sensations, expressed the invariantive, deterministic aspect of our knowledge to which we ourselves are subject, independently of the different modalities of our sensations. They then clearly derive from the outer world, as being independent of us. On the contrary, our different subjective representations, sounds, colors, smells, flavours, are interpretable as only secondary symbolic effects, though supplied with specific and mutually irreducible physiological modalities (according to J. Müller's physiological law of specific energies of sensory nerves), of the actions of the things in themselves on our senses, actions primarily producing, as we have seen, dynamical and geometrical symbols as starting points of our unconscious inferences from the variety of our experiences to the lawlike objectivity of the outer world.[7] But the physical interpretation of these primary and general, that is transcendental and a priori, dynamical and geometrical physiological symbols of the outer world in terms of precise physical objects, clearly implied different possibilities all along Helmholtz's scientific career, as they were, contrary to Kant, almost devoid of any a priori content, apart from their a priori invariantive deterministic lawlike form, as such compatible with different possible modes of representation. In particular, they did not demonstrate a priori the validity of Euclidean geometry as the geometry of the physical world. Then Helmholtz spaced all along his life from discrete atomism and action at a distance to continuity and vortex atoms theory of matter,[9] apart from the general a priori lawlike geometrical and dynamical nature of the base of all our experience, in the sense of mere spatiality and physical dynamism, that is mutual dynamical interactions among parts of space (contrary to Kant, no longer necessarily Euclidean, though of constant curvature). The precise structure of physical space and dynamics, beyond mere 'parallelism of laws' and formal invariance, had then to be apprehended through experience, which in particular confirmed the physical validity of Euclid's axioms,[9] and could then be modeled through different possible modes of representation, corresponding to different possible interpretations of our experience, however conceived as a physiological symbolization of the outer world. At any rate, it was the transcendental physiological, at least in part subjective basis of our experience that gave coherence and stability to our knowledge of the physical world, then making it possible at all in Kant's sense. If this one was Helmholtz's physiological reduction of Kant's a priori, reducing it to minimum, in fact neo-Kantian rather than just Kantian, terms, E. Mach tried to dispel at all this component of Kant's epistemology, just saving Kant's psychophysiological characterisation of knowledge.[10] Thus, for him the sense-elements are actively connected among themselves through sensory experiences and then elaborated into concepts and models, but the connections no longer have any a priori validity and the concepts and models even have no autonomous cognitive content. In fact these ones only have a practical and economic, 'survival' value, to be finally, hopefully, superseded by a complete purely empirical description of facts through mathematical equations. So, only these will regulate experience in order to have a more efficient comparison and unification of facts, from a practical point of view, than visual concepts and models can, much more ambiguously, provide. In this sense, such concepts as atoms and ether, poles and continua are only quite provisional constructions of our mind, no longer explications of facts just to be confronted with them in order to be accepted, provided their compatibility with universal physical laws, as they were still in Helmholtz's view. Quite on the contrary, they are for Mach only literally fictional instruments aimed at connecting psychologically our experiences without any a priori or even empirical validity, in view of subsequent mere mathematical correlation of facts. For Mach, indeed, this is a completion of previous attempts to unify science through visual concepts, by showing their ambiguity and partiality, in their claim to grasp an ontological reality precluded to us.[11] Then he tried to realize Kant's unification program, in terms of a mere mathematical regulation of experience, at the cost of dropping out Kant's transcendental a priori conception of an elementary base of construction of experience as a structural reality underlying appearances. If this one is in fact conceived as a necessary and universal, then a not purely empirical, perhaps a physiological as in Helmholtz's view, and in any case not purely ontological precondition of experience, then the purely sensorial, always changing and instable, only mathematically regulated elements of Mach's theory are not at all comparable with it. In particular, Mach's elements do not comprehend dynamical sources or causes of phenomena, the forces that Kant and Helmholtz put at the base of their 'reconstruction' of physics. They only contain the dynamical empirical effects or accelerations, only mathematically correlated among themselves.[12]

3. THE ACME OF KANTIANISM IN PHYSICS: H. HERTZ

This criticism of the concept of force as a metaphysical entity was anyway incorporated into a Kantian view of physics more ortodox than Mach's, the one held by the great physicist H. Hertz, Helmholtz's most celebrated pupil.[13] In fact, Hertz reinstated the a priori component of Kantian view, but renouncing the Newtonian and Kantian primary concept of force. However, in order to give a sufficient transcendental base to physics, another Newtonian concept, that of mass, had to be expanded at its place for Hertz, together with purely Kantian conceptions of space and time, conceived by him, contrary to his teacher Helmholtz, in more ortodox Kantian terms, as a priori justified Euclidean and Kantian ones. In fact, Hertz substituted a dynamical material continuum, filling up all physical space, at the base of his Kantian conception, to Kant's and Helmholtz's elementary dynamical interactions among parts of space.[14] Thus he explicitely incorporated into a Kantian, transcendental view, Maxwell's mathematical and mechanical re-elaboration of Faraday's lines of force, which had already substituted a material ethereal structure, underlying appearances and mathematically defined, to immaterial pure lines of force of Faradayan physics. Then, Hertz's overcoming of Helmholtz's polarization theory by the conception, derived from Faraday and Maxwell, of a universal field as a material, dynamical continuum filling Euclidean and Kantian space, served as a transcendental base unifying mechanics and electromagnetism in terms of hidden masses and ondulatory motions of ether, without any need to specify any longer, in more traditional terms, the internal mechanical composition of that continuum in terms of elements and their elementary interactions.[15] Maxwell, from his part, had already considered the particular mechanical models he introduced to explain electromagnetic phenomena as quite provisional, endowed as they were with an essentially heuristic role, instrumental to the application of mathematical algorithms, applied until then to mechanics, also to electromagnetism, and not as realistic images of things, as they were even contradictory among themselves. Provided they were just compatible with his mechanical continuum view of the world, in terms of matter and motion, as a quite general dynamical and mathematical theory of the material continuum, they did not claim to have any truth value other than an heuristic function as a provisional bridge between different branches of physics, such as mechanics and electromagnetism.[16] Hertz perfected this view just underlining the necessary transcendental character of the conception of the continuum ether, linked to experiments but underlying them as their unifying conception overcoming particular branches of physics, even at the cost of leaving some physical effects, though acknowledged, unexplained in the light of his otherwise (as in the case of the discovery of Hertzian waves) enormously fertile a priori physical view,[17] and also at the cost, as we have already seen, of dropping out the constructive aspects of the Kantian view, that is the representations of elements and interactions or elementary forces that were then instead still developed by other contemporary interprets and followers of that view in physics.

4. KANTIANISM WITHOUT DOGMA IN PHYSICS: POINCARÉ TO EINSTEIN

In particular, H. Poincaré accepted, in his critical view of physics, both components of Kantianism: 1) the transcendental a priori component, aimed at unifying the various domains of phenomena through physico-mathematical principles as group invariance, relativity, symmetry and conservation principles, endowed with an everlasting validity, at least for what we can foresee (such principles were unfalsifiable for him, and at most to be left aside, not invalidated, if revealed, contrary to what is known up to now, unfertile),[18] and 2) the constructive intuitive component, as physical hypotheses linking conjecturally those general principles with experience, without claiming any necessary validity nor a complete transparency and a priori deducibility, but only aspiring to an intuitive fecundity and an effective constructive capacity to link principles to experience. So, Euclidean geometry is conceived by Poincaré as a mere hypothesis on empirical space, neither induced from phenomena nor deduced a priori from more general a priori principles as group invariance, though being compatible with and partially derivable from them. Then it is only partially justified a priori, but in the last instance only intuitively and empirically, not logically, justified as useful and comfortable in the scientific 'struggle for survival', rather than really true.[19] And so the concepts of force, ether, electron, and so on, neither deduced totally a priori from general physico-mathematical principles, nor directly induced from pure facts of experience, were conceived as hypothetical intuitive links between the two requiring, for their acceptance, together with an agreement with physico-mathematical principles and a confirmed empirical fecundity, which were both necessary but not sufficient, also a certain intuitive conviction, not devoid of obscurity, to serve as constructive starting points of physical explanation.[20] Then the goal of explanation was pursued with a certain flexibility in the construction, through the possible revision of the specific details of the hypotheses when confronted with phenomena. This point is well exemplified by Poincaré's attempts at constructing a more and more adequate explanation of the absence of an ether wind, revealed by the Michelson and Morley experiment, through ether models, trying both to reinstate the relativity principle and to save his personal conviction on the existence of ether as a unifying, though irreducibly hypothetical, explanation of phenomena.[21] If the 'classical' and Kantian framework of Poincaré's physics of principles is evident when it required that the physico-mathematical principles were unfalsifiable, then a priori, but to be supplemented, in order to explain facts, with constructive hypotheses as intuitive models trying to guarantee the unity and efficacity of the principles themselves as stable bases of scientific knowledge, Einstein, on the contrary, thought that the principles as bases of physical explanation could be applied directly to experience, without 'constructive' preoccupations of the sort of Poincaré's. However he accepted, contrary to Hertz who, also doing without constructive hypotheses, included Euclidean geometry and continuity of physical matter among the principles themselves, that they had a more abstract and general physico-mathematical meaning, as they indeed already had in Poincaré's writings. So, he explicitely purified Kant's heritage from the excesses both of Poincaré's 'constructivism' and of Hertz's 'mechanism', to go back to what he considered the core of Kant's criticism, the so called 'Copernican revolution'.[22] Kant used this expression to qualify the strict dependence of our knowledge on a priori forms of our rationality, that is on transcendental structures of our self-conscious mind rather than on the outer world, just as in Copernican theory the visible motions of the celestial bodies largely depended on those of the earth itself. But, while those rational structures were identified by Kant with absolute forms of intuition and categories, such as space, time and causality, identified at their turn with historical conceptions such as Euclid's and Newton's, in Einstein's view they were identified with much more general conceptions of space-time localization and determinism. Then they were no longer reified in terms of specific intuitions and cathegories, but corresponded to most general coherence, completeness, relativity and symmetry requirements.[23] In any case, the novelty of Einstein's approach consisted, in front of Poincaré's comparable identification of Kant's a priori with most abstract physico-mathematical principles, in the proposal of directly submitting phenomena to those principles without the mediation of constructive hypotheses, and of developing only afterwards, on the base of a preliminary mathematical treatment of facts on the base of the principles, interpretative models of the facts themselves as their physical reconstructions, no longer presupposed, as in 'constructive' physics, but subsequent to equations. In this new process of physical explanation, aiming at a direct physico-mathematical treatment of phenomena, whence models could be derived only after this treatment had been completed, the demand of giving an explanation of facts beyond experience was indeed still maintained, in terms of concrete physical interpretations depending on the equations of phenomena. However, as these interpretations, as we have seen, followed and no longer preceded mathematical treatment in terms of principles, these ones sometimes ran the risk of being too much detached from the objects to which they were applied, and then of being almost void, as in some developments of General Relativity and Unified Field Theory.[25] Whence Einstein's lifelong search for a correspondence between principles and facts which were as most complete and realistic as possible, also in his subsequent physical models and interpretations. So, trying to reach a full comprehension of physical reality, he developed his famous criticism of Quantum Mechanics in the thirties just for its indeterminacy and incompleteness, in fact developing, in new, no longer dogmatic terms, Kant's originary program of unification of physics through most general concepts, which however strictly adhered, with maximum universality and flexibility, to the multiplicity of experience.[26]

5. THE DISSOLUTION OF KANTIANISM IN PHYSICS: N. BOHR

If Einstein still retained, on Kant's wake, a demand for a strict correspondence between a priori principles and facts, even before subsequent hypotheses and models took place, quite different appairs by the way N. Bohr's attitude. In fact, in his construction of the new Quantum Mechanics, Bohr reinterpreted a priori principles in purely linguistic terms, thus reducing their meaning to mere necessary limitations in our expression of facts, no longer implying their unitary and coherent comprehension, as in the Kantian sense of a priori principles. Thus, in Bohr's quantum view, space, time and causality, instead of being the universal necessary components of the principles themselves, either to grasp directly physical reality, as for Einstein, or to be developed into 'constructive' unitary though tentative images of phenomena, as for Poincaré, reduced to mere useful codes and symbols for dealing with experience. Their function was indeed for him to help one framing by models and analogies a reality, as the quantum one, which was in fact irreducible to them, by trying to reconstruct it in the only terms we can use unambiguously, the classical ones, in order to give an objective stance to it, but with the consciousness that we can apply those terms to it only partially and according to reciprocal strict limitations, that is taking them not literally but in a symbolic sense. Consequently, the principles are no longer Kantian principles of coherent and unitary comprehension of reality, but only the necessary application rules of symbolic images of it, establishing precise limitations to their application.[27] The last Kantian element present in Bohr is then the necessity to have recourse to classical images, though according to strict reciprocal limitations established by the Complementarity and Correspondence principles, meant as mere rules of their use, even in the study of quantum phenomena, still depending on our universal subjective, that is transcendental limitations. But those images and their limitations are now reduced to mere linguistic, even if still necessary, partial and fragmentary constructions, as they were never in Kant, to the effect of being unable to generate objects in classical sense, and of eliminating, due to their asserted necessity in this fragmentary form, that sort of unification of physics which even the purely phenomenistic interpretation of Kantianism held by Mach still maintained in Kantian sense, as we have seen before, in terms of a universal and intuitively coherent correlation of all phenomena.[28]

Department of Physics
University of Lecce

NOTES

[1] Engl.transl. by J. Ellington (Indianapolis & New York, Library of Liberal Arts, Bobs Merrill,1970).
[2] Cp. L. Pearce Williams, *The Origins of Field Theory* (Lanham, N. York, London, University Press of America, 1980), ch. 2, pp. 32–43.
[3] Of course, in the negative sense of the *First Critique*, as unknowable, not in the positive sense of the *Second Critique*, as supra-sensorial object of our free will, that is as value. See G. Buchdahl, *Kant and the Dynamics of Reason* (Oxford & Cambridge, Mass., Blackwell, 1992), p. 43.
[4] Cp. L. Pearce Williams,*Michael Faraday* (London, Chapman & Hall, 1965). But C. Truesdell, in *Essays in the History of Mechanics* (Berlin, Heidelberg, N. York, Springer Verlag, 1968), pp. 180–182, stresses the fact that Faraday, though mathematically illitterate, had a certain mathematical instinct in treating field phenomena as sets and structures, only not yet formalized in algebraic and topological terms, as the appropriate mathematical language had not been yet discovered.
[5] Cp. R.B. Lindsay, *Julius Robert Mayer Prophet of Energy* (Oxford, New York, Toronto, Pergamon Press, 1973).
[6] Cp. H. von Helmholtz, 'On the Conservation of Energy: a Physical Memoir'(1847), in Taylor, *Scientific Memoirs*, (Nat. Phil.), 1853.
[7] Maybe the best exposition of Helmholtz's physiological translation of Kant's 'transzendentale' on Lockian lines is 'The Facts in Perception'(1878), in R.S.Cohen, Y. Elkana, eds., H. von Helmholtz, *Epistemological Writings*, Boston Studies in the Philosophy of Science, vol. XXXVII (Dordrecht and Boston, Reidel, 1977), pp. 115–85.
[8] Cp. H. Helmholtz, 'On Integrals of the Hydrodynamic Equations which express Vortex-Motions', Phil. Mag. [4], XXXIII (1858).
[9] Cp. H. Helmholtz, 'Ueber den Ursprung und die Bedeutung der geometrischen Axiome', in *Populaere wissenschaftliche Vortraege*, 2 Heft (Braunschweig, 1870).
[10] Cp. especially *The Analysis of Sensations and the Relation of the Physical to the Psychical* (New York, 1959).
[11] Cp. especially *The Science of Mechanics: a Critical and Historical Account of Its Development* (La Salle, Ill., Open Court 1960).
[12] Cp. *ibidem*, ch. 2, par. 8.
[13] Helmholtz highly praised his former pupil Hertz in his preface to Hertz's postumous *Principles of Mechanics* (Engl. ed. New York, Dover, 1956), even appreciating his a priori Kantian ortodox

treatment of phenomena, in terms of absolute space and time, continuity of matter and hidden masses, but underlining its lack of precise constructive explanatory hypotheses of detail, that were especially developed by British physicists as Kelvin and Maxwell.

[14] Notwithstanding his formal agreement with Hertz's elimination of the concept of force as a great simplification in physical explanation, also Mach in *The Science of Mechanics* quoted, ch.2, par. 9, as Helmholtz already did in his preface to Hertz's postumous *Principles of Mechanics* quoted above, holds that the main defect of Hertz's treatment of particular, gravitational as electromagnetic phenomena, is the excess of apriorism, leading him to apply without reservation and specification an absolute image of physics, in terms of pure masses, to phenomena, at the cost of artifices and exaggerations, which do not duly consider data of fact, as empirical dynamical accelerations, and their most direct confrontation.

[15] Cp. H. Hertz, *Electric Waves* (N. York, Dover, 1962).

[16] Cp. J.C. Maxwell, *A Treatise on Electricity and Magnetism*, ch. V (N. York, Dover, 1954).

[17] Cp. S.D' Agostino, 'The Physics of the 'Effects': a Gate to Atomism', in D. Hoffmann, F. Bevilacqua and R. H. Stuewer (eds.), *The Emergence of Modern Physics* (Pavia, 1996), pp. 27–43.

[18] Cp. J.H. Poincaré, *The Value of Science* (N. York, Dover, 1958), ch. VII–IX.

[19] Cp. J.H. Poincaré, 'On the Foundations of Geometry', *The Monist,* **9** (1898), pp. 1–43.

[20] Cp. J.H. Poincaré, *Science and Hypothesis* (N. York, Dover, 1952), ch. VI, where he stresses the difference between Hertz's doubtless logical derivation of hypotheses from principles and their confirmation by facts, and his own only fiduciary, revisible and tentative acceptation of them.

[21] Cp. A. I. Miller, *Imagery in Scientific Thought*(Cambridge Mass. and London, MIT Press, 1986), ch. 1, 'Poincaré and Einstein', which insists on Poincaré's continuity through local changes in his faith on ether (and electrons).

[22] Cp. A. Einstein, 'Physics and Reality'(1936), in *Essays in Physics* (N. York, Philosophical Library, 1936), p. 18.

[23] In the last instance, Einstein refuses to start from particular intuitions and models because he searches for the most general structures of objective reality inside and outside ourselves, in Spinozian sense, not identifiable with peculiar images, but only with the most highly invariantive characters of the whole, conceived in deterministic, symmetric, complete and unitary terms, so enlarging and purifying Kantian heritage in neo-Kantian sense. Cp. 'Physics and Reality' quoted above.

[24] The distinction between 'constructive physics' and 'physics of principles', that is his own, is clearly outlined by Einstein in 'Time, Space and Gravitation' (1948), in *Out of My Later Years* (Phil. Libr. Repr., N. York, 1950).

[25] Cp. S. D' Agostino, 'Einstein's Lifelong Doubts on the Physical Foundations of the General Relativity and Unified Field Theories', in C. Garola, A. Rossi (eds.), *The Foundations of Quantum Mechanics. Historical Analysis and Open Questions* (Dordrecht, Kluwer, 1995), pp.167–78.

[26] Cp. D. Howard, 'Nicht Sein Kann Was Nicht Sein Darf', in A. Miller (ed.), *Sixty-Two Years of Uncertainty: Historical, Philosophical and Physical Inquires into the Foundations of Quantum Mechanics* (N. York, Plenum Press, 1990), pp. 61–112.

[27] Cp. C. Chevalley, 'Niels Bohr's words and the Atlantis of Kantianism', in J. Faye and H.J. Folse (eds.), *Niels Bohr and contemporary philosophy* (Dordrecht, Kluwer, 1994), pp. 33–55. But, at variance with our interpretation, Chevalley traces back her view of Bohr's positions to 19th century physicists, such as Helmholtz, Hertz, etc., already attributing them a linguistic conception of principles as metaphors, inspired by Kant's *Critique of Judgement* and Goethe, which is indeed incoherent with Helmholtz's physiological view of a priori and Hertz's absolute logico-transcendental view of it.

[28] It is just the permanence in Bohr of non-empirical elements in terms of complementary linguistic images of quantum phenomena, contradicting each other according to principles meant as their application rules and in contrast with a purely phenomenistic unification of physics, which made P.K. Feyerabend speak of a 'positivism of a higher order' with regard to Bohr's view in 'Complementarity', Supplementary Volume 23 of *The Proceedings of Aristotelian Society* (1958), pp. 75–104, as still maintaining the necessity of a sort of a priori, though no longer in Kantian terms.

FABIO BELLISSIMA

EPIMORIC RATIOS AND GREEK MUSICAL THEORY

1. INTRODUCTION

If one landmark overtops all others in the evolution of science, it is the discovery by Pythagoras of the connection between musical harmonies and numbers.[1]

The discovery to which the historian of mathematics Eric Temple Bell refers consisted in observing that the three fundamental kinds of consonances, the *octave* (for instance, the interval *C-c)*, the *fifth* (as the interval *C-G*) and the *fourth* (as the interval *C-F*) were obtained by plunking two sections of a string whose lengths were in the ratio of 2 to 1 for the octave, of 3 to 2 for the fifth and of 4 to 3 for the fourth. So, the three most natural consonances could be expressed by means of the numbers 1, 2, 3 and 4 (the numbers which form the *tetraktys* of the decad!), and the following figure (in which 6, 8, 9, 12 are the smallest numbers producing these ratios) became the manifesto of the Greek music theory. (We observe that, if we set the numbers in incresing order and we consider them as representing the lenghts of the strings, then the pitches of the corresponding notes are decreasing. And actually the Greek scales were descending.)

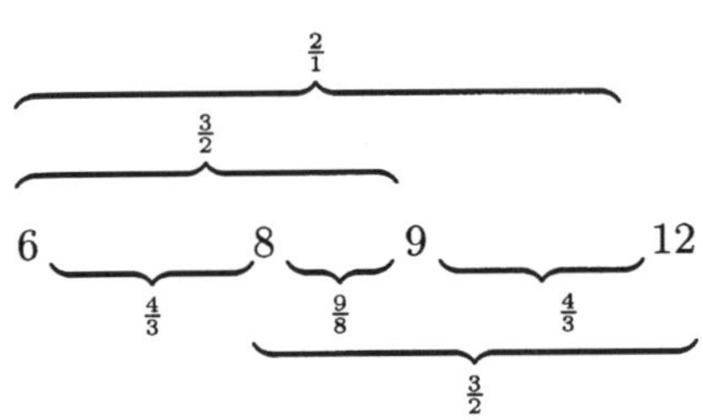

Figure 1.

So, by means of the fundamental consonances, it was possible to fix four notes in the ambit of an octave, at intervals of a fourth, a tone and a fourth (usually, in modern literature, they are denoted by e_2, b_1, a_1 and e_1). Therefore four notes were still needed to complete the octave, two inside each interval of fourth (i.e. d_2, c_2 and g_1, f_1).

M.L. Dalla Chiara et al. (eds.), Language, Quantum, Music, 303–326.

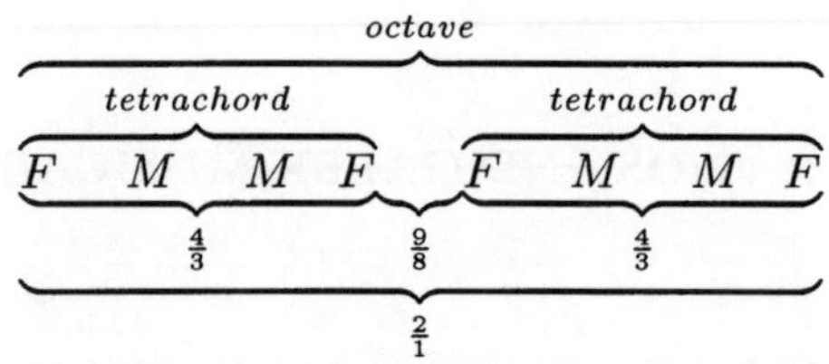

Figure 2.

Also for the Greeks, in fact, the octave (*diapason*) was constituted by eight notes (and according to this, the names given to the intervals of fifth and fourth were *diapente* and *diatessaron*).[2] But, while on one hand the ratios among the four preceding notes remained unchanged during the whole of the Greek and Medieval period, thanks to their connection with those clear numerical ratios, on the other hand it was not possible to extend this situation to the remaining four notes. For them there have been in fact so many proposals, and so different between them, that they were called *moveable notes*, in oppositions to the old four notes that were called *fixed notes* (see Figure 2). But, which were the causes of such a differentiation, and which the criteria employed in the various cases? Now, one element which simplifies our investigation is constituted by the fact that for the Greeks the nucleus constituting the harmonic space was the *tetrachord*, i.e. a group of four notes whose extremes produce an interval of fourth; the octave was obtained by two tetrachords separated by a tone (*disjoint tetrachords*, as in Figure 2) and it was not the only way to combine tetrachords. From all this it follows that:

> *defining an attunement reduced to determining the position of the two moveable notes inside the tetrachord.*

First of all, the attunements were divided in three sets, the *genera*. If the two moveable notes were very close to each other, we had an *enharmonic tetrachord*; if the two moveable notes were a little more distant we had a *chromatic tetrachord*; finally, if the two moveable notes were still more distant we had a *diatonic tetrachord*. So, considering that a fourth is, more or less, two tone and a half, we have that the three genera were variations on these themes:

> *enharmonic tetrachord*: ditone[3] + quarter of tone + quarter of tone;
> *chromatic tetrachord*: semiditone[4]+ semitone + semitone;
> *diatonic tetrachord*: tone + tone + semitone.[5]

But the choice among the three genera was not the point of the theoretical disputation, because it was mainly due to aesthetical and historical reasons; in fact, almost each theorist proposed his own attunements for each of the three genera. Because of the weakening, for the moveable notes, of the relationship between sounds and numbers (or rather between fine sounds and fine numbers), a first splitting among the theorists happened; on one side there were those who attributed more weight to the sounds (the most important of which was Aristoxenus), and on the other side there were those who attributed more weight to the numbers. The former ones were called *Harmonicists*, the latter ones *Pythagoreans* but, of course, the last term has in this context a much wider meaning than that of 'followers of Pythagoras'.

Before concluding the introduction we want to clarify a problem, both mathematical and linguistic, which has caused (and sometimes still does) many misunderstandings. All the musical terminology is additive (the term *interval* is emblematic), while the underlying 'reality' is multiplicative. For instance: in musical language we say that the octave is the *sum* of a fifth and a fourth, but $\frac{2}{1}$, which is the ratio corresponding to the octave, is the *product* of $\frac{3}{2}$ and $\frac{4}{3}$, which are the ratios corresponding to the fifth and to the fourth. This passage from the product to the sum drags all the other operations in a logarithmical correspondence, as shown in Table 3:

Musical image:	*Arithmetical operation:*
interval	ratio
sum (the octave is a fourth plus a fifth)	product $(\frac{2}{1} = \frac{3}{2} \cdot \frac{4}{3})$
difference (the tone is the difference between a fifth and a fourth)	quotient $(\frac{9}{8} = \frac{3}{2} / \frac{4}{3})$
multiplication by n (triple tone)	n-th power $((\frac{9}{8})^3)$
division in n equal parts (quart of tone)	n-th root $(\sqrt[4]{\frac{9}{8}})$
arithmetical mean	geometrical mean.

Table 3.

Now we can redefine our problem. We have seen that, for the Greeks, to determine an attunement meant to determine the position of the two moveable notes inside the tetrachord. And since the interval of fourth corresponds to the fraction $\frac{4}{3}$, and since the Pythagoreans avoided irrational numbers, we have, in accordance with Table 3, that:

> *defining an attunement reduced to determining three fractions* $\frac{p_1}{q_1}, \frac{p_2}{q_2}, \frac{p_3}{q_3}$ *whose product is* $\frac{4}{3}$.

It is to the mathematical aspects of the criteria employed by the Pythagoreans in defining these fractions that this article in mainly devoted. But the prevalence of the numerical considerations over those connected with the musical practice, prevalence that some *Pythagoreans* carried to radical extremes (cf. for instance, Plato's *Republic* 531 a-c), must not lead us to believe that their research was alien to what we mean today by the term 'music'. The sing-song *do-re-mi-fa-sol-la-si-do*, which describes our division of the harmonic space and represents the means by which we receive innumerably many messages, is rooted in our way of considering the musical phenomena to such a point that we are induced to think of it as an *a priori* or, at most, as a partition of a *continuum* in intervals of 'reasonable' width. But the choice of the about 85 sounds of a keyboard, which in their disposition and in their names are repeated in groups of 12, is not analogous to a choice of 85 colors for a box of pastels. It represents the solution, given by our civilization, to the problem of finding, in the greatest possible number of configurations (and sacrificing a little bit of precision in order to increase such a number) those ratios that the Greeks considered as consonant together with others that they refused. And this solution contains a lot of mathematics; the music that we would obtain from a piano in which we had increased the numbers of keys in order to reduce the intervals among them would be, contrary to the case of a richer box of pastels, radically different and unrecognizable. From this point of view, the numerical research, even the most abstract of it, that the *Pythagoreans* assigned to the musical ambit is connected with the contemporary meaning of the term 'music' since it concerns the fundamental aspect of the music language: the choice of the sounds, that is, the choice of its alphabet.

2. EPIMORIC ATTUNEMENTS

In this section we consider the criteria followed by the *Pythagoreans* in determining their attunements. The most immediate of them, in the sense that it was the most direct consequence of the disposition of the fixed notes, consisted in selecting as interval of tone between the moveable notes the same one existing between the fixed notes, i.e. the ratio $\frac{9}{8}$. In such a way, since the difference between a fourth and two of such tones is $\frac{4}{3}/\left(\frac{9}{8}\right)^2 = \frac{256}{243}$ (recall Table 3!), then the diatonic divisions of the tetrachord and consequently that of the octave became as in Figure 4:

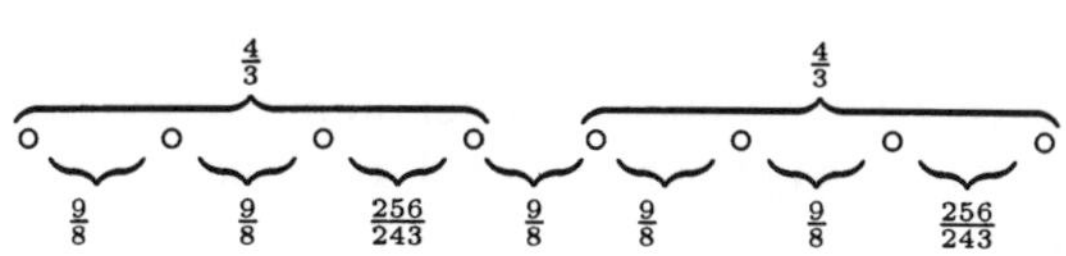

Figure 4.

This attunement is usually called *Pythagorean diatonic* (where the term *Pythagorean* must this time be intended in a strict sense), and it was without doubt the most important among the Greek attunements.[6] Since the tone is the difference between a fifth and a fourth, and since the fourth is the difference between an octave and a fifth, each note of this scale may be obtained by going back and forth by intervals of octave and fifth. This musical property is the counterpart of the principal mathematical characteristic of the Pythagorean diatonic, very Pythagorean indeed, constituted by the fact that each interval of this scale is expressed by ratios of type $\frac{2^m}{3^n}$ or $\frac{3^m}{2^n}$.

But the ratios $\frac{2}{1}$, $\frac{3}{2}$ and $\frac{4}{3}$, which represent the only consonances accepted by the Greeks inside the octave, have another common characteristic: they are of type $\frac{n+1}{n}$, a kind of ratios that the Greeks called *epimoric*, excepting $\frac{2}{1}$ which was considered among the *multiple* ratios (cf. §3). This fact, together with some theoretical reasons that we analyze in the next section, led to the formulation of the following law, which we shall call *Principle of Consonance*:

> *If an interval is consonant then it is expressed by a multiple or an epimoric ratio.*

This law appears explicitly in the Introduction of Euclid's *Sectio Canonis*, and it is repeated by almost all other theorists. Having an implicative form, with the musical property as antecedent and the arithmetical one as consequent, the connection between sounds and numbers expressed by this principle could have only a 'negative' function from a musical point of view: it did not allow to attribute but only to negate the *status* of consonance to an interval.

The Principle of Consonance was modified and extended from the consonances to the *melodic intervals*, that is, the intervals between two *consecutive* notes of a scale, providing in such a way a criterion for the division of the tetrachord. We find this new law explicitly stated by Ptolemy in his *Harmonics*,[7] and we call it *Principle of Melodic Intervals*:

> *Melodic magnitudes must be in epimoric ratios.*

This principle has the same implicative form as the above one, and therefore has only a negative value in order to establish new melodic intervals. Actually, Ptolemy's insistence in attempting to justify his attunements in an acoustic way serves the purpose to confirm the fact that the condition of being epimoric is necessary but not sufficient in order to be melodic. But we shall show that, on the contrary, the principle was employed, together with some adjunctive hypotheses, as a sufficient condition.

For the sake of simplicity we call *epimoric interval* an interval expressed by an epimoric ratio, and *epimoric attunement* a division of the tetrachord in three epimoric intervals. Now, the epimoric attunements quoted in the *Harmonics* of Ptolemy are in Table 5.

Without doubt they are many, even taking the multiplicity of the genera into account. They are so many that Manuel Bryenne, a Byzanthine theorist of the

Diatonic tetrachords

Ptolemy's even $\frac{10}{9} \cdot \frac{11}{10} \cdot \frac{12}{11} = \frac{4}{3}$ Ptolemy's tense: $\frac{10}{9} \cdot \frac{9}{8} \cdot \frac{16}{15} = \frac{4}{3}$

Didymus': $\frac{9}{8} \cdot \frac{10}{9} \cdot \frac{16}{15} = \frac{4}{3}$ Archytas': $\frac{9}{8} \cdot \frac{8}{7} \cdot \frac{28}{27} = \frac{4}{3}$
(Ptolemy's tonic)

Ptolemy's soft: $\frac{8}{7} \cdot \frac{10}{9} \cdot \frac{21}{20} = \frac{4}{3}$

Chromatic tetrachords

Ptolemy's tense: $\frac{7}{6} \cdot \frac{12}{11} \cdot \frac{22}{21} = \frac{4}{3}$ Didymus': $\frac{6}{5} \cdot \frac{25}{24} \cdot \frac{16}{15} = \frac{4}{3}$

Eratosthenes': $\frac{6}{5} \cdot \frac{19}{18} \cdot \frac{20}{19} = \frac{4}{3}$ Ptolemy's soft : $\frac{6}{5} \cdot \frac{15}{14} \cdot \frac{28}{27} = \frac{4}{3}$

Enharmonic tetrachords

Archytas': $\frac{5}{4} \cdot \frac{36}{35} \cdot \frac{28}{27} = \frac{4}{3}$ *Didymus':* $\frac{5}{4} \cdot \frac{31}{30} \cdot \frac{32}{31} = \frac{4}{3}$

Ptolemy's: $\frac{5}{4} \cdot \frac{24}{23} \cdot \frac{46}{45} = \frac{4}{3}$

Table 5.

XIV Century, in his comment on Ptolemy's *Harmonics*, asserted that they were all the possible epimoric attunements.[8] This assertion is wrong, as Pierre de Fermat observed, in turn, commenting Bryenne. Fermat, in fact, presented, just as examples, two new decompositions of $\frac{4}{3}$, namely $\frac{5}{4} \cdot \frac{17}{16} \cdot \frac{256}{255}$ and $\frac{8}{7} \cdot \frac{13}{12} \cdot \frac{14}{13}$, which were absent in Ptolemy's list.[9] At such a point Fermat put also the problem concerning the number of decomposition of any epimoric fraction into a given number of epimoric fractions. Now, in the specific case of the decompositions of $\frac{4}{3}$ in three epimoric fractions, we have the 26 decompositions listed in Table 6 below.[10] The list of all the theoretical possibilities will play a fundamental role in our attempt of better explaining the choices of Greek theorists.

1)	4	16	255		14)	5	14	27	Pt. soft chrom.
2)	4	17	135		15)	5	15	24	Did. chrom.
3)	4	18	95		16)	5	18	19	Er. chrom.
4)	4	19	75		17)	6	8	63	
5)	4	20	63		18)	6	9	35	
6)	4	21	55		19)	6	11	21	Pt. tense chrom.
7)	4	23	45	Pto. en.	20)	6	14	15	
8)	4	25	39		21)	7	7	48	
9)	4	27	35	Arc. en.	22)	7	8	27	Arc. (Pt.tonic) diat.
10)	4	30	31	Did. en.	23)	7	9	20	Pt. soft diat.
11)	5	10	99		24)	7	12	13	
12)	5	11	54		25)	8	9	15	Did. (Pt. tense) diat.
13)	5	12	39		26)	9	10	11	Pt. even diat.

Table 6.

In Table 6 we have written m instead of $\frac{m+1}{m}$; so, for instance, the triple 4-16-255 denote the decomposition $\frac{4}{3} = \frac{5}{4} \cdot \frac{17}{16} \cdot \frac{256}{255}$. The factors of each decomposition are in decreasing order and so, since $m > t$ implies that $\frac{m+1}{m} < \frac{t+1}{t}$, the denominators increase. Since we now identify two decompositions if they differs only by the order of the factors, the 12 diatonic attunements quoted by Ptolemy reduces to 11, because Ptolemy's diatonic and Didymus' diatonic have the same factors (see Table 5).

The 11 decompositions which are present in Ptolemy's *Harmonics* constitute, up to the order, less than half of the possible decompositions, and this might seem a proof of the fact that the Principle of Melodic Intervals was used as necessary but not sufficient condition. But consider the following.

As we have seen, the Greek theory admitted three *genera* of tetrachords, qualitatively different between them: the enharmonic, the chromatic and the diatonic. And each of them represented a model, even if quantitatively not coercive, for the attunements. The minimum interval considered was the quarter of tone, and considering $\frac{9}{8}$ as the standard ratio for the tone we have that $\sqrt[4]{\frac{9}{8}}$ is between $\frac{35}{34}$ and $\frac{34}{33}$. Therefore, intervals smaller than $\frac{35}{34}$, i.e. epimoric ratios $\frac{n+1}{n}$ with $n > 34$, would be outside of the theory. Moreover, the ditone, which is the largest interval in the enharmonic, must be around $\left(\frac{9}{8}\right)^2 \left(= \frac{81}{64} = 1.265625...\right)$; therefore it can be reasonably expressed, among the epimoric ratios, only by $\frac{5}{4} \left(= \frac{80}{64} = 1.25\right)$, because $\frac{6}{5}$ is too far and qualitatively different.[11] Analogously, the semiditone of the chromatic genus, which must be around $\left(\sqrt[2]{\frac{9}{8}}\right)^3 = 1.193242....$, can be reasonably expressed, always among the epimoric ratios, only by $\frac{6}{5} (= 1.2)$, because $\frac{7}{6} \left(= 1.1\bar{6}\right)$ is again too far. For the intervals of tone in the diatonic genus, one cannot use both times the ratio $\frac{9}{8}$, because the interval of semitone that we obtain, that is $\frac{4}{3} / \left(\frac{9}{8}\right)^2 = \frac{256}{243}$, is not an epimoric ratio (actually, what we obtain is the Pythagorean diatonic – see Figure 3 – which is not an epimoric attunement); so one may employ, beside $\frac{9}{8}$, also the two epimoric ratios closest to it, i.e. $\frac{8}{7}$ and $\frac{10}{9}$. Now, these conditions all together are equivalent to this one:

> (2.1) *Each tetrachord must be composed by three epimoric ratios (whose product is $\frac{4}{3}$) which are greater or equal to $\frac{35}{34}$, and must contain either exactly one among $\frac{5}{4}$ and $\frac{6}{5}$ or at least two among $\frac{8}{7}$, $\frac{9}{8}$ and $\frac{10}{9}$.*

This condition expresses therefore the conjunction of the Principle of Melodic Intervals with some conditions which naturally arise from the traditional definitions of the *genera*. And if we look at Table 6 we can realize that (2.1) has been used as a *sufficient* condition. In fact there are exactly nine epimoric attunements satisfying it, namely the numbers 9, 10, 14, 15, 16, 22, 23 and 25, and *all* of them have been employed by some Greek theorist. Furthermore, among the epimoric attunements occurring in the *Harmonics* of Ptolemy, the only three which do not satisfy (2.1), that are the numbers 7, 19 and 26 of the table, are of Ptolemy himself. In fact, his enharmonic presents the interval $\frac{46}{45}$, his tense chromatic contains none of $\frac{5}{4}, \frac{6}{5}, \frac{8}{7}, \frac{9}{8}$ and $\frac{10}{9}$, and his even diatonic presents only one among $\frac{8}{7}, \frac{9}{8}$ and $\frac{10}{9}$. But, and this is probably the most interesting fact, before violating (2.1) Ptolemy, by means of his soft chromatic and soft diatonic, exhausts all the epimoric possibilities which are in accord to (2.1). So, notwithstanding the frequent recalls to the role of experience in music, Ptolemy acts here on a purely arithmetical level.[12] The mathematical criteria by which Ptolemy and the others Greek theorists found their attunements will be analyzed in §5.

3. DEBATE ON THE PRINCIPLE OF CONSONANCE.

In the preceding section we described the use that multiple ratios and, most of all, epimoric ratios had in the problem of tuning the scale. Now we want to investigate the arguments that were employed to found this fact. The most frequent justification of the Principle of Consonance was based on the analogies between ratios among sounds and ratios among numbers (and the importance that the inferences by analogy had in the Pythagorean philosophy is well known). In order to define these analogies we must therefore examine the Greek classification of the ratios among natural numbers. First we describe it in modern terms.

Given two natural numbers p, q with $p > q \geq 1$, let $\frac{m}{n}$ be the ratio $\frac{p}{q}$ reduced to the lowest terms, and let $m = sn + r$ be the euclidean division of m by n.

1. If $s = 0$ then $\frac{p}{q}$ is called a *multiple ratio.*

2. If $s = 1, r = 1$ (i.e. if $m = n + 1$) and $n \neq 1$ then $\frac{p}{q}$ is called an *epimoric ratio* (in Latin, *superparticular ratio*).

3. If $s = 1$ and $r > 1$ then $\frac{p}{q}$ is called an *epimeric ratio* (in Latin, *superpartient ratio*).

4. If $s > 1$ and $r = 1$ then $\frac{p}{q}$ is called a *multiple-epimoric ratio.*

5. If $s > 1$ and $r > 1$ then $\frac{p}{q}$ is called a *multiple-epimeric ratio.*

The above cases exhaust all the possibilities if $p > q$. The inverse ratios were denoted by the prefix *hypo*.

The preceding definitions are of immediate comprehension for a modern reader, but are somewhat far from the spirit of the original ones, since those did not employ the concept of reduction to the lowest terms. For instance, the original definition of epimoric ratio was as follows:

> the ratio $\frac{p}{q}$ is epimoric if $p = q + d$, where d is a divisor of q.[13]

In fact, the term 'epimoric' derives from *epi* (over) and *morion* (divisor), and the Latin term 'superparticular' is a literal translation. For instance, if we set $q = 6$ then the numbers which are in epimoric ratio with it are $7(= 6 + 1)$, $8(= 6 + 2)$ and $9(= 6 + 3)$. The fact that if $\frac{p}{q}$ is an epimoric ratio then it can be expressed in the form $\frac{n+1}{n}$ was a theorem, due to Archytas (see §4). In general, the greek classification of the ratios was based on the way by which the numerator overcomes the denominator. Under this point of view (so far from ours) multiple and epimoric ratios turn out to have a common characteristic: they are the only kinds of ratios in which p can be expressed in terms of q by using only one number besides q. In fact, if $\frac{p}{q}$ is a multiple ratio then $p = q + kq$, and if $\frac{p}{q}$ is epimoric then $p = q + \frac{q}{k}$ (where, of course, k is a divisor of q). On the other hand, in the remaining cases we need either two or three new numbers: if $\frac{p}{q}$ is epimeric then $p = q + h(\frac{q}{k})$, if it is multiple-epimoric then $p = hq + \frac{q}{k}$, and if it is multiple-epimeric then $p = hq + j\frac{q}{k}$. The simplicity in denoting multiple and epimoric ratios was interpreted as a mark of distinction of these class of numbers, which therefore were worth to be chosen as counterpart for a privileged class of music intervals, the consonances. This was the fundamental analogy. Ptolemy reports it in these terms:

> They (the Pythagoreans) laid down a first principle of their method that was entirely appropriate, according to which equal numbers should be associated with equal-toned notes, and unequal numbers with unequal-toned notes; and from this they argued that just as there are two primary classes of unequal-toned notes, that of the concords and that of the discords, and that of the concords is finer, so there are also two primary distinct classes of ratios between unequal numbers, one being that of what are called 'epimeric' or 'number to number' ratios, the other being that of the epimorics and multiples; and of these the latter is better than the former on account of the simplicity of the comparison, since in this class the difference, in the case of epimorics, is a simple part, while in the multiples the smaller is a simple part of the greater. For this reason they fit the epimorics and multiple ratios to the concords, and link the octave to the duple ratio [2:1], the fifth to hemiolic [3:2], the fourth to the epitritic [4:3]. Their procedure here is very rational, since the octave is the finest of the concords and the duple is the best of the ratios, the former because it is nearest to the equal-toned, the latter because it alone makes an excess equal to that which is exceeded; and again, because the octave consists of the two first concords taken successively, and the duple consists of the two first epimorics taken successively, the hemiolic and the epitritic; and while in the latter case the hemiolic ratio is greater than the epitritic, in the former the concord of the fifth is grater than that of the fourth, so that the difference between them – that is, the tone, is assigned to epogdoic ratio [9:8], by which the hemiolic is greater than the epitritic.[14]

According to Ptolemy, the Pythagoreans used therefore the analogy not only to found the Principle of Consonance, but also to put each consonance in relation to its proper ratio. And if, from one hand, the justification of the Principle of Consonance can be summarized, without peculiar loss of content, in the slogan 'good ratios for good sounds', on the other hand the justification for the assignment of the octave to the ratio $\frac{2}{1}$, which is the keystone of the second part, is based on the argument 'the best ratio for the best consonance'. Clearly, the tentative was that of ennobling the original empiric discovery by a 'rational' justification, in view of its insertion into a global conception of the universe.

In this direction, and again among the Pythagoreans, there was another, more sophisticated, attempt. Instead of using the method of analogy twice, both for the general case (the Principle of Consonance) and for the particular attributions of certain ratios to certain consonances, it happened that the general case, still based on analogy, was employed to *demonstrate*, in Euclidean style, the particular attributions. So the principle, obtained by extrapolation from some empirical facts, was assumed to prove these facts themselves. The inversion of perspective with respect to the original Pythagorean discovery is complete, and had the evident aim of freeing the knowledge from experience: it was a pure emanation of the Platonic ideology!

This proof appeared first in Euclid's *Sectio Canonis*, and then in Ptolemy, Porphyry and Boethius.[15] A problem, first observed by Tannery in [13], is constituted by the fact that the proof employes, besides the Principle of Consonance (*If an interval is consonant then it is expressed by a multiple or an epimoric ratio*), also the following implication: *If an interval is not consonant then it is not expressed by a multiple or an epimoric ratio*. Clearly, this implication is not a logical consequence of the Principle of Consonance, but is equivalent to its converse: *if an interval is expressed by a multiple or an epimoric ratio then it is consonant*. This assertion, under a 'reasonable' interpretation of the term 'consonant', is false and in contrast with every Greek tradition. A hypothesis proposed to justify this fact is that all the *Sectio Canonis* 'tacitly relies in the quaternary $\{1, 2, 3, 4\}$'.[16] Under this restriction also the converse of the Principle of Consonance is true.[17] But even if the use of the converse of the principle had been an error, there would be many possible justifications (and so we would not consider it a sufficient reason to deny the attribution of the *Sectio Canonis* to Euclid). Ptolemy, who is in general a harsh critic, reports the proof accurately without noticing any error. In fact, the most efficient antidote to this kind of logical errors is a good semantic interpretation of the terms involved. And the link between sounds and numbers, which constitutes the semantic of these reasoning, was sometimes so weak that the purely arithmetical aspect had assumed a remarkable self-sufficiency. For instance, we have shown in the preceding section that another harmonic criterion, even if enounced in the following implicative form – *If an interval is melodic then it is expressed by an epimoric ratio* –, it has been employed also in its converse form – *If an interval is expressed by an epimoric ratio then it is melodic* –. The symmetry between this situation and that of the *Sectio Canonis* is evident. At any rate, if we assume the new axiom: *An interval is consonant if and only if it is expressed by a multiple or an epimoric ratio* (and we can do this without contradicting the text because, contrary to the *Elements*, in the *Sectio Canonis* the axioms are not explicitly stated) the proof is correct and very interesting. The musical properties employed (again not explicitly stated) are:

1. The double octave is a concord.

2. The double fifth and the double fourth are discords.

3. The fifth and the fourth are consonant.

4. The octave is composed by a fifth and a fourth.

The arithmetical properties employed (and previously proved) are:

5. If m and n are in epimoric ratio, then there is no geometrical mean between them (i.e. the geometrical mean is not a natural number; this result, due to Archytas, constitutes Proposition 3 of the *Sectio Canonis*).

6. If p is the geometrical mean between m and n, then n is multiple of m if and only if p is multiple of n (Propositions 1, 2, 4 and 5 of the *Sectio Canonis*).

7. The double ratio is composed by the two maximal epimoric ratios (i.e. $\frac{2}{1} = \frac{3}{2} \cdot \frac{4}{3}$; recall that the Greeks considered $\frac{2}{1}$ among multiple ratios, and therefore $\frac{3}{2}$ is the maximal epimoric ratio).

The only bridges between this two groups of properties, the former without numbers, the latter without sounds, are the conjunction of the Principle of Consonance with its converse (Axiom 8) and a part of the logarithmic relationship between musical intuition and arithmetical operations (Axiom 9).

8. An interval is consonant if and only if it is expressed by a multiple or a epimoric ratio.

9. Each interval corresponds to the geometrical mean of its double (i.e. if $\frac{p}{n}$ corresponds to a certain interval, then $\frac{m}{n}$ corresponds to the doubled interval if and only if p is the geometrical mean of m and n).

Now there are all the elements to prove that:

The octave corresponds to $\frac{2}{1}$, *the fifth to* $\frac{3}{2}$ *and the fourth to* $\frac{4}{3}$.

The Euclidean proof develops as follows. The double octave is consonant (by 1). Therefore its ratio is either multiple or epimoric (by 8). If it were epimoric then it could not have a geometrical mean (by 5). But the octave is the geometrical mean of the duble octave (by 9), which therefore cannot be an epimoric ratio. So it is a multiple ratio. Therefore (by 9 and 6) also its geometrical mean, the octave, is a multiple ratio. Moreover, the double fifth is dissonant (by 2); therefore it is neither multiple nor epimoric (by 8), and its geometrical mean, the fifth, is not a multiple ratio (by 9 and 6). But it is consonant (by 3), and therefore (by 5) it corresponds to an epimoric ratio. By an identical proof one obtains that also the fourth corresponds to an epimoric ratio. Now, it has been proved that the octave corresponds to a multiple ratio. Therefore it is double ($\frac{2}{1}$) or more than double. But the octave is composed by a fifth and a fourth (4). And the double ratio is composed by the two maximal epimoric ratios (7). Therefore the octave cannot correspond to a multiple ratio more than double, and therefore corresponds to the double ratio. And since it has been shown that the fifth and the fourth correspond to epimoric ratios, and since the fifth is bigger than the fourth, from 4 and 7 it follows that the fifth corresponds to $\frac{3}{2}$ and the fourth to $\frac{4}{3}$.[18]

Finally, there was also a third way that the Pythagorean attempted to prove the correspondence between consonances and their corresponding ratios.[19] They attributed a 'degree of dissonance', say $d(\frac{m}{n})$, to each ratio $\frac{m}{n}$, as follows: $d(\frac{m}{n}) = (m-1) + (n-1)$. Notwithstanding the fact that Ptolemy defined this formula 'utterly ridiculous', it is not meaningless. If we imagine a constant sound as a sequence of impulses at regular intervals, and a variation of its pitch as a variation of the frequency of such impulses, then the previous formula has this interesting interpretation: if we have two sources S_1 and S_2 which start simultaneously and produce respectively m and n dots for time unit, then the number of unsyncronized dots per time unit is exactly $d(\frac{m}{n})$. If we remain in the ambit of the octave and take m and n coprime, this formula produces results which are substantially correct (in 1770 Euler, in the fourth of his *Lettres a une Princesse d'Allemagne* will justify the consonances in the same way); but, above all, it is important because it overcomes the borders of the method of analogy, trying a way which could be defined 'physical-psychological'.

On the opposite side of all these justifications, also the limits of the Principle of Consonance were evidentiated by some Greek theorists. A precise, criticism is reported by Ptolemy. According to the principle, the orthodox Pythagoreans did not consider the eleventh among the consonances because, being an octave plus a fourth, it corresponds to the ratio $\frac{2}{1} \cdot \frac{4}{3} = \frac{8}{3}$, which is neither multiple nor epimoric, while they accepted the twelfth, which is an octave plus a fifth and therefore corresponds to the multiple ratio $\frac{2}{1} \cdot \frac{3}{2} = \frac{3}{1}$. This discrimination does not have any musical consistence because, as Ptolemy himself observes, the addition of an octave cannot change the *status* of consonance.[20] Another difficulty that was evidentiated brings us back to the 'error' in the proof of the *Sectio Canonis*: why the Principle of Consonance expresses only a necessary and not also a sufficient condition? If the connection between the epimorics and the concords is due to the fact that the epimorics 'are spoken of under a single name in relation to one another',[21] why only some of them correspond to concords, while all of them have this property? A tentative solution was proposed by some Hellenistic theorists, and consisted in enriching the distinctions among sounds, in order to have many degrees of goodness to put in correspondence with the epimoric ratios in function of the distance from $\frac{2}{1}$. Nevertheless, this way was not firmly followed, and did not weaken the distinction between the concords and the discords. For instance, it was necessary to wait until the XIII century to find a theorist, Walter Odington, who accepted the major third and the minor third among the consonances, notwithstanding their corresponding ratios, $\frac{5}{4}$ and $\frac{6}{5}$, i.e. the first epimorics following $\frac{2}{1}$, $\frac{3}{2}$ and $\frac{4}{3}$, should have forced the Greeks in this direction.

4. SOME ARITHMETICAL PROPERTIES OF EPIMORIC RATIOS.

As seen in the previous section, the fundamental support of the Principle of Consonance can be synthesized in the slogan 'good ratios for good sounds'. Now, the fact that the concords were considered good sounds gives no problems: they are good by definition. In the same way, it is perfectly imaginable that, among the ratios, the Greeks considered the multiple ones good: in fact a multiple ratio can be expressed by a natural number. But, why should one consider the epimoric ratios as good ratios? As seen, the standard justification was that they are 'among those numbers which are spoken of under a single name in relation to one another', i.e. the class of epimoric ratios can be described as $\{\frac{p}{q} : p = q + \frac{q}{k}, \textit{for k divisor of q}\}$, using only one number ('*name*') besides p and q. But, even conceding that this justification were more significant for the Greeks than it is for us, it is not conceivable that this property alone could have rendered the epimoric ratios good and therefore worth joining the concords. Actually, there were other mathematical properties (some of them discovered by the Greeks, some others probably only perceived) that, making these numbers more interesting, strenghtened this joining; this last section is devoted to these properties.

The relationship between epimoric ratios and the means was very important. In is well known, in fact, that problems regarding the means were among the oldest treated by the Greeks, and had a fundamental role in the Pythagorean mathematics and in its arithmo-geometrical characterization. The tradition says that Pythagoras learned these concept in the Orient, but the first definition that we have is due to Archytas and regards three means:

There are three means in music. One is arithmetic, the second geometric, the third subcontrary, which they call 'harmonic'. There is an arithmetic mean when there are three terms, proportional in that they exceed one another in the following way: the second exceeds the third by the same amount as that by which the first exceeds the second.[...] There is a geometric mean when they are such that as the first is to the second, so is the second to the third.[...] There is a subcontrary mean, which we call 'harmonic', when they are such that the part of the third by which the middle term exceed the third is the same as the part of the first by which the first exceeds the second.[22]

Given two numbers n, m such that $n > m$, if we denote by $A(n,m)$, $G(n,m)$ and $H(n,m)$ the *arithmetic*, *geometric* and *harmonic mean* respectively then, following the above definitions, we have that $n - A(n,m) = A(n,m) - m$, $\frac{n}{G(n,m)} = \frac{G(n,m)}{m}$ and $\frac{m}{H(n,m)-m} = \frac{n}{n-H(n,m)}$, and therefore:

$$A(n,m) = \frac{n+m}{2}; \quad G(n,m) = \sqrt{nm}, \quad H(n,m) = \frac{2nm}{n+m}$$

An important result, that we have already mentioned and which regards epimoric ratios and means, is due to Archytas himself, and it is a 'negative' result:

> (Theorem of Archytas) *If two natural numbers n and m are in epimoric ratio then there is no natural z such that $z = G(n,m)$.*[23]

The proof employs the fact that if m and n are in epimoric ratio than $m : n = t+1 : t$ for a certain t. From a historical point of view this result is important because it presupposes several other results that we shall find in the arithmetical books of the *Elements* (as, for instance, Propositions VII.20,22,33 and VIII.8), and therefore it allows us to presume that there already existed, at least as early as the time of Archytas, treatises on the Elements of Arithmetic in a form similar to the Euclidean one, and containing many of the results afterwards embodied by Euclid in his arithmetical books.[24] But this result has also a musical consequence. It shows in fact that, given n, m such that $\frac{n}{m} = \frac{9}{8}$ then there is no natural z such that $\frac{n}{z} = \frac{z}{m}$, and therefore it is not possible to express the interval of the exact semitone by means of natural numbers. We find this result in the *Sectio Canonis* Proposition 16: *The tone will not be divided into two or more equal intervals*), and it was one of the principal elements of distinction between the *Harmonicists* and the *Pythagoreans;* the former ones contented themselves by finding the semitone by ear, while the latter ones, according to the arithmo-geometrical philosophy, interpreted this result as the proof of the inexistence of the semitone.[25]

Another important (and this time not negative) result about the connection between epimoric ratios and means is expressed by the so called *Musical Proportion*, considered by the Greeks the most perfect proportion. According to Nicomachus and Iamblichus, it was discovered by the Babylonians and was first introduced into Greece by Pythagoras. It is as follows:

$$n : A(n,m) = H(n,m) : m.$$

(and therefore $n : H(n,m) = A(n,m) : m$). Now, if $\frac{n}{m} = \frac{2}{1}$, i.e. if n and m produce an interval of octave, then $\frac{n}{H(n,m)} = \frac{A(n,m)}{m} = \frac{3}{2}$ (the fifth), $\frac{n}{A(n,m)} = H(n,m)m = \frac{4}{3}$ (the fourth), and $\frac{A(n,m)}{H(n,m)} = \frac{9}{8}$ (the tone). This is the reason why Nicomachus asserted that the name of 'harmony' was due to this proportion only. We can generalize this by taking two numbers n, m such that $\frac{n}{m} = \frac{t+1}{t}$ for some t, instead of $\frac{n}{m} = \frac{2}{1}$ as above, and obtain that:

> (4.1) *If* $\frac{n}{m} = \frac{t+1}{t}$ *then* $\frac{n}{H(n,m)} = \frac{A(n,m)}{m} = \frac{2t+1}{2t}$, $\frac{n}{A(n,m)} = \frac{H(n,m)}{m} = \frac{2t+2}{2t+1}$ and $\frac{A(n,m)}{H(n,m)} = \frac{2t^2+4t+1}{2t^2+4t}$.

This property has not been explicitly stated by the Greeks, but it was probably employed to find epimoric attunements; for instance, if we assume $\frac{n}{m} = \frac{3}{2}$, then $\frac{n}{H(n,m)} = \frac{A(n,m)}{m} = \frac{5}{4}$, $\frac{n}{A(n,m)} = \frac{H(n,m)}{m} = \frac{6}{5}$, and $\frac{A(n,m)}{H(n,m)} = \frac{25}{24}$; and these ratios were very frequently employed in the attunements (see §2). Because of the obvious proportion $n : m = \frac{n}{m} : 1$, Proposition 4.1 has the following consequence:

(4.2) *If x is an epimoric ratio, then $A(x,1)$, $H(x,1)$ and $\frac{A(x,1)}{H(x,1)}$ are epimoric ratios.*

This form makes evident that we are in front of a closure result, but also suggests a new result, which expresses a generation property:

(4.3) *Let us consider the operators $f_A(x) = A(x,1)$ and $f_H(x) = H(x,1)$. The set of the epimoric ratios can be generated from the set $\{\frac{2}{1}\}$ by means of $f_A(x)$ and $f_H(x)$.*[26]

Leaving the means, we pass to a second group of results, of which we can sometime see some partial employments in music, and which can be obtained by this single immediate property of the epimoric ratios:

(4.4) *For every natural number $n > 1$, $n = \frac{2}{1} \cdot \frac{3}{2} \cdot \ldots \cdot \frac{n}{n-1}$.*

Now, let p, q be natural numbers such that $p > q$. Considering that $p = \frac{2}{1} \cdot \frac{3}{2} \cdot \ldots \cdot \frac{p}{p-1}$ and $q = \frac{2}{1} \cdot \frac{3}{2} \cdot \ldots \cdot \frac{q}{q-1}$ we obtain that:

(4.5) *For every rational number $\frac{p}{q} > 1$, $\frac{p}{q} = \frac{q+1}{q} \cdot \frac{q+2}{q+1} \cdot \ldots \cdot \frac{p}{p-1}$.*

So, each rational number bigger than 1 is the product of epimoric ratios. If we connect this result with (4.3) we obtain a way to generate the set of the rational numbers bigger than 1 from the set $\{\frac{2}{1}\}$ remaining in the ambit of epimoric ratios. Moreover, an immediate consequence of 4.5 is that we can also decompose any epimoric ratio $\frac{n+1}{n}$ in m epimoric ratios, for every m. In fact, if we set $p = m(n+1) = mn + m$ and $q = mn$ we obtain that:

(4.6) $\frac{n+1}{n} = \frac{mn+m}{mn} = \frac{mn+1}{mn} \cdot \frac{mn+2}{mn+1} \cdot \ldots \cdot \frac{mn+m}{mn+m-1}$.

As we shall see in the next section, this equation had an important role in the research of the epimoric attunements. For now we want just to observe a fact which is connected with its geometrical interpretation. Suppose that OA and OB are two segments such that $\frac{\overline{OB}}{\overline{OA}} = \frac{n+1}{n}$, and choose $m-1$ points $A_1, A_2, \ldots, A_{m-1}$ on AB, thus dividing AB in the m segments $AA_1, A_1A_2, \ldots, A_{m-1}B$. We obtain that $\frac{n+1}{n} = \frac{\overline{OB}}{\overline{OA}} = \frac{\overline{OA_1}}{\overline{OA}} \cdot \frac{\overline{OA_2}}{\overline{OA_1}} \cdot \ldots \cdot \frac{\overline{OB}}{\overline{OA_{m-1}}}$, and this identity coincides with 4.6 if and only if the segments $AA_1, A_1A_2, \ldots, A_{m-1}B$ are equal to each other, i.e. if we have divided AB in m equal parts. Now, if in 4.6 we set $n = m = 3$, we obtain the even diatonic of Ptolemy: $\frac{4}{3} = \frac{10}{9} \cdot \frac{11}{10} \cdot \frac{12}{11}$. So the preceding observation implies that the term 'even' (*omolos equabilis* in the latin translation) have been used by Ptolemy in an additive sense, because $\overline{OA}, \overline{OA}_1, \overline{OA}_2, \ldots, \overline{OB}$ form an arithmetic progression. The even diatonic corresponds therefore to those attunements obtained, for instance, by those primitive flutes in which the holes were equidistant.[27]

The epimoric ratios were involved also in problems of arithmo-geometry. Let us consider the following numerical sequences:

n	1	2	3	4	5	...
n^2	1	4	9	16	25	...
$n(n+1)$	2	6	12	20	30	...

The first one is that of the natural numbers, the second one that of the square numbers and the third that of the 'oblong' numbers.[28] For the Pythagoreans this third sequence was almost as important as that of the square numbers, for this reason: if, from one hand, n^2 is the sum of the first n odd numbers (i.e. $n^2 = 1 + 3 + 5 + \ldots + o_n$), on the other hand $n(n+1)$ represents the sum of the first n even numbers (i.e. $n(n+1) = 2 + 4 + 6 + \ldots + e_n$). The epimoric ratios are strictly connected with these sequences. Since $(n+1)^2 : n(n+1) = n(n+1) : n^2 = n+1 : n$, then each epimoric ratio can be generated by comparing the sequences vertically or diagonally (see the following figure):

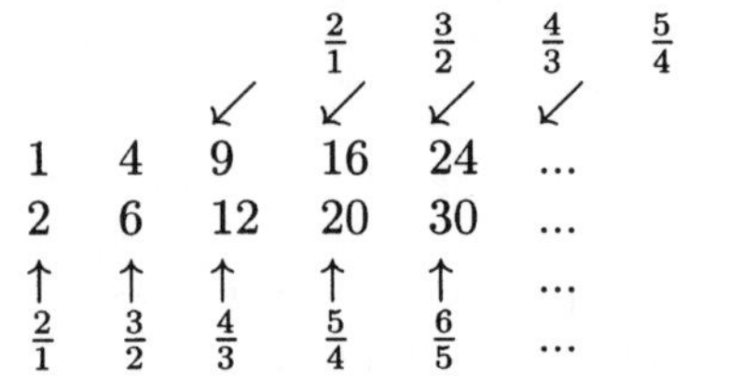

Figure 7:

Furthermore, the three numbers $(n+1)^2, n(n+1)$ and n^2, i.e. the sum of the first $n+1$ odds, the sum of the first n evens and the sum of the first n odds, are the smallest natural numbers a, b, c such that $\frac{a}{b} = \frac{b}{c} = \frac{n+1}{n}$.[29] As we can see, no reference to music is present in the above result; and in fact Boethius inserts it in his *De Institutione Arithmetica* (II.33) and not in his *De Institutione Musica.* But results like this, making the epimoric ratios important, contributed to strenghten the law 'good ratios for good sounds', and thus the correspondence of these numerical ratios with the class of consonances and of melodic intervals.

But the property of epimoric ratios that we think was the basis for the formulation of the concept of epimoric attunement is the following 4.7. We define $\frac{t+1}{t}$ a *good divisor* of $\frac{n+1}{n}$ if $\frac{n+1}{n} / \frac{t+1}{t} = \frac{s+1}{s}$ and $\frac{s+1}{s} \leq \frac{t+1}{t}$. In other words, a good divisor of $\frac{n+1}{n}$ is the biggest element of a decomposition $\frac{t+1}{t} \cdot \frac{s+1}{s}$ of $\frac{n+1}{n}$, and therefore (since the two elements cannot be equal because the square of an epimoric ratio cannot be an epimoric ratio) the good divisors of $\frac{n+1}{n}$ are as many as the decompositions of $\frac{n+1}{n}$ in two epimoric ratios. Since, obviously, $\frac{t+1}{t} < \frac{n+1}{n}$, we may represent $\frac{t+1}{t}$ in the form $\frac{n+q+1}{n+q}$, q a natural number. In [3] we have proved that:

(4.7) $\frac{n+q+1}{n+q}$ *is a good divisor of* $\frac{n+1}{n}$ *if and only if* $q \leq n$ *and* q *is a divisor of* $n \cdot (n+1)$.

Now, Property 4.7 implies, roughly speaking, that if n is small then there is a good probability that, dividing $\frac{n+1}{n}$ by an epimoric ratio not too far from it, the rest is still epimoric. As we shall see in §5, this fact was very relevant in the first searches of epimoric attunements. The Greeks did not known Property 4.7 but, without doubt, they observed the frequency of the divisions of two epimoric ratios having an epimoric result. This fact interacted with some other situations that, even if not generalizable under a theorem because they are false in general, are always, or almost always, satisfied among numbers which are not too big. All this generated the impression, among people dealing with epimoric numbers, that they are omnipresent and still richer of properties than they actually are. We give two examples.

First example. From 4.7 it follows that, if n is small, *many* decompositions of $\frac{n+1}{n}$ in 3 epimoric factors can be obtained *in two times*, i.e. by decomposing first $\frac{n+1}{n}$ in two epimoric factors and then by decomposing one of them in two epimoric factors again. Of course, it is false, in general, that all the decompositions of $\frac{n+1}{n}$ can be obtained in such a way. For instance, in the case $\frac{n+1}{n} = \frac{5}{4}$, we have that the decomposition $\frac{5}{4} = \frac{8}{7} \cdot \frac{17}{16} \cdot \frac{35}{34}$ is such that no among the products of two of these factors is epimoric. But, in case of $\frac{4}{3}$, we have that *all* the decompositions in three epimoric factors, i.e. all the attunements, can be obtained in two times, because from Table 6 easily follows the following statements:

(4.8) *In any decomposition of* $\frac{4}{3}$ *in three epimoric factors the product of two of them is epimoric.*

Second example. Let us consider the set X of the rational numbers between 1 and 2, i.e. in the Greek terminology, the set of the epimoric and the epimeric ratios (cf. §2). As we have seen in (4.5), each rational number greater than 1 is the product of epimoric ratios, and, *a fortiori*, so is each element of X. On the other hand it is false that each element of X is the product of two epimoric ratios.[30] But this, besides beeing true for the epimoric themselves, is true also for the numbers of type $\frac{n+2}{n}$ and $\frac{n+3}{n}$, two important classes of ratios under the Greek classification of the rational numbers,[31] and it is frequent in many other cases. Therefore, if for instance we consider the ratios $\frac{m}{n} \leq \frac{3}{2}$, where $n \leq 20$ and $m \leq 30$, then we have that among the 100 possible pairs (m, n) which satisfy these conditions (64 of which are not epimoric ratios) only 6 are such that $\frac{m}{n}$ cannot be decomposed in two epimoric ratios (namely $\frac{17}{13}, \frac{19}{13}, \frac{23}{16}, \frac{22}{17}, \frac{28}{19}$ and $\frac{29}{19}$). And this few cases can be not sufficient, under a superficial analisys, to avoid an at least hypotethical generalization.

5. HOW THE EPIMORIC ATTUNEMENTS WERE OBTAINED

In the research of epimoric attunements Greek theorists employed methods having the common characteristic of beein 'in two times"; in fact the factors were found by decomposing first $\frac{4}{3}$ in two epimoric factors and then by decomposing one of them in two epimoric factors again. Clearly, this kind of methods are simpler because reduce the problem of decomposing in three factor to that of decomposing in two factors; incidentally, Proposition 4.8 tell us that, in the case of $\frac{4}{3}$, all the decompositions can be obtained by means of them. The ways employed to decompose in two factors are linked to Proposition 4.6.

If in 4.6 we set $m = 2$, we obtain (5.1) $\frac{n+1}{n} = \frac{2n+1}{2n} \cdot \frac{2n+2}{2n+1}$, which is the decomposition where the two epimoric factors are closest between them, and it is still linked with the means because (cf. 4.1 and 4.2) it can be expressed in the form $\frac{n+1}{n} = A(\frac{n+1}{n}, 1) \cdot H(\frac{n+1}{n}, 1)$. Five among the twelve attunements of Table 5 can be obtained by decomposing first $\frac{4}{3}$ in two epimoric ratios (and the Greeks knew that $(\frac{5}{4}, \frac{16}{15})$, $(\frac{6}{5}, \frac{8}{9})$ and $(\frac{7}{6}, \frac{8}{7})$ are the only possible ways), and then decomposing one of them by 5.1. A property that can be immediately noticed about the attunements obtained by this method is that two epimoric ratios are consecutive and the smaller of them has an even denominator. We refer to it as *Method A*.[32]

If, again in 4.6, we set $m = 3$ we obtain (5.2) $\frac{n+1}{n} = \frac{3n+1}{3n} \cdot \frac{3n+2}{3n+1} \cdot \frac{3n+3}{3n+2}$, and this identity is involved in *Ptolemy's Method*, an algorithm explicitly stated in [12] Book 1,Ch.15. Ptolemy realized, even if without showing it in general, that in decomposition 5.2 one among $\frac{3n+1}{3n} \cdot \frac{3n+2}{3n+1}$ and $\frac{3n+2}{3n+1} \cdot \frac{3n+3}{3n+2}$ is epimoric.[33] Thus he obtained two new kinds of decompositions of an epimoric fraction in two epimoric fractions, which are $(\frac{3n+1}{3n} \cdot \frac{3n+2}{3n+1}, \frac{3n+3}{3n+2})$ and $(\frac{3n+1}{3n}, \frac{3n+2}{3n+1} \cdot \frac{3n+3}{3n+2})$. So he modified the Method A in the sense that he still considered the decomposition of $\frac{4}{3}$ in two epimoric ratios, but he divided one of them in two epimoric ratios following his new way. The property that can be noticed about the attunements obtained by this *method* is that the numerator of one factor is the half of the denominator of the other factor. All the attunements of Ptolemy can be obtained by this method, excepting the tonic diatonic (which coincides with Archytas' diatonic).[34]

Three important epimoric attunements are still absent, in the sense that they cannot be obtained by the two methods of above; they are Architas' diatonic, Architas' enharmonic and Didymus' chromatic. Probably, they were obtained through attempts. For instance: for the ratio to be put together with $\frac{9}{8}$ in his diatonic, Architas choose the smallest among those bigger than $\frac{9}{8}$, i.e. $\frac{8}{7}$; and he was lucky because what he obtained, by dividing $\frac{4}{3}$ by $\frac{9}{8} \cdot \frac{8}{7}$, was still an epimoric ratio, $\frac{28}{27}$. Then he decided to insert this ratio also in his enharmonic and, taking the natural ratio $\frac{5}{4}$ for the ditone, divided $\frac{4}{3}$ by $\frac{5}{4} \cdot \frac{28}{27}$, thus being lucky again because he obtained an epimoric ratio, $\frac{36}{35}$. Didymus, in turn, choose the standard $\frac{6}{5}$ for the semiditone of his chromatic, and the ratio $\frac{16}{15}$, the same imploied in his diatonic, for the interval of semitone. Then he divided $\frac{4}{3}$ by $\frac{6}{5} \cdot \frac{16}{15}$ and he was as lucky as Archytas, because what he obtained was once more an epimoric ratio, $\frac{25}{24}$.

So, it might seem that luck had a very determinant role in finding epimoric attunements. But we want to show that it was not luck but favourable probability. As an effect of (4.7), which determines a high density of epimoric decompositions among the epimoric ratios involving small numbers, we obtain that 'often' the product of two of such epimoric ratios is epimoric, and therefore 'often' the division of an epimoric ratio by a product of two of such epimoric ratios is epimoric itself. In fact:

The probability of obtaining an epimoric attunement by proceeding randomly is more than $\frac{6}{10}$.

Clearly, the above proposition depends on the meaning of the expression 'randomly'. As we shall see, by this we mean that the only considerations that are made in trying an element simply regard the lower and the upper bounds among which this element is taken. Under these assumptions the probability will $\frac{26}{43}$. As in Table 6, we consider the factors of a decomposition in decreasing order, and then we order the set of the decompositions lexicographically. In other words, by an attunement we mean a triple (x, y, z) of epimoric fractions such that $x \cdot y \cdot z = \frac{4}{3}$ and $x \geq y \geq z$, and we say that (x, y, z) precedes (x', y', z') if either $x > x'$, or $x = x'$ and $y > y'$, or $x = x', y = y'$ and $z > z'$. We consider in the same way the decompositions (x, y) of an epimoric fraction $\frac{n+1}{n}$ in two epimoric elements. Now, consider the element x (i.e. the biggest) of a decomposition (x, y, z) of $\frac{4}{3}$. Clearly, since $y \cdot z > 1$ and $x \cdot y \cdot z = \frac{4}{3}$, we have that $x < \frac{4}{3}$, and thus, since it must be epimoric, we obtain $x \leq \frac{5}{4}$. Moreover, the last (still under the lexicographic order) decomposition of $\frac{4}{3}$ is $(\frac{10}{9}, \frac{11}{10}, \frac{12}{11})$, and in general $(\frac{3n+1}{3n}, \frac{3n+2}{3n+1}, \frac{3n+3}{3n+2})$ is the last decomposition of $\frac{n+1}{n}$ in three factors (the Greeks knew this kind of decompositions well, and the fact that they are the last ones is rather intuitive). Thus the bounds that we naturally obtain for the first factor x of a decomposition are $\frac{5}{4} \geq x \geq \frac{10}{9}$, i.e. we have 6 possible first factors. Now, from $x \cdot y \cdot z = \frac{4}{3}$ it follows $y \cdot z = \frac{4}{3}/x$, and (Property 4.7 in action!) we have that in 5 among this 6 possible cases $\frac{4}{3}/x$ is epimoric (the only exception is constituted by $x = \frac{9}{8}$, because $\frac{4}{3}/\frac{9}{8} = \frac{32}{27}$ which is not epimoric). So for $x \neq \frac{9}{8}$ we can repeat the above considerations in choosing, for a given x, a possible second factor y which follows x. Consider for instance $x = \frac{5}{4}$. In this case $y \cdot z = \frac{4}{3}/\frac{5}{4} = \frac{16}{17}$. A second factor y must therefore satisfy these conditions: $y \leq \frac{5}{4}$ (by the general condition $x \geq y$), $y \leq \frac{17}{16}$ (because $y \cdot z = \frac{16}{15}$ and $z > 1$) and this condition annuls the previous one, and $y \geq \frac{31}{30}$ (because $(\frac{31}{30}, \frac{32}{31})$ is the last decomposition of $\frac{16}{15}$, and, in general, $(\frac{2n+1}{2n}, \frac{2n+2}{2n+1})$ is the last decomposition of $\frac{n+1}{n}$ in two factors). Now among the 15 possible epimoric ratios y such that $\frac{17}{16} \geq y \geq \frac{31}{30}$, 10 are favourable, in the sense that the rational number $z = \frac{4}{3}/(\frac{5}{4} \cdot y)$ is still epimoric (again an effect of Property 4.7). For $x = \frac{6}{5}$ we obtain, by the same considerations, that $\frac{11}{10} \geq y \geq \frac{19}{18}$, and 6 among these 9 cases are favourable. For $x = \frac{7}{6}$ the possible cases are 7, and 4 of them are favourable. For $x = \frac{8}{7}$ the possible cases are 6 and 4 of them are favourable. For $x = \frac{10}{9}$ there is only 1 possible case that is $y = \frac{11}{10}$ (in fact y must satisfies these conditions: $y \leq x = 109$, $y < \frac{4}{3}/109 = \frac{6}{5}$, i.e. $y \leq \frac{7}{6}$, and this condition is annulled by the previous one, and $y \geq \frac{11}{10}$, because, $(\frac{11}{10}, \frac{12}{11})$ is the last decomposition of $\frac{6}{5}$) and it is favourable. Up to now we have that among the 38 possible cases, 25 have been favourable. Finally, we consider the case $x = \frac{9}{8}$. For the upper bound of y we have, as in the previous case, the conditions $y \leq x = \frac{9}{8}$ and $y < \frac{4}{3}/\frac{9}{8} = \frac{32}{27}$, and this second condition is annulled by the first. But for the lower bound we cannot refer to the usual last decomposition in two epimoric ratios of $\frac{32}{27}$, because it is not epimoric. But, considering that $y \geq z$ and $y \cdot z = \frac{32}{27}$, we obtain $y^2 \geq \frac{32}{27}$, and the least epimoric ratio which satisfies this condition is $\frac{13}{12}$. So the range for y we obtain from this elementary considerations is $\frac{9}{8} \geq y \geq \frac{13}{12}$, i.e. 5 cases among which only one is favourable. Therefore, altogether we have 43 possible cases and 26 favourable which, assuming that they are equiprobable, give a probability of $\frac{26}{43}$, bigger than $\frac{6}{10}$

We think that the role of this high probability was not limited to the attunements that cannot be obtained either by Method A or by Ptolemy's Method, i.e. Architas' diatonic, Architas' enharmonic and Didymus' chromatic. In fact, we can note this peculiar situation. As observed, the other two Dydimus' attunements can be obtained by Method A. But the way might have been as follows. Also for his diatonic he proceeded by attempts: at first he chose, together with $\frac{9}{8}$, the ratio $\frac{10}{9}$ which is the closest to it, and then, dividing $\frac{4}{3}$ by $\frac{9}{8} \cdot \frac{10}{9}$, he obtained an epimoric ratio, $\frac{16}{15}$. At this point, from one hand he tried his luck again taking $\frac{16}{15}$ together with $\frac{6}{5}$ for his chromatic; and, as we have seen, the result was positive. On the other hand, by observing the characteristics of the identity $\frac{9}{8} \cdot \frac{10}{9} = \frac{5}{4}$, he discovered Method A by which he obtained his enharmonic. The same did not happen to Archytas for his third attunement. The ratio $\frac{28}{27}$ appears in both his epimoric attunements, the diatonic and the enharmonic. If he had employed $\frac{28}{27}$ with the ratio $\frac{6}{5}$ as semiditone also for his chromatic, he would have been lucky too, because $\frac{4}{3}/(\frac{6}{5} \cdot \frac{28}{27}) = \frac{15}{14}$. Nevertheless this way was not followed by him (and in fact Architas' chromatis is $(\frac{32}{27}, \frac{243}{224}, \frac{28}{27})$, and it is not an epimoric attunement), but by Ptolemy (his chromatic is in fact $(\frac{6}{5}, \frac{15}{14}, \frac{28}{27})$). And Ptolemy, observing the characteristics of the identity $\frac{28}{27} \cdot \frac{15}{14} = 109$ (which has the numerator of one factor double than the denominator of the other, and therefore can be transformed in $\frac{28}{27} \cdot \frac{29}{28} \cdot \frac{30}{29}$ by the well known identity 5.1) formulated the method by which he obtained his other attunements. Various elements seem to support this hypothesis, first of all the description given by Ptolemy of his tonic diatonic, and the fact that this attunement, which is the only one among his attunements which cannot be obtained by his method, coincides with Archytas' diatonic. And if this reconstruction is correct, then Property 4.7 would have had a still more important role in determining epimoric attunements.

Department of Mathematics
University of Siena

NOTES

1 E.T. Bell, *The magic of numbers*, 1946.

2 We can observe that the custom of counting both extremes of an interval dates back at least to the Greeks. In such a way the sum of an n-interval with an m-interval gives an $(n + m - 1)$-interval.

3 A *ditone*, i.e. a double tone, corresponds to a major third in our terminology.

4 A *semiditone*, i.e. a tone and a half, corresponds to a minor third.

5 The scales presented by the Greek theorists were descending; therefore the order of the intervals in the original definitions was inverted with respect to that we have given. For instance, the division of the diatonic tetrachord was *semitone* + *tone* + *tone*. But, since this fact is not relevant for our discussion we have preferred to translate the original definitions in ascending order. So the diatonic genus, expanded to the octave, produces our diatonic scale.

6 The first witnessing of this attunement is in a fragment of Philolaus; it was also employed in Plato's *Timaeus* and in Euclid's *Sectio Canonis*. Ptolemy called it *ditonic diatonic* and attributed it to Eratosthenes. In almost all the Medieval treatises it is the only attunement for which the instructions to divide the monochord are given.

7 See [12] , Book 1, end of Ch.5 (in [2] at p.286). Above in Ptolemy's book, melodic intervals are defined in a wider sense, as intervals 'acceptable to the hearing'. In this sense, every interval between two notes of the scale, even if non-consecutive, would be melodic. But it is not in this wider sense that the term has been used in saying that the melodic intervals must be epimoric, because no scale could satisfy this property.

8 See [6], p.403.

9 See [8], p.396. Fermat proposed also the decomposition $\frac{6}{5} \cdot \frac{19}{18} \cdot \frac{20}{19}$. But this decomposition, absent in the comment of Bryenne, was present in Ptolemy.

10 Beside our [3], the only reference to this problem that we have found is that of Tannery [13] who, nevertheless, asserts that the decompositions are 24.

11 The *difference* between $\frac{5}{4}$, which represents the major third in the natural scale, and the ditone of $\left(\frac{9}{8}\right)^2$ is $\frac{81}{64} / \frac{80}{64} = \frac{81}{80}$, and was called *lemma* by the Greeks. This value corresponds also to the difference between $\frac{6}{5}$, that is the minor third in the natural scale, and the pythagorean semiditone of $\frac{9}{8} \cdot \frac{256}{243}$. In the literature this value is often quoted as *Didymus' Lemma*.

12 Looking at Table 6, a question arises: why Ptolemy did not chosen, for his enharmonic, the triple (4,25,39), i.e. $(\frac{5}{4}, \frac{26}{25}, \frac{40}{39})$, or, for his tense chromatic, the triple 6-14-15? This choise in fact would have been closer to the definitions of the genera. Note 34 contains an answer to this question.

13 In this definition q is not considered a divisor of itself. In fact $\frac{2}{1}$ was considered a multiple and not an epimoric ratio.

14 [12], Book 1, Ch.5 (in [2] at p. 285).

15 See [7] Propositions 10–11, [12] Book 1, Ch.5, [11] Book 1, Ch.5, and [5] Book 2, Chs. 22–25. In the *Sectio Canonis* the justification of the Principle of Consonance is a little more elaborated, even if substantially equivalent, to that presented by Ptolemy, and is as follows: "Among notes we also recognize some as concordant, others as discordant, the concordant making a single blend out of the two, while the discordant do not. In view of this, it is to be expected that the concordant notes,

since they make a single blend of sound out of the two, are among those numbers which are spoken of under a single name in relation to one another, being either multiple or epimoric."

16 See [2] p.17.

17 In fact, beside the well known $\frac{2}{1}(=\frac{4}{2})$, $\frac{3}{2}$ and $\frac{4}{3}$, we have $\frac{3}{1}$ (octave + fifth) and $\frac{4}{1}$ (double octave).

18 Probably, the structure of the Euclidean proof may be more appreciated is it is expressed in form of an algebraic exercise, without any interpretation of the terms. In such a way one can better realize what this proof represents, that is the research of the solutions of a set of conditions. *Exercise.* Let M be the set of natural numbers bigger than 1 and $E = \{\frac{n+1}{n} : n > 1\}$ (consider M and E as subset of the set Q of the rational numbers). Let μ be a one-to-one function from Q into Q such that (i) $D(\mu) \cap E = \emptyset$, where $D(\mu)$ is the domain of μ, and (ii) $x \in M$ iff $\mu(x) \in M$. Let a, b, c, d, e, f be real numbers. Then the following conditions have one and only one solution:

1. $a, b, c, f \in M \cup E$,
2. $d, e \notin M \cup E$,
3. $d, e, f \in D(\mu)$ and $\mu(d) = a$, $\mu(e) = b$, $\mu(f) = c$,
4. $c = a \cdot b$ and $a < b$

Proof. Since $f \in M \cup E$ and $f \in D(\mu)$ we obtain, by (i), that $f \in M$ and by (ii) that $c \in M$. On the other hand, from $d, e \notin M$ we obtain, by (3) and (ii), that $a, b \notin M$. So, by (1), $a, b \in E$. But the only triple a, b, c which satisfies (4) and such that $a, b \in E$ and $c \in M$ is $(2, \frac{3}{2}, \frac{4}{3})$; in fact $2 = \frac{3}{2} \cdot \frac{4}{3}$, and $\frac{3}{2}$ and $\frac{4}{3}$ are the two biggest elements of E while 2 is the smallest element of M. So $a = \frac{3}{2}$, $b = \frac{4}{3}$, $c = 2$, $d = \mu^{-1}(a)$, $e = \mu^{-1}(b)$, $f = \mu^{-1}(c)$.

Clearly, by interpreting the terms of the exercise to obtain the proof of the *Sectio Canonis*, we have that M and E stand for the multiple and epimoric ratios, a, b, c, d, e, f for the ratios corresponding respectively to the fourth, the fifth, the octave, the double fourth, the double fifth and the double octave, and μ is the function whose domain is the set of rational numbers $\frac{m}{n}$ such that $m > n$ and $G(m, n)$ is a natural number, and such that $\mu(x) = \frac{G(m,n)}{n}$ (observe that if $\frac{m}{n} = \frac{m'}{n'}$ then $G(m, n)$ is a natural number iff $G(m', n')$ is a natural number and $G(m, n) = \frac{G(m',n')}{n'}$; so μ is well defined). In the exercise we have limited ourselves in giving the properties of μ that are necessary for the proof instead of defining μ explicitly. The converse of the Principle of Consonance is expressed in particular by condition (2). If we erase this condition the system of the remaining conditions admits infinitely many solutions, even if we define the function μ explicitly, as before. In fact, for any pair $a, b \in M$, the six-tuple $(a, b, a \cdot b, \mu^{-1}(a), \mu^{-1}(b), \mu^{-1}(a \cdot b))$ satisfies this reduced set of conditions. So, the converse of the Axiom of consonance is necessary to obtain the uniqueness, and therefore to prove the statement of the *Sectio Canonis*.

19 See [12] Book 1, Ch.6, and [11] Book 1, Ch.6.

20 See [12] Book 1, Ch.5.

21 Cf. [7], Introduction.

22 Cf, [11]. In [2] at p.42.

23 We found it in [7] Proposition 3, in [11] Book 1, Ch. 5, and also in [5] Book 3, Ch.11.

24 Cf. T. Heath, *A History of Greek Mathematics*, Oxford, 1960, I, p.216.

25 The already quoted Plato's *Republic* 531 a-c refers to this fact. We can also observe that, since $\sqrt{\frac{9}{8}} = \frac{3}{2\sqrt{2}}$, the problem of the semitone coincides with that of $\sqrt{2}$.

26 We give an informal proof. Let $x = \frac{n+1}{n}$. Then $f_A(\frac{n+1}{n}) = \frac{2n+1}{2n}$ and $f_H(\frac{n+1}{n}) = \frac{2n+2}{2n+1}$. So, if we consider only the denominators, by f_A we pass from n to $2n$ and by f_H from n to $2n + 1$. Therefore if we start from 1 (which is the denominator of our generator $\frac{2}{1}$) we obtain a binary tree which contains all the natural numbers (one can show by induction that at level s of the tree we have all the numbers n such that $2^s \leq n < 2^{s+1}$).

27 Clearly, when one says that the semitones of our tempered scale are equal between them, the meaning of the term 'equal' is not additive but multiplicative, and the measures $\overline{OA}, \overline{OA}_1, \overline{OA}_2, \ldots; \overline{OB}$ form a geometric progression, like the keys of a guitar. Coming back to the even diatonic, if we extend this attunement from the tetrachord to the octave in the standard way, we have a peculiar alternance of different progressions. This fact appears clearly if we consider the minimal natural numbers that produces all the ratios of the even diatonic octave. This numbers, given by Ptolemy himself, are 18,20,22,24,27,30,33,36. The numbers 18,24,27,36 (that are the

triples of the numbers 6,8,9,12, so popular among Greek and Medieval theorists) express the two disjoint tetrachords forming the octave (in fact $\frac{24}{18} = \frac{4}{3}, \frac{27}{24} = \frac{9}{8}$ and $\frac{36}{27} = \frac{4}{3}$), and thus 18,24 and 27,36 are two geometrical progression of order $\frac{4}{3}$. But the division of these tetrachords in three equal parts produces two different arithmetical progressions, one of order 2 (18,20,22,24) and the other of order 3 (27,30,33,36). Incidentally, since $\frac{9}{8}$ is the epimoric ratio which 'precedes' $\frac{10}{9}, \frac{11}{10}$ and $\frac{12}{11}$, in this second progression also number 24 can be embodied. Thus the intere sequence reduces to the conjunct union of two different arithmetical progressions.

28 The greek term is *eteromekes*, and the latin term is *parte altera longior*.

29 This result is a consequence of Proposition VIII.2 of the *Elements*, which asserts that, given the ratio $\frac{m}{n}$, the smallest s natural numbers $a_1, ..., a_s$ such that, for $1 \leq i \leq s$, $\frac{a_{i+1}}{a_i} = \frac{m}{n}$ are n^{s-1}, $n^{s-2}m$, $n^{s-3}m^2, ..., nm^{s-2}, m^{s-1}$.

30 Let us denote by E^n the set of the rational number which are product of n epimoric ratios. Then, for each n, $E^n \subset E^{n+1}$ and $E^n \cup \{0\}$ is the set of the limit points of E^{n+1}(cf. [3]).

31 *Case* $\frac{n+2}{n}$: If n is even then $\frac{n+2}{n} = \frac{m+1}{m}$ for $m = \frac{n}{2}$, and so we are in the preceding case. If n is odd, we consider the identity $\frac{n+2}{n} = \frac{n+3}{n+1} \cdot \frac{n^2+3n+2}{n^2+3n}$. Since in this case $n+1$ and n^2+3n are even, we have $\frac{n+2}{n} = \frac{m+1}{m} \cdot \frac{s+1}{s}$ for $m = \frac{n+1}{2}$ and $s = \frac{n^2+3n}{2}$.
Case $\frac{n+3}{n}$. If n is a multiple of 3, then $\frac{n+3}{n} = \frac{m+1}{m}$ for $m = \frac{n}{3}$, and so it is an epimoric fraction. Otherwise one among $n+1$ and $n+2$ is a multiple of 3. In the first case we consider the identity $\frac{n+3}{n} = \frac{n+4}{n+1} \cdot \frac{n^2+4n+3}{n^2+4n}$. Since $n+1 = m \cdot 3$, we obtain that $n+4 = (m+1) \cdot 3$, $n^2+4n = n(m+1) \cdot 3$, and $n^2+4n+3 = (n(m+1)+1) \cdot 3$. Therefore $\frac{n+3}{n} = \frac{m+1}{m} \cdot \frac{n(m+1)+1}{n(m+1)}$. In the second case we consider the identity $\frac{n+3}{n} = \frac{n+5}{n+2} \cdot \frac{n^2+5n+6}{n^2+5n}$. Since $n+2 = m \cdot 3$, we obtain that $n+5 = (m+1) \cdot 3$. Moreover, since $n^2+5n+6 = (n+2)(n+3)$, we obtain that $n^2+5n+6 = s \cdot 6$, because $n+2$ is a multiple of 3 and one among $(n+2)$ and $(n+3)$ is a multiple of 2. Therefore $\frac{n+3}{n} = \frac{m+1}{m} \cdot \frac{s}{s-1}$.

32 We show in details the attunements that can be obtained by Method A. At first, consider $\frac{4}{3} = \frac{5}{4} \cdot \frac{16}{15}$. If we decompose $\frac{5}{4}$ by 5.1 we obtain $\frac{9}{8}$ and $\frac{10}{9}$ which, together with $\frac{16}{15}$ are the factors of Didymus' diatonic and Ptolemy's tense diatonic; if we decompose $\frac{16}{15}$ we obtain $\frac{31}{30}$ and $\frac{32}{31}$ which, together with $\frac{5}{4}$, are the factors of Didymus' enharmonic. Then, consider $\frac{4}{3} = \frac{6}{5} \cdot \frac{10}{9}$. If we decompose $\frac{6}{5}$ we obtain $\frac{11}{10}$ and $\frac{12}{11}$ which, together with $\frac{10}{9}$ are the factors of Ptolemy's even diatonic; if we decompose $\frac{10}{9}$ we obtain $\frac{19}{18}$ and $\frac{20}{19}$ which, together with $\frac{6}{5}$ are the factors of Eratosthenes' chromatic. Finally, the two epimoric attunements that could be obtained by $\frac{4}{3} = \frac{7}{6} \cdot \frac{8}{7}$, that is, $(\frac{8}{7}, \frac{13}{12}, \frac{14}{13})$ and $(\frac{7}{6}, \frac{15}{14}, \frac{16}{15})$, were not employed by the Greeks. But it is interesting to observe that they both were employed by Fermat to falsify the hypothesis of Bryenne (cf. Notes 6 and 7, and also [3]).

33 In fact, if n is even then $\frac{3n+1}{3n} \cdot \frac{3n+2}{3n+1} = \frac{3n+2}{3n} = \frac{3n21}{3n2}$, while if n is odd then $\frac{3n+2}{3n+1} \cdot \frac{3n+3}{3n+2} = \frac{3n+3}{3n+1} = \frac{3n+12+1}{3n+12}$.

34 We analyze in details the attunements that can be obtained by Ptolemy's method. First consider $\frac{4}{3} = \frac{5}{4} \cdot \frac{16}{15}$. From $\frac{5}{4} = \frac{13}{12} \cdot \frac{14}{13} \cdot \frac{15}{14}$, since $\frac{13}{12} \cdot \frac{14}{13} = \frac{7}{6}$ we obtain the two epimorics $\frac{7}{6}$ and $\frac{15}{14}$ and thus the decomposition of $\frac{4}{3}$ in $(\frac{7}{6}, \frac{15}{14}, \frac{16}{15})$. We shall see at the end of this note why this attunement was not employed by Ptolemy. If we consider $\frac{16}{15} = \frac{46}{45} \cdot \frac{47}{46} \cdot \frac{48}{47}$, since $\frac{47}{46} \cdot \frac{48}{47} = \frac{24}{23}$ we obtain the decomposition of $\frac{4}{3}$ in $(\frac{5}{4}, \frac{24}{23}, \frac{46}{45})$, i.e. Ptolemy's enharmonic. Now consider $\frac{4}{3} = \frac{6}{5} \cdot \frac{10}{9}$. From $\frac{6}{5} = \frac{16}{15} \cdot \frac{17}{16} \cdot \frac{18}{17}$, since $\frac{17}{16} \cdot \frac{18}{17} = \frac{9}{8}$, we obtain the decomposition $(\frac{9}{8}, \frac{10}{9}, \frac{16}{15})$, which gives the factors of Ptolemy's tense diatonic. From $\frac{10}{9} = \frac{28}{27} \cdot \frac{29}{28} \cdot \frac{30}{29}$, since $\frac{29}{28} \cdot \frac{30}{29} = \frac{15}{14}$ we obtain $(\frac{6}{5}, \frac{15}{14}, \frac{28}{27})$, i.e. Ptolemy's soft chromatic. Finally consider $\frac{4}{3} = \frac{7}{6} \cdot \frac{8}{7}$. From $\frac{7}{6} = \frac{19}{18} \cdot \frac{20}{19} \cdot \frac{21}{20}$, since $\frac{19}{18} \cdot \frac{20}{19} = \frac{10}{9}$ we obtain $(\frac{8}{7}, \frac{10}{9}, \frac{21}{20})$, i.e.Ptolemy's soft diatonic. From $\frac{8}{7} = \frac{22}{21} \cdot \frac{23}{22} \cdot \frac{24}{23}$, since $\frac{23}{22} \cdot \frac{24}{23} = \frac{12}{11}$ we obtain $(\frac{7}{6}, \frac{12}{11}, \frac{22}{21})$, which is the Ptolemy's tense chromatic.

Now we can answer the question, formulated in Note 12, on the reasons for which Ptolemy, after exhausting all the possible orthodox cases (i.e. in accord to our 2.1), choose $(\frac{5}{4}, \frac{24}{23}, \frac{46}{45})$ instead of the less heterodox $(\frac{5}{4}, \frac{26}{25}, \frac{40}{39})$ for his enharmonic, and $(\frac{7}{6}, \frac{12}{11}, \frac{22}{21})$ instead of $(\frac{7}{6}, \frac{15}{14}, \frac{16}{15})$ for his tense chromatic (see Table 6). As regards this last case, Ptolemy helps us in finding the answer. As he explains (cf.[12], I.15), he always considered, for applying his method, the undecomposed fraction as the first interval of the tetrachord. Therefore, after having obtained the triple $(\frac{16}{15}, \frac{7}{6}, \frac{15}{14})$, in this order, he refuted it instead of commuting the intervals. Why he limited the possibilities of his method in such a way, it is difficult to say. On the contrary, as regards the first case, we think that the answer is simply that he did not found the decomposition $(\frac{5}{4}, \frac{26}{25}, \frac{40}{39})$, because it was not obtainable by his method, and he contented himself with the triple $(\frac{5}{4}, \frac{24}{23}, \frac{46}{45})$. But the fraction $\frac{46}{45}$, which is between a fifth and a sixth of tone, is really too small to represent a musical interval, even if in an enharmonic attunement.

REFERENCES

[1] Barbera André: 1991, *The Euclidean Division of the Canon.* University of Nebraska Press.

[2] Barker Andrew: 1989, *Greek Musical Writings. II.* Cambridge University Press.

[3] Bellissima Fabio: 1995, *A Problem of Fermat Relative to the Decomposition of Epimoric Fractions.* Rapporto Matematico dell'Universitá di Siena.

[4] Boethius: 1867, *De Istitutione Arithmetica.* Teubner.

[5] Boethius: 1867, *De Institutione Musica.* Teubner.

[6] Bryennius Manuel: *Harmonics*, in [14].

[7] Euclid, *Sectio Canonis*: text [9], text and and Latin translation in Heiberg-Menge, *Euclidis opera omnia*VIII, Lipsiae 1916; English translation in [1] and [2], Italian translation [15].

[8] Fermat Pierre: Notes Critiques sur les Harmoniques de Manuel Bryenne, in *Ouvres de Fermat*, Appendice.

[9] Jan Carl von; 1895, *Musici scriptores graeci.* Leipzig.

[10] Nicomachus, *Enchiridion*, in [9], English translation in [2], Italian translation in [15].

[11] Porphyry, *Commentary on Ptolemy's Harmonics*, in [14].

[12] Ptolemy, *Harmonics*, in [14], English translation in [2].

[13] Tannery Paul: 'Du Rôle de la Musique Grecque dans le Développment de la Mathématique Pure', *Bibliotheca Mathematica*, 3. Folge, t. III, 1902, 161–175 (also in *Mémories Scientifiques de Paul Tannery*, III, 68–89).

[14] Wallis John: 1972, *Opera Mathematica, III.* Georg Olms Verlag.

[15] Zanoncelli Luisa: 1990, *La Manualistica Musicale Greca.* Guerini Studio.

GIULIANO TORALDO DI FRANCIA

MUSIC AND SCIENCE

1. A (TOO) GOOD QUESTION

What is music? That's a **good question**.

The usual – albeit trivial – way to tackle such problems is to have recourse to a good dictionary. Let us take, for example, the Oxford Dictionary. Under the entry 'music' one finds the definition: "*the art of arranging the sounds of voices or instruments or both in a pleasing sequence or combination*" Is this an **adequate** definition? One may doubt.

Notice, for instance, that the sounds of voices coming from the kitchen, plus the noises made by pots and pans being handled by the cook, may represent a *pleasing combination* to the hungry person; but one would hardly attribute the effect to musical reasons. Moreover, a number of noted musicologists and philosophers have insisted that, though music can cause *pleasure* (or *sorrow*), pleasure and sorrow are not the **essence** of music.

In a not much different way, one may note that, although a mathematician may derive a lot of aesthetic pleasure from his work, *aesthetics* is not the essence of mathematics.

A young physicist from Paris (E. Klein) has recently written: "*Une question n'est vraiment une bonne question que si elle est plus juste que toute réponse qu'on lui connaît*" (No question is really a good question, unless it is sounder than any answer we know to it). Very well put, indeed.

That tenet, so far as we can see, may pretty well apply to our inquiry, inasmuch as no one has ever been able to find a satisfactory definition, by actually *telling* us what is music. In this connection, an alarming suspicion might arise, too. A good question that does not admit of a satisfactory answer, may simply have **no answer** at all; in which case, it is a **bad question**!

On what grounds may we be led to surmise that our question about music **cannot** be answered? It should be sufficient to observe that any *human* answer is necessarily conceived and expressed in a *language,* i.e. in words. But, strictly speaking, music can neither be *defined* nor *explained* by means of mere **words**. In this connection, it would even be nonsense to criticise the authors of dictionaries, on account of their not being able to do what is impossible.

If you are to learn the meaning of the term 'music', no *verbal* explanation can really teach you. You have first to **listen** to music. Thus, the words of the would-be teacher must at least be accompanied by a sort of **ostension.**

M.L. Dalla Chiara et al. (eds.), Language, Quantum, Music, 327–338.

Try and teach what is music to a deaf-born person! You will never succeed, no matter whether the learner is familiar, say, with *sign language* or can fluently read *written language.* Of course, you will be likely to make the unlucky person understand that normal people have, besides *sight, smell, taste* and *touch,* a fifth sense, termed *hearing,* that gives them a special kind of additional **information** about their surroundings.

For instance, you may state that a sound, called 'barking', informs *normal* people of the presence of a dog. You will also state that hearing enables one to perceive the **words** uttered by other people, without any need to **read** them in the sign language, from lip movements, or in a printed book.

In this respect, the situation presents some analogy to the one obtaining when you want to tell a blind person that there is a sense, called *sight*, capable of informing us about the state of the outer world. That information can be checked to be correct in different ways, but mainly by *touching* the objects around us.

All this is easily understood, when one is dealing with a sense that supplies useful information about the outer world. However, we may be tempted to ask: **what kind of information is supplied by music?** And about what? Does it perhaps inform us that a stretched string is being made to vibrate or that a horn is blown or that the parchment of a kettledrum has been struck? That kind of *material* information is certainly supplied to the knowledgeable listener, but it is rather irrelevant to **understanding** music.

What is hardly believable, is that you may ever be able to explain to the deaf person how and why music, through the sense of hearing, can convey **feelings and emotions** of a kind **otherwise unknown** to the listener. Incidentally, we may recall that there are also quite a few people with *normal* hearing, who seem to be absolutely insensitive to music.

Be that as it may, we shall not attempt here to answer the original question: what is music? Ironically, we shall rather try and clarify **what music is not**, by drawing attention to a number of factors, which though inseparable from one another, are often treated in a disjointed fashion.

Music is the end product of at least four kinds of interwoven phenomena, namely: 1) **physical**, 2) **physiological**, 3) **neurological**, 4) **cortical**. Concentrating the attention on only one of these aspects has frequently led students to misunderstandings and wrong conclusions.

2. VIBRATIONS

In order to introduce the physics of vibrations, it will be expedient to start from **Hooke's law**, which governs the response of an elastic body to deformation, by stating that: *ut tensio, sic vis.* The deformation of the body is proportional to the force that causes it.

Thus, for example, if you want to stretch or compress a spring by an amount x with respect to its normal length at rest, you must apply to the end of the spring a force F given by:

$$F = kx$$

where k represents a constant of the spring.

As a consequence of this law being **linear**, if you attach a mass m to the end of the stretched spring and then release the spring, it starts oscillating with a **harmonic motion:**

$$x = a \sin 2\pi f t = a \sin 2\pi t/T$$

t represents time, a is the *amplitude* – an arbitrary constant depending on the initial conditions – f the *frequency* – being given by $f = (1/2\pi)\sqrt{(k/m)}$ – and $T = 1/f$ the *period.* If t and T are measured in *seconds*, f is measured in *hertz* (*Hz*).

A great deal of what we believe to know about the material world and its oscillations is based on the above simple equations. Of course, when using them, we rely on the daring assumption that the deformation of an elastic body can be described by a *single coordinate* x. But real bodies are three-dimensional, and strictly speaking we ought to take into account two additional coordinates y and z. It turns out, however, that in a number of cases of interest the assumption of a predominant parameter x represents *a good approximation*; accordingly we shall start by ignoring y and z.

Hooke's law is linear; but, ironically, **there are no true linear relations in nature!** In the case of elasticity, even setting apart *dissipation* and *hysteresis*, Hooke's law should be replaced by a more realistic series expansion:

$$F = k_1 x + k_2 x^2 + k_3 x^3 \ldots$$

Nevertheless, since we are living in a comparatively *cold* world, where the intermolecular cohesion forces are rather strong with respect to the forces we normally put in action, the first term of the expansion – i.e. the original Hooke law – often yields a very good approximation to the real phenomenon. As a consequence, harmonic vibration has taken a great deal of importance in human evolution, and our ear is particularly accustomed and *fitted* to it.

Moreover, there is a *mathematical* reason why harmonic motion should hold a privileged place in the physics of vibrations: according to the well-known *Fourier theorem*, any periodic motion results from the sum of a (theoretically infinite) number of harmonic motions, whose frequencies are all integral multiples of the *fundamental frequency* of the oscillation.

3. A SIMPLE SCHOOL-BOOK THEORY

A number of old-fashioned school-books introduce the science of musical acoustics in the following way.

A **sound** is produced by a **periodic** variation of the air pressure striking the ear-drum. It is fully qualified by three characteristic features, namely: 1) **pitch**, 2) **intensity**, and 3) **timbre**.

The pitch of the sound is determined by the frequency of the periodic motion (i.e. by its *fundamental* frequency). The greater the frequency, the *higher* (or more *acute*) is the the sound (or the *note*) perceived by the listener.

The intensity (or *loudness,* or *volume*) of the sound is determined by the mechanical *power* transported by the air wave across a unit area perpendicular to its motion. The power, in turn, is proportional to the *square* of the amplitude of the oscillation.

The timbre (or *quality)* of the sound results from its *harmonic structure,* i.e. from the intensity of the different higher harmonics as compared with the intensity of the fundamental oscillation. A pure sound is a simple harmonic motion – or sine wave – not accompanied by any higher harmonic. Allegedly, it corresponds to the **poorest** timbre the ear can perceive. A (nearly) perfect pure sound is produced by a tuning fork. The listener should recognise the timbre of a periodic motion, by (subconsciously) making the *harmonic analysis* of the sound.

So much for periodic vibrations. If the variation of the air pressure is **not periodic**, the result is generally called **noise**, instead of **sound**. It does not have a well-defined pitch. Percussion of a material body generally produces a noise.

All these school-book definitions may well have some *didactical* value as a starting point; provided they are immediately followed by a lot of criticism. Reality is different, and (luckily) not so simple. In order to arouse the first doubts in a pupil, it might be sufficient to note that even the *timbre of an instrument* is to a large extent a sort of fiction or convention. Strike, on a piano, first a bass note, then a high-pitch note: do they have any likeness? Not at all; only by habit do we recognise the same instrument.

4. THE IMPORTANCE OF BEING NON-LINEAR

As was already stated, nothing in nature is really linear. In a number of applications of science, this is reckoned as a nuisance we would gladly dispense with. Not so in music! It is indisputable that much of the richness and fascination of music derives exactly from a number of essential non-linearities, **both of the instruments and of the ear**.

To start with, let us emphasise that the organ of hearing does not transmit linearly the form of the air vibration to the brain of the listener. Some **distortion** already occurs at the eardrum, at the little bones of the middle ear, and at the oval window that passes the stimulation to the inner ear. But non-linear effects seem to arise mainly in the *cochlea* – i.e. in the spiral cavity, along which the sensitive *basilar membrane* runs – and beyond it.

By monitoring the vibration with a microelectrode implanted in the acoustic nerve, it has been established that even an allegedly *pure sound* – especially if very loud – is not transmitted as a simple sine wave. In a sense, **pure sounds do**

not exist, and nobody knows what they would really sound like!

Moreover, it has long been known that when two frequencies f_1 and f_2 are sounded at the same time, the listener perceives also the **combination** frequencies:

$$f_1 - f_2,\ 2f_2 - f_1\ ,3f_2 - 2f_1,\ \text{and so on.}$$

There is also something else of great interest to be added to the simple school-book theory of **harmonic analysis**. Fourier's theorem holds only for a **true periodic motion**. In order for the vibration to be such, it should go on and on unaltered from $t = -\infty$ to $t = +\infty$, a condition absolutely impossible for a real instrument. Any note must have a beginning and an end. In such situation, a generalisation of the simple Fourier theorem holds, stating that the vibration can still result from the sum of harmonic motions, provided their frequencies are **continuously** distributed from zero to infinity. In mathematical terms, this means that instead of a Fourier **series**, we must apply a Fourier **integral**.

What is of paramount importance for the determination of timbre is indeed **the way the note starts and stops**, thereby exhibiting a continuous distribution of frequencies. These **transient** frequencies are generally very rapidly attenuated, and die out, leaving in between what is a good approximation to a true periodic motion, namely to a musical **note**. Nevertheless, the transient states of the vibration are characteristic of each instrument and of the way it is played. They are indispensable to the recognition of timbre!

Let us take an example. If you don't hear the attack or the ending of a sustained note, which, as far as you know, is played either on a cello or on a bassoon, you may have some difficulty telling which instrument is playing. Mind you, the attack and the ending of the note **are not interchangeable**. If you record a *presto* piece of music on a tape, and then run the tape backwards, you are likely to hear a medley of sounds, where no clear timbre can be recognised.

The most striking example is given by the piano. The piano is both a **percussion** and an **harmonic** instrument! In fact, when the hammer strikes the strings, *all frequencies* – in theory from zero to infinity – are aroused, and very soon die out. There remain the fundamental frequency and the higher harmonics, which in turn die out, starting from the higher ones (for the sake of simplicity, let us ignore the pedal). This structure of the piano sound – elaborated and refined over a couple of centuries – has resulted in perhaps the most brilliant and fascinating timbre of western music. Playing a piano note *in reverse* would make no sense.

But, apart from the initial and final transients of a note, should a musical instrument be as *harmonic* as possible? Not at all. Indeed, small departures from that condition are not so much inevitable, as even **desirable**, in order to achieve a not too *poor* or *dull* timbre. For instance, the three strings of a piano note should not be exactly tuned at unison, as every good piano-tuner knows. A violin string has a thickness and some rigidity, which cause the higher frequencies not to be exact multiples of the fundamental one; in addition to the **transversal** modes of vibration, there are the **longitudinal** and the **torsional** modes. Any resonant cavity – from the body of a piano to that of a violin – is three-dimensional; consequently its frequencies do not form an arithmetic series. We could go on with a detailed and lengthy discussion of all musical instruments; but the above hints should be sufficient to get the gist of the argument.

5. THE INTENSITY OF SOUND

The *physical* definition of the intensity of sound – as the mechanical power conveyed by the air wave across a unit surface perpendicular to its direction – is all right for many applications. But it is utterly inadequate to describe the intensity of the *sensation* perceived by the listener. And anyhow, is it possible to **measure a sensation?**

There is no denying that the latter endeavour may appear devoid of any sense. In spite of this skepticism, some moderately good results have been obtained by relying on the Weber-Fechner psychophysical law, which states that in order for two barely different stimuli p and $p + \Delta p$ to give rise to two recognisably different sensations s and $s + \Delta s$, their difference Δp should be a quantity proportional both to p and to Δs. In mathematical terms, we can write:

$$\Delta s = k\Delta p/p.$$

By making the daring assumption that both Δs and Δp can be treated as *differentials* (ds and dp), and making an integration, we arrive at the well-known equation:

$$s = K \log p$$

K is a new constant, depending on the units that are used.

Now, it is customary to adopt as unit for p the **threshold** power p_0, under which no sound is perceived. Moreover the logarithm is taken to the base 10, and $K = 10$. It is said then that the intensity of the sound is measured in dB (decibels).

Please, remember that **convention** represents only the best we can do, in order to move around in a quantitative fashion in the realm of acoustics; but do not yield to the temptation of thinking you have found the *proper and exact* way to measure a sensation!

The first snag is encountered when we realise that the threshold power p_0 does not have the same value for all individuals and all ages. We must adopt for it an *average value* determined for *young* listeners. Moreover, even for one and the same individual, **the threshold varies as a function of frequency**. It is customary to choose for p_0 the average value of the threshold at the frequency of 1000Hz. This is close to the region of maximum sensitivity of the human ear, which is about 3000Hz. Sensitivity starts at a few tens of Hz reaches its maximum at $3000Hz$, then declines and ceases altogether at about $10,000Hz$ (but children can go much beyond that limit).

The different sensitivity of the ear to different frequencies may turn out to be of great importance in music, when it is desired that a particular timbre should not be **masked** by the sound of other instruments. For instance, remember the G_6 note ($3,136Hz$), played by the piccolo and perfectly audible during the *storm* episode in Beethoven's sixth symphony. Again, a trained singer is accustomed to give his or her voice a *singing formant* (or *ring*) in the range of $2000 \leftrightarrow 3000Hz$, which stands out beautifully even against a full orchestra playing more or less in the vicinity of the standard ($440Hz$)

Finally, it is interesting to note that the sensitive cells of the ear are subject to a rapid *saturation,* when exposed to a steady frequency. In this connection, a singer's *vibrato* – consisting in a pitch oscillation of $1/4$ to $3/4$ of a tone, about 5 to 7 times a second – may help give great prominence to the voice.

6. THE MYSTERY OF CONSONANCE

We have tried to *justify* the acoustical preference of humans for harmonic (or more generally periodic) oscillations of the air pressure by a philogenetic argument based on Hooke's law. But we do not claim for a moment that that *justification* should have the value of an *explanation.* The deep reason of our preference for well-defined frequencies lies undoubtedly in the structure of the human brain, whatever the philogenetic – or ontogenetic – derivation of that preference may be.

Even much harder is to find the reason why **intervals** (i.e. *ratios* of frequencies) are more important and naturally recognised than single **frequencies**. As a matter of fact, very few (and well trained) people can exhibit an *absolute ear* and tell right away, say, an *A* from an *A sharp*. Nevertheless, most people can easily perceive the passage from *A* to *A sharp*, when sounded in succession, and be trained to recognise that it is the same interval obtaining between *G* and *G sharp*.

One might easily be led to surmise that something similar happens with colors: we can indeed recognise two slightly different hues, when shown together, or in rapid succession. But no one can be trained to assess in a **quantitative** way whether or not the interval between two hues of blue is the same as that between two hues of red! Such an assessment does not seem to make any sense.

Well, the most striking fact in the field of acoustics is the existence of some **privileged** intervals that are cherished and immediately recognised as being **consonant** by virtually all people in the world. There are *at least* two such intervals, namely the **octave** (double frequency) and the **fifth** (frequency multiplied by $3/2$). There follow a succession of other fundamental intervals, that are gradually less and less universally recognised as consonant: namely the **fourth** (ratio $4/3$), the **major third** (ratio $5/4$), the **minor third** (ratio $6/5$), and so forth.

Ever since the time of Pythagoras it was realised that consonance has to do with (small) **natural numbers** and their ratios. However, a complete and convincing explanation of this finding seems still rather hard to formulate. In the remote past, people frequently had recourse to mystical speculations about the alleged **divine** nature of numbers (but remember that even in recent times – albeit in a quite different context – Kronecker asserted that natural numbers were created by God, whereas all other numbers were constructed by man).

A much more scientific explanation – substantially due to H. Helmholtz – relies on the phenomenon of **beats**. When two notes of different frequencies, say f and $f + \Delta f$, are superimposed, the **intensity** of the resulting sound oscillates with a frequency equal to Δf. That phenomenon gives rise to a *roughness* of the sound and seems rather *unpleasant* to the human ear, whenever the beats repeat a few times per second, namely when the difference Δf is comparatively small.

According to the formulation recently elaborated by E. Zwicker, R. Plomp and W. Levelt, there is a *critical band* of frequencies around any sound; roughness is perceived whenever two sounds are superimposed and a harmonic of the first falls within the critical band of a harmonic of the second.

When two frequencies are substantially different, there should be no problem in hearing them at the same time (apart from perceiving, as already noted, an additional frequency equal to their difference). But mind you! Some higher harmonics of the two notes might happen to have very close values, giving rise to the disturbing roughness. Now, consider any fundamental note N. There is no beat danger for the **octave**, due to the fact that the octave itself and all its higher harmonics are *exactly* higher harmonics of N and do not beat with them. Passing to the **fifth**, we recognise that all its *even* harmonics are also harmonics of N, while all the others are too remote to be disturbing. Similarly, in the case of the **fourth**, we find that within any *triple* of successive harmonics there is a harmonic of N. And so on.

Is this sufficient to *explain* why those notes are consonant? Perhaps so, but with a word of caution! As a matter of fact, consonant notes sound *pleasant* not only when superimposed, but also when played in succession in a melody. This can tentatively be accounted for by the **reverberation** of the room, or by a sort of *persistence* within the ear, or by something else. Whatever the explanation may be, the fact remains that a number of notes seem to us *to go well together*, and this is at the base of the musical scales adopted at different times in different places. Whether or not the beat theory can **explain** everything, is still under debate; as a consequence, we will take it only as a viable *working hypothesis*.

7. GOD IS SUBTLE, BUT NOT MALICIOUS

The main problem in establishing a scale is represented by the requirement of being able to **transpose** a melody, or in general any piece of music, from a given pitch-range to another. For instance, everybody knows that one and the same tune can be sung by a soprano, by a baritone, or by a bass. Can the same instrument, or orchestra accompany each one of them? Clearly, the problem consists in having at our disposal **exactly the same intervals** in the different pitch-ranges.

There would be no point to tackle here the ungainly task of presenting and analysing all the musical scales that have been conceived or adopted at different times in different places. We will limit ourselves to recalling a couple of particularly important cases.

The first case is that of the **natural scale** *of the physicists.* This is the scale generated by a (perfect) stretched string (a *monochord*), whose length is successively divided into a number of equal parts. If the original length vibrates at a fundamental frequency f_0, the string divided into n parts will vibrate at the frequency$f_n = nf_0$.

Let us call C_1 the fundamental note we start from. For $n = 2$, we obtain $f_2 = 2f_0$, namely the **octave** C_2, of C_1, while for $n = 3$, we pass to $f_3 = 3f_0 = (3/2)f_2$, or to the note G_2, which is the **fifth** of C_2. For $n = 4$, we find, of course, the **second octave** C_3 of C_1. For $n = 5$, we find $f_5 = (5/4)f_4$, namely the **major third** E_3 of C_3. For $n = 6$, we reach $f_6 = (3/2)f_4$, namely G_3 , or the fifth of C_3. So far, so good!

Later, however, we get into some trouble. For instance, by going on in the same way, we are forced to admit for the interval of what we call **one tone** two different values: namely 9/8 (**major second**) between C_4 and D_4, and 10/9 (**minor second**) between D_4 and E_4. This discrepancy leads to an interval $9/8 : 10/9 = 81/80$, which is termed **Zarlino's comma** (after the name of the 16th century scholar who proposed this diatonic scale).

Another scale of interest is **Pythagoras' scale**, which proceeds by *ascending fifths* (i.e. by multiplying at every step by 3/2), and *descending octaves* (i.e. by dividing by 2 whenever necessary, in order to always return to the original octave). It turns out that the first operation has to be repeated twelve times, and the second operation seven times. The trouble is that one does not return exactly to the starting point, because the two numbers, $(3/2)^{12} = 129.75$ and $2^7 = 128$ are different. Their ratio is termed **Pythagoras' comma.**

Thus, any effort to build a completely satisfactory scale seems to be desperate. Yet, as Einstein once put it: *God is subtle, but not malicious.* In fact, every time the solution of a scientific problem seems to be hopeless, we are likely, by stubbornly studying the situation, to end up finding that God has provided some way out.

In our case, the way out is represented by the *miraculous* number a $\alpha = \sqrt[12]{2}$, which in decimal representation turns out $\alpha = 1.0594\ldots$ This number, if taken to represent invariably a **semitone**, engenders the **equably tempered scale**, which is often said to have been introduced into western music in 1691 by the organist and composer A. Werckmeister (but see more details in A. Frova's book).

In this scale, twelve semitones give obviously $\alpha^{12} = 2$, or exactly an **octave**. Seven semitones give $\alpha^7 = 1.498$, or (almost exactly) a **fifth**. Five semitones yield $\alpha^5 = 1.335$, or (almost exactly) a **fourth**. Four semitones yield $\alpha^4 = 1.260$, or something very close to a **major third**. Three semitones result in $\alpha^3 = 1.188$, or something very close to a **minor third**. Two semitones give $\alpha^2 = 1.122$, or something quite acceptable for **one tone**. Eleven semitones yield for the **seventh** $\alpha^{11} = 1.875$, which, of course, needs an additional semitone to reach the octave.

All in all, the tempered scale allows one to pass from any bass to any acute note by always adding semitones of the same value. The very small discrepancies from an *ideal* scale do not appreciably disturb, and anyway are readily learned and accepted by a Western listener. With that scale it is an easy matter to transpose any piece of music, to modulate, to pass from one key to another within the same piece.

That scale, put in the hands of J. S. Bach, soon gave birth to the *Well tempered clavier*, one of the greatest monuments of human civilisation.

8. TOO MUCH MATHEMATICS?

Numbers, numbers, numbers: is that really the **essence** of music? Such a question prompts the present author to set forth a few *personal* considerations, without claiming any general agreement on them. Perhaps, they may easily be contradicted by knowledgeable musicologists; anyway, let them merely serve as a stimulus for further discussion.

A sort of *pythagoricist*, as well as *platonicist*, tradition of very long standing has tended to place music in the same realm as *hard sciences*. Remember that music was one of the four disciplines of the so-called **quadrivium**, next to arithmetic, geometry and astronomy, the *arts* of rational thought. It was taken for granted that they were all based upon numbers, even before Galileo asserted that the *book of nature* is written in mathematical language.

Kepler himself, who put the laws of the heliocentric system in mathematical form, still believed in the **harmony** of the celestial bodies, regulated by numbers. Even in recent times, quite a few physicists have insisted in trying to guess the profound significance of the *dimensionless constants*, i.e. of the **pure numbers** appearing in the theory. (Incidentally, such an enterprise seems not to make much sense to me; but that is immaterial).

We have already recognised the immense advantage that Western civilisation has derived from having introduced the *right* (or should we say *convenient* numbers into music. Bach's genius did a lot to persuade composers that music had to be *mathematised.* Following the same way, the *right* laws of harmony were firmly established. In a century or so the apex of what could be done in that direction had been reached. What else, what more sublime was possible?

Perhaps the spell began to break down with the advent of romanticism. Composers started to rebel. The (more or less mathematical) **rules** were challenged, and gradually ceased to dominate musical activity. Beethoven's fifth symphony marks a turning point. Following a long, difficult, but irreversible path, first *traditional* harmony dwindled and collapsed, then new rules (such as twelve-tone music) were introduced, and finally **all a priori rules** were discarded.

Meanwhile, new technical possibilities were provided by science. Electronic music, stochastic music, computer music, concrete music, and what not, were introduced. A new, immense horizon was disclosed. Perhaps, it is too early to assess and to synthesise the results of such a revolution, though a number of first-class achievements might already be mentioned. Posterity will judge; the present author does not feel himself in a position to formulate an honest verdict on such a crowded and diversified field.

Let us instead speculate a little on what had happened at the time when music had tended to be based almost entirely on mathematical rules. We know today that the two *hemispheres* of the human brain – though not at all isolated from one another – tend to some extent to perform two somewhat different functions. Putting it in a very simplistic way, one can say the following. The left side of the cortex is mainly responsible for rational thought, the use of language, calculus, and so on. The right side instead is mainly responsible for pattern recognition, artistic creation, music, and so on.

Ever since Pythagoras' time, scholars have tended – as already noted – to think of music as a kind of rational language. This *logocentrism* has necessarily relied much more on the left than on the right capabilities of the brain. As J. Derrida would put it, a language is a sort of **writing** (*écriture*) of our thought. There is always a danger to confuse the **significant** (i.e. the language) with the **signifed** (i.e. the thought).

Incidentally, it may be interesting to remark that something pretty similar to a gradual substitution of the significant for the signified also occurred in poetry. For a long time people made a lot of confusion between poetry and **rhymes**, or **verses**, or **metre**. Even today it is common to talk, say, of 'Petrarca's rhymes'. But rhymes are by no means the essence of poetry, as any contemporary poem can show.

Perhaps **musical thought** (whatever that term may mean) has been for a long time confused with its writing in notes; consequently the operation of the right cortex has been unduly dwarfed. As is well known, both sides of the brain are connected by the *corpus callosum*; this bridge has recently been found (by G. Schlaug et al.) to be considerably thicker and more developed in musicians than in non-musicians, as though the right side had been forced by training to depend on the operation of left side, perhaps much more substantially than was planned by mere nature.

Music should, at long last, be recognised to be simply what it is: it is not a language, it is not a combination of symbols or numbers, it is not geometry. It is just music.

Department of Physics
University of Florence

REFERENCES

Assayag, G., Cholleton, J.P.: 1995, 'Musique, Nombres et Ordinateurs', *La Recherche*, 804.

Bellissima, F.: 1995, *La Sezione del Monocordo di Euclide*. preprint.

Cremer, L.: 1984, *The Physics of the Violin*. MIT Press.

Derrida, J.: 1967, *De la Grammatologie*. Ed. de Minuit.

Fletcher, N.H., Rossing, T.D.: 1991, *The Physics of Musical Instruments*. Springer.

Frova, A.: 1997, *Musica e Fisica*. Zanichelli.

Habib, M., Robichon, F., De monet, J.: 1996, 'Le Singulier Cerveau des Dyslexyques', *La Recherche*, **289**.

Helmholtz, H.: 1954, *On the Sensation of Tone* . Dover.

Knobloch, E.: 1992, *Rapports Historiques entre Musique, Mathématique et Cosmologie*. IPCM.

Leipp, E.: 1989, *Acoustique et Musique*. Masson.

Plomp, R., Levelt, W.: 1965, 'Tonal Consonance and Critical Bandwidth' *J Acoust Soc Am*.

Roederer, J.G.: 1979, *Introduction to the Physics and Psychophysics of Music*. Springer.

Rossing, T.D.: 1989, *The Science of Sound*. Addison-Wesley.

Schlaug, G. et al.: 1995, *Science*, **267**, 699.

Zwicker, X. et al.: 1957, Critical Bandwidth, *J Acoust Soc Am*.

139. L. Nowak, *The Structure of Idealization.* Towards a Systematic Interpretation of the Marxian Idea of Science. 1980 ISBN 90-277-1014-7
140. C. Perelman, *The New Rhetoric and the Humanities.* Essays on Rhetoric and Its Applications. Translated from French and German. With an Introduction by H. Zyskind. 1979 ISBN 90-277-1018-X; Pb 90-277-1019-8
141. W. Rabinowicz, *Universalizability.* A Study in Morals and Metaphysics. 1979 ISBN 90-277-1020-2
142. C. Perelman, *Justice, Law and Argument.* Essays on Moral and Legal Reasoning. Translated from French and German. With an Introduction by H.J. Berman. 1980 ISBN 90-277-1089-9; Pb 90-277-1090-2
143. S. Kanger and S. Öhman (eds.), *Philosophy and Grammar.* Papers on the Occasion of the Quincentennial of Uppsala University. 1981 ISBN 90-277-1091-0
144. T. Pawlowski, *Concept Formation in the Humanities and the Social Sciences.* 1980 ISBN 90-277-1096-1
145. J. Hintikka, D. Gruender and E. Agazzi (eds.), *Theory Change, Ancient Axiomatics and Galileo's Methodology.* Proceedings of the 1978 Pisa Conference on the History and Philosophy of Science, Volume I. 1981 ISBN 90-277-1126-7
146. J. Hintikka, D. Gruender and E. Agazzi (eds.), *Probabilistic Thinking, Thermodynamics, and the Interaction of the History and Philosophy of Science.* Proceedings of the 1978 Pisa Conference on the History and Philosophy of Science, Volume II. 1981 ISBN 90-277-1127-5
147. U. Mönnich (ed.), *Aspects of Philosophical Logic.* Some Logical Forays into Central Notions of Linguistics and Philosophy. 1981 ISBN 90-277-1201-8
148. D. M. Gabbay, *Semantical Investigations in Heyting's Intuitionistic Logic.* 1981 ISBN 90-277-1202-6
149. E. Agazzi (ed.), *Modern Logic – A Survey.* Historical, Philosophical, and Mathematical Aspects of Modern Logic and Its Applications. 1981 ISBN 90-277-1137-2
150. A. F. Parker-Rhodes, *The Theory of Indistinguishables.* A Search for Explanatory Principles below the Level of Physics. 1981 ISBN 90-277-1214-X
151. J. C. Pitt, *Pictures, Images, and Conceptual Change.* An Analysis of Wilfrid Sellars' Philosophy of Science. 1981 ISBN 90-277-1276-X; Pb 90-277-1277-8
152. R. Hilpinen (ed.), *New Studies in Deontic Logic.* Norms, Actions, and the Foundations of Ethics. 1981 ISBN 90-277-1278-6; Pb 90-277-1346-4
153. C. Dilworth, *Scientific Progress.* A Study Concerning the Nature of the Relation between Successive Scientific Theories. 3rd rev. ed., 1994 ISBN 0-7923-2487-0; Pb 0-7923-2488-9
154. D. Woodruff Smith and R. McIntyre, *Husserl and Intentionality.* A Study of Mind, Meaning, and Language. 1982 ISBN 90-277-1392-8; Pb 90-277-1730-3
155. R. J. Nelson, *The Logic of Mind.* 2nd. ed., 1989 ISBN 90-277-2819-4; Pb 90-277-2822-4
156. J. F. A. K. van Benthem, *The Logic of Time.* A Model-Theoretic Investigation into the Varieties of Temporal Ontology, and Temporal Discourse. 1983; 2nd ed., 1991 ISBN 0-7923-1081-0
157. R. Swinburne (ed.), *Space, Time and Causality.* 1983 ISBN 90-277-1437-1
158. E. T. Jaynes, *Papers on Probability, Statistics and Statistical Physics.* Ed. by R. D. Rozenkrantz. 1983 ISBN 90-277-1448-7; Pb (1989) 0-7923-0213-3
159. T. Chapman, *Time: A Philosophical Analysis.* 1982 ISBN 90-277-1465-7
160. E. N. Zalta, *Abstract Objects.* An Introduction to Axiomatic Metaphysics. 1983 ISBN 90-277-1474-6
161. S. Harding and M. B. Hintikka (eds.), *Discovering Reality.* Feminist Perspectives on Epistemology, Metaphysics, Methodology, and Philosophy of Science. 1983 ISBN 90-277-1496-7; Pb 90-277-1538-6

162. M. A. Stewart (ed.), *Law, Morality and Rights.* 1983 ISBN 90-277-1519-X
163. D. Mayr and G. Süssmann (eds.), *Space, Time, and Mechanics.* Basic Structures of a Physical Theory. 1983 ISBN 90-277-1525-4
164. D. Gabbay and F. Guenthner (eds.), *Handbook of Philosophical Logic.* Vol. I: Elements of Classical Logic. 1983 ISBN 90-277-1542-4
165. D. Gabbay and F. Guenthner (eds.), *Handbook of Philosophical Logic.* Vol. II: Extensions of Classical Logic. 1984 ISBN 90-277-1604-8
166. D. Gabbay and F. Guenthner (eds.), *Handbook of Philosophical Logic.* Vol. III: Alternative to Classical Logic. 1986 ISBN 90-277-1605-6
167. D. Gabbay and F. Guenthner (eds.), *Handbook of Philosophical Logic.* Vol. IV: Topics in the Philosophy of Language. 1989 ISBN 90-277-1606-4
168. A. J. I. Jones, *Communication and Meaning.* An Essay in Applied Modal Logic. 1983 ISBN 90-277-1543-2
169. M. Fitting, *Proof Methods for Modal and Intuitionistic Logics.* 1983 ISBN 90-277-1573-4
170. J. Margolis, *Culture and Cultural Entities.* Toward a New Unity of Science. 1984 ISBN 90-277-1574-2
171. R. Tuomela, *A Theory of Social Action.* 1984 ISBN 90-277-1703-6
172. J. J. E. Gracia, E. Rabossi, E. Villanueva and M. Dascal (eds.), *Philosophical Analysis in Latin America.* 1984 ISBN 90-277-1749-4
173. P. Ziff, *Epistemic Analysis.* A Coherence Theory of Knowledge. 1984 ISBN 90-277-1751-7
174. P. Ziff, *Antiaesthetics.* An Appreciation of the Cow with the Subtile Nose. 1984 ISBN 90-277-1773-7
175. W. Balzer, D. A. Pearce, and H.-J. Schmidt (eds.), *Reduction in Science.* Structure, Examples, Philosophical Problems. 1984 ISBN 90-277-1811-3
176. A. Peczenik, L. Lindahl and B. van Roermund (eds.), *Theory of Legal Science.* Proceedings of the Conference on Legal Theory and Philosophy of Science (Lund, Sweden, December 1983). 1984 ISBN 90-277-1834-2
177. I. Niiniluoto, *Is Science Progressive?* 1984 ISBN 90-277-1835-0
178. B. K. Matilal and J. L. Shaw (eds.), *Analytical Philosophy in Comparative Perspective.* Exploratory Essays in Current Theories and Classical Indian Theories of Meaning and Reference. 1985 ISBN 90-277-1870-9
179. P. Kroes, *Time: Its Structure and Role in Physical Theories.* 1985 ISBN 90-277-1894-6
180. J. H. Fetzer, *Sociobiology and Epistemology.* 1985 ISBN 90-277-2005-3; Pb 90-277-2006-1
181. L. Haaparanta and J. Hintikka (eds.), *Frege Synthesized.* Essays on the Philosophical and Foundational Work of Gottlob Frege. 1986 ISBN 90-277-2126-2
182. M. Detlefsen, *Hilbert's Program.* An Essay on Mathematical Instrumentalism. 1986 ISBN 90-277-2151-3
183. J. L. Golden and J. J. Pilotta (eds.), *Practical Reasoning in Human Affairs.* Studies in Honor of Chaim Perelman. 1986 ISBN 90-277-2255-2
184. H. Zandvoort, *Models of Scientific Development and the Case of Nuclear Magnetic Resonance.* 1986 ISBN 90-277-2351-6
185. I. Niiniluoto, *Truthlikeness.* 1987 ISBN 90-277-2354-0
186. W. Balzer, C. U. Moulines and J. D. Sneed, *An Architectonic for Science.* The Structuralist Program. 1987 ISBN 90-277-2403-2
187. D. Pearce, *Roads to Commensurability.* 1987 ISBN 90-277-2414-8
188. L. M. Vaina (ed.), *Matters of Intelligence.* Conceptual Structures in Cognitive Neuroscience. 1987 ISBN 90-277-2460-1

189. H. Siegel, *Relativism Refuted*. A Critique of Contemporary Epistemological Relativism. 1987 ISBN 90-277-2469-5
190. W. Callebaut and R. Pinxten, *Evolutionary Epistemology*. A Multiparadigm Program, with a Complete Evolutionary Epistemology Bibliograph. 1987 ISBN 90-277-2582-9
191. J. Kmita, *Problems in Historical Epistemology*. 1988 ISBN 90-277-2199-8
192. J. H. Fetzer (ed.), *Probability and Causality*. Essays in Honor of Wesley C. Salmon, with an Annotated Bibliography. 1988 ISBN 90-277-2607-8; Pb 1-5560-8052-2
193. A. Donovan, L. Laudan and R. Laudan (eds.), *Scrutinizing Science*. Empirical Studies of Scientific Change. 1988 ISBN 90-277-2608-6
194. H.R. Otto and J.A. Tuedio (eds.), *Perspectives on Mind*. 1988 ISBN 90-277-2640-X
195. D. Batens and J.P. van Bendegem (eds.), *Theory and Experiment*. Recent Insights and New Perspectives on Their Relation. 1988 ISBN 90-277-2645-0
196. J. Österberg, *Self and Others*. A Study of Ethical Egoism. 1988 ISBN 90-277-2648-5
197. D.H. Helman (ed.), *Analogical Reasoning*. Perspectives of Artificial Intelligence, Cognitive Science, and Philosophy. 1988 ISBN 90-277-2711-2
198. J. Woleński, *Logic and Philosophy in the Lvov-Warsaw School*. 1989 ISBN 90-277-2749-X
199. R. Wójcicki, *Theory of Logical Calculi*. Basic Theory of Consequence Operations. 1988 ISBN 90-277-2785-6
200. J. Hintikka and M.B. Hintikka, *The Logic of Epistemology and the Epistemology of Logic*. Selected Essays. 1989 ISBN 0-7923-0040-8; Pb 0-7923-0041-6
201. E. Agazzi (ed.), *Probability in the Sciences*. 1988 ISBN 90-277-2808-9
202. M. Meyer (ed.), *From Metaphysics to Rhetoric*. 1989 ISBN 90-277-2814-3
203. R.L. Tieszen, *Mathematical Intuition*. Phenomenology and Mathematical Knowledge. 1989 ISBN 0-7923-0131-5
204. A. Melnick, *Space, Time, and Thought in Kant*. 1989 ISBN 0-7923-0135-8
205. D.W. Smith, *The Circle of Acquaintance*. Perception, Consciousness, and Empathy. 1989 ISBN 0-7923-0252-4
206. M.H. Salmon (ed.), *The Philosophy of Logical Mechanism*. Essays in Honor of Arthur W. Burks. With his Responses, and with a Bibliography of Burk's Work. 1990 ISBN 0-7923-0325-3
207. M. Kusch, *Language as Calculus vs. Language as Universal Medium*. A Study in Husserl, Heidegger, and Gadamer. 1989 ISBN 0-7923-0333-4
208. T.C. Meyering, *Historical Roots of Cognitive Science*. The Rise of a Cognitive Theory of Perception from Antiquity to the Nineteenth Century. 1989 ISBN 0-7923-0349-0
209. P. Kosso, *Observability and Observation in Physical Science*. 1989 ISBN 0-7923-0389-X
210. J. Kmita, *Essays on the Theory of Scientific Cognition*. 1990 ISBN 0-7923-0441-1
211. W. Sieg (ed.), *Acting and Reflecting*. The Interdisciplinary Turn in Philosophy. 1990 ISBN 0-7923-0512-4
212. J. Karpiński, *Causality in Sociological Research*. 1990 ISBN 0-7923-0546-9
213. H.A. Lewis (ed.), *Peter Geach: Philosophical Encounters*. 1991 ISBN 0-7923-0823-9
214. M. Ter Hark, *Beyond the Inner and the Outer*. Wittgenstein's Philosophy of Psychology. 1990 ISBN 0-7923-0850-6
215. M. Gosselin, *Nominalism and Contemporary Nominalism*. Ontological and Epistemological Implications of the Work of W.V.O. Quine and of N. Goodman. 1990 ISBN 0-7923-0904-9
216. J.H. Fetzer, D. Shatz and G. Schlesinger (eds.), *Definitions and Definability*. Philosophical Perspectives. 1991 ISBN 0-7923-1046-2
217. E. Agazzi and A. Cordero (eds.), *Philosophy and the Origin and Evolution of the Universe*. 1991 ISBN 0-7923-1322-4

218. M. Kusch, *Foucault's Strata and Fields.* An Investigation into Archaeological and Genealogical Science Studies. 1991 ISBN 0-7923-1462-X
219. C.J. Posy, *Kant's Philosophy of Mathematics.* Modern Essays. 1992 ISBN 0-7923-1495-6
220. G. Van de Vijver, *New Perspectives on Cybernetics.* Self-Organization, Autonomy and Connectionism. 1992 ISBN 0-7923-1519-7
221. J.C. Nyíri, *Tradition and Individuality.* Essays. 1992 ISBN 0-7923-1566-9
222. R. Howell, *Kant's Transcendental Deduction.* An Analysis of Main Themes in His Critical Philosophy. 1992 ISBN 0-7923-1571-5
223. A. García de la Sienra, *The Logical Foundations of the Marxian Theory of Value.* 1992 ISBN 0-7923-1778-5
224. D.S. Shwayder, *Statement and Referent.* An Inquiry into the Foundations of Our Conceptual Order. 1992 ISBN 0-7923-1803-X
225. M. Rosen, *Problems of the Hegelian Dialectic.* Dialectic Reconstructed as a Logic of Human Reality. 1993 ISBN 0-7923-2047-6
226. P. Suppes, *Models and Methods in the Philosophy of Science: Selected Essays.* 1993 ISBN 0-7923-2211-8
227. R. M. Dancy (ed.), *Kant and Critique: New Essays in Honor of W. H. Werkmeister.* 1993 ISBN 0-7923-2244-4
228. J. Woleński (ed.), *Philosophical Logic in Poland.* 1993 ISBN 0-7923-2293-2
229. M. De Rijke (ed.), *Diamonds and Defaults.* Studies in Pure and Applied Intensional Logic. 1993 ISBN 0-7923-2342-4
230. B.K. Matilal and A. Chakrabarti (eds.), *Knowing from Words.* Western and Indian Philosophical Analysis of Understanding and Testimony. 1994 ISBN 0-7923-2345-9
231. S.A. Kleiner, *The Logic of Discovery.* A Theory of the Rationality of Scientific Research. 1993 ISBN 0-7923-2371-8
232. R. Festa, *Optimum Inductive Methods.* A Study in Inductive Probability, Bayesian Statistics, and Verisimilitude. 1993 ISBN 0-7923-2460-9
233. P. Humphreys (ed.), *Patrick Suppes: Scientific Philosopher.* Vol. 1: Probability and Probabilistic Causality. 1994 ISBN 0-7923-2552-4
234. P. Humphreys (ed.), *Patrick Suppes: Scientific Philosopher.* Vol. 2: Philosophy of Physics, Theory Structure, and Measurement Theory. 1994 ISBN 0-7923-2553-2
235. P. Humphreys (ed.), *Patrick Suppes: Scientific Philosopher.* Vol. 3: Language, Logic, and Psychology. 1994 ISBN 0-7923-2862-0
     Set ISBN (Vols 233–235) 0-7923-2554-0
236. D. Prawitz and D. Westerståhl (eds.), *Logic and Philosophy of Science in Uppsala.* Papers from the 9th International Congress of Logic, Methodology, and Philosophy of Science. 1994 ISBN 0-7923-2702-0
237. L. Haaparanta (ed.), *Mind, Meaning and Mathematics.* Essays on the Philosophical Views of Husserl and Frege. 1994 ISBN 0-7923-2703-9
238. J. Hintikka (ed.), *Aspects of Metaphor.* 1994 ISBN 0-7923-2786-1
239. B. McGuinness and G. Oliveri (eds.), *The Philosophy of Michael Dummett.* With Replies from Michael Dummett. 1994 ISBN 0-7923-2804-3
240. D. Jamieson (ed.), *Language, Mind, and Art.* Essays in Appreciation and Analysis, In Honor of Paul Ziff. 1994 ISBN 0-7923-2810-8
241. G. Preyer, F. Siebelt and A. Ulfig (eds.), *Language, Mind and Epistemology.* On Donald Davidson's Philosophy. 1994 ISBN 0-7923-2811-6
242. P. Ehrlich (ed.), *Real Numbers, Generalizations of the Reals, and Theories of Continua.* 1994 ISBN 0-7923-2689-X

243. G. Debrock and M. Hulswit (eds.), *Living Doubt*. Essays concerning the epistemology of Charles Sanders Peirce. 1994 ISBN 0-7923-2898-1
244. J. Srzednicki, *To Know or Not to Know*. Beyond Realism and Anti-Realism. 1994 ISBN 0-7923-2909-0
245. R. Egidi (ed.), *Wittgenstein: Mind and Language*. 1995 ISBN 0-7923-3171-0
246. A. Hyslop, *Other Minds*. 1995 ISBN 0-7923-3245-8
247. L. Pólos and M. Masuch (eds.), *Applied Logic: How, What and Why*. Logical Approaches to Natural Language. 1995 ISBN 0-7923-3432-9
248. M. Krynicki, M. Mostowski and L.M. Szczerba (eds.), *Quantifiers: Logics, Models and Computation*. Volume One: Surveys. 1995 ISBN 0-7923-3448-5
249. M. Krynicki, M. Mostowski and L.M. Szczerba (eds.), *Quantifiers: Logics, Models and Computation*. Volume Two: Contributions. 1995 ISBN 0-7923-3449-3
Set ISBN (Vols 248 + 249) 0-7923-3450-7
250. R.A. Watson, *Representational Ideas from Plato to Patricia Churchland*. 1995 ISBN 0-7923-3453-1
251. J. Hintikka (ed.), *From Dedekind to Gödel*. Essays on the Development of the Foundations of Mathematics. 1995 ISBN 0-7923-3484-1
252. A. Wiśniewski, *The Posing of Questions*. Logical Foundations of Erotetic Inferences. 1995 ISBN 0-7923-3637-2
253. J. Peregrin, *Doing Worlds with Words*. Formal Semantics without Formal Metaphysics. 1995 ISBN 0-7923-3742-5
254. I.A. Kieseppä, *Truthlikeness for Multidimensional, Quantitative Cognitive Problems*. 1996 ISBN 0-7923-4005-1
255. P. Hugly and C. Sayward: *Intensionality and Truth*. An Essay on the Philosophy of A.N. Prior. 1996 ISBN 0-7923-4119-8
256. L. Hankinson Nelson and J. Nelson (eds.): *Feminism, Science, and the Philosophy of Science*. 1997 ISBN 0-7923-4162-7
257. P.I. Bystrov and V.N. Sadovsky (eds.): *Philosophical Logic and Logical Philosophy*. Essays in Honour of Vladimir A. Smirnov. 1996 ISBN 0-7923-4270-4
258. Å.E. Andersson and N-E. Sahlin (eds.): *The Complexity of Creativity*. 1996 ISBN 0-7923-4346-8
259. M.L. Dalla Chiara, K. Doets, D. Mundici and J. van Benthem (eds.): *Logic and Scientific Methods*. Volume One of the Tenth International Congress of Logic, Methodology and Philosophy of Science, Florence, August 1995. 1997 ISBN 0-7923-4383-2
260. M.L. Dalla Chiara, K. Doets, D. Mundici and J. van Benthem (eds.): *Structures and Norms in Science*. Volume Two of the Tenth International Congress of Logic, Methodology and Philosophy of Science, Florence, August 1995. 1997 ISBN 0-7923-4384-0
Set ISBN (Vols 259 + 260) 0-7923-4385-9
261. A. Chakrabarti: *Denying Existence*. The Logic, Epistemology and Pragmatics of Negative Existentials and Fictional Discourse. 1997 ISBN 0-7923-4388-3
262. A. Biletzki: *Talking Wolves*. Thomas Hobbes on the Language of Politics and the Politics of Language. 1997 ISBN 0-7923-4425-1
263. D. Nute (ed.): *Defeasible Deontic Logic*. 1997 ISBN 0-7923-4630-0
264. U. Meixner: *Axiomatic Formal Ontology*. 1997 ISBN 0-7923-4747-X
265. I. Brinck: *The Indexical 'I'*. The First Person in Thought and Language. 1997 ISBN 0-7923-4741-2
266. G. Hölmström-Hintikka and R. Tuomela (eds.): *Contemporary Action Theory*. Volume 1: Individual Action. 1997 ISBN 0-7923-4753-6; Set: 0-7923-4754-4

267. G. Hölmström-Hintikka and R. Tuomela (eds.): *Contemporary Action Theory.* Volume 2: Social Action. 1997 ISBN 0-7923-4752-8; Set: 0-7923-4754-4
268. B.-C. Park: *Phenomenological Aspects of Wittgenstein's Philosophy.* 1998 ISBN 0-7923-4813-3
269. J. Paśniczek: *The Logic of Intentional Objects.* A Meinongian Version of Classical Logic. 1998 ISBN 0-7923-4880-X
270. P.W. Humphreys and J.H. Fetzer (eds.): *The New Theory of Reference.* Kripke, Marcus, and Its Origins. 1998 ISBN 0-7923-4898-2
271. K. Szaniawski, A. Chmielewski and J. Woleński (eds.): *On Science, Inference, Information and Decision Making.* Selected Essays in the Philosophy of Science. 1998 ISBN 0-7923-4922-9
272. G.H. von Wright: *In the Shadow of Descartes.* Essays in the Philosophy of Mind. 1998 ISBN 0-7923-4992-X
273. K. Kijania-Placek and J. Woleński (eds.): *The Lvov–Warsaw School and Contemporary Philosophy.* 1998 ISBN 0-7923-5105-3
274. D. Dedrick: *Naming the Rainbow.* Colour Language, Colour Science, and Culture. 1998 ISBN 0-7923-5239-4
275. L. Albertazzi (ed.): *Shapes of Forms.* From Gestalt Psychology and Phenomenology to Ontology and Mathematics. 1999 ISBN 0-7923-5246-7
276. P. Fletcher: *Truth, Proof and Infinity.* A Theory of Constructions and Constructive Reasoning. 1998 ISBN 0-7923-5262-9
277. M. Fitting and R.L. Mendelsohn (eds.): *First-Order Modal Logic.* 1998 ISBN 0-7923-5334-X
278. J.N. Mohanty: *Logic, Truth and Modalities.* From a Phenomenological Perspective. 1999 ISBN 0-7923-5550-4
279. T. Placek: *Mathematical Intuitionism and Intersubjectivity.* A Critical Exposition of Arguments for Intuitionism. 1999 ISBN 0-7923-5630-6
280. A. Cantini, E. Casari and P. Minari (eds.): *Logic and Foundations of Mathematics.* 1999 ISBN 0-7923-5659-4
281. M.L. Dalla Chiara, R. Giuntini and F. Laudisa (eds.): *Language, Quantum, Music.* 1999 ISBN 0-7923-5727-2

Previous volumes are still available.

KLUWER ACADEMIC PUBLISHERS – DORDRECHT / BOSTON / LONDON